# Student Solutions Manual

**Tamas Wiandt**
*Rochester Institute of Technology*

to accompany

---

# Calculus for the Life Sciences

**Sebastian J. Schreiber**
*University of California, Davis*

**Karl J. Smith**
*Santa Rosa Junior College*

**Wayne M. Getz**
*University of California, Berkeley*

WILEY

| | |
|---|---|
| PUBLISHER | Laurie Rosatone |
| ACQUISITIONS EDITOR | Shannon Corliss |
| FREELANCE PROJECT EDITOR | Anne Scanlan-Rohrer |
| ASSISTANT EDITOR | Jacqueline Sinacori |
| EDITORIAL ASSISTANT | Michael O'Neal |
| SENIOR CONTENT MANAGER | Karoline Luciano |
| SENIOR PRODUCTION EDITOR | Kerry Weinstein |

Founded in 1807, John Wiley & Sons, Inc. has been a valued source of knowledge and understanding for more than 200 years, helping people around the world meet their needs and fulfill their aspirations. Our company is built on a foundation of principles that include responsibility to the communities we serve and where we live and work. In 2008, we launched a Corporate Citizenship Initiative, a global effort to address the environmental, social, economic, and ethical challenges we face in our business. Among the issues we are addressing are carbon impact, paper specifications and procurement, ethical conduct within our business and among our vendors, and community and charitable support. For more information, please visit our website: www.wiley.com/go/citizenship.



ISBN 978-1-118-64559-8

10 9 8 7 6 5 4 3 2 1

# Contents

**Problem Set 1.1 - Real Numbers and Functions**

**1. a.** Function; $D : \{3, 4, 5, 6\}$, $R : \{4, 7, 9\}$.

**b.** Not a function, just a set of numbers.

**3. a.** Not a function; multiple values for all $x$ values.

**b.** Function; $D : x \neq 0$, $R : \{-1, 1\}$.

**5. a.** Not a function; two values for certain $x$ values.

**b.** Function; $D : \mathbb{R}$, $R : y \geq -4$.

**7. a.** Function; $D : \mathbb{R}$, $R : y \geq 0$.

**b.** Function; $D : \mathbb{R}$, $R : \mathbb{R}$.

**9. a.** Not a function; two values for $-2 < x < 2$.

**b.** Function; $D : x \neq \pm 2$, $R : y \neq 1$.

**11. a.** Not a function; two values for $x > -3$.

**b.** Function; $D : \mathbb{R}$, $R : y \geq -80$.

**13.** $D : \mathbb{R}$, $f(0) = 3$, $f(1) = 4$, $f(-2) = -5$.

**15.** $D : x \neq -3$, $f(2) = 0$, $f(0) = -2$, $-3$ is not in $D$.

**17.** $D : \mathbb{R}$, $f(3) = 4$, $f(1) = 2$, $f(0) = 4$.

**19.** $F(x) = x^2$.

**21.** $M(x) = 3x - 7$, $M(5) = 8$, $M(0) = -7$, $M(-3) = -16$.

**23.** $D : [-6, 6]$, $R : [-6, 5]$, increasing: $[0, 6]$, decreasing: $[-6, 0]$.

**25.** $D : x \neq 2$, $R : \{5\}$, constant: on $D$.

**27.** $D : [-5, 3) \cup (3, \infty)$, $R : [-3, 6) \cup (6, \infty)$, increasing: $[0, 3)$ and $(3, \infty)$, decreasing: $[-2, 0]$, constant: $[-5, -2]$.

**29.** $f(x) = 3x - 5$, $D : \mathbb{R}$.

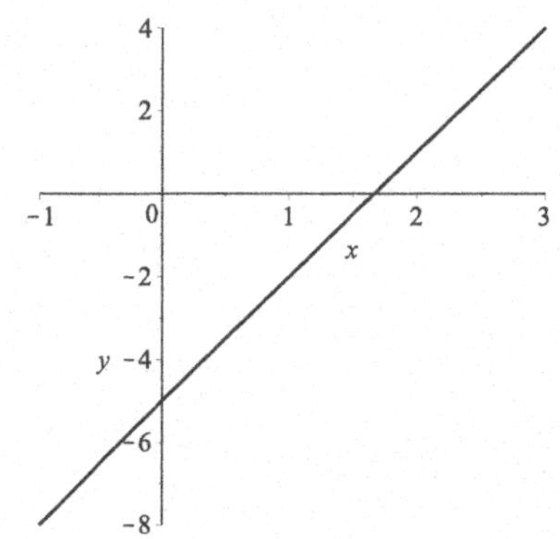


**31.** $f(x) = \sqrt{5 - x}$, $D : (-\infty, 5]$.

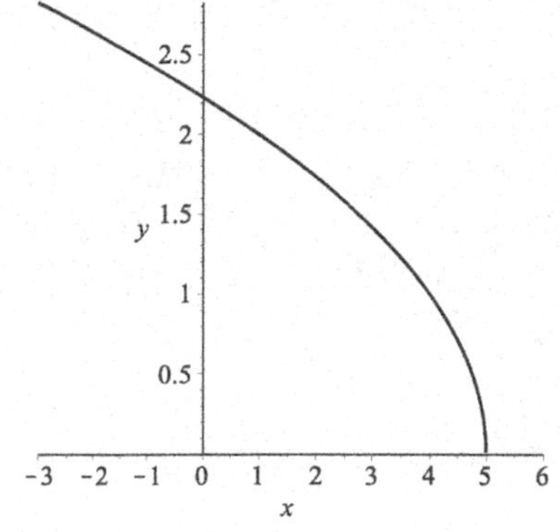


**33.** $f(x) = (5 + x)^2 - x^2 = 25 + 10x$.

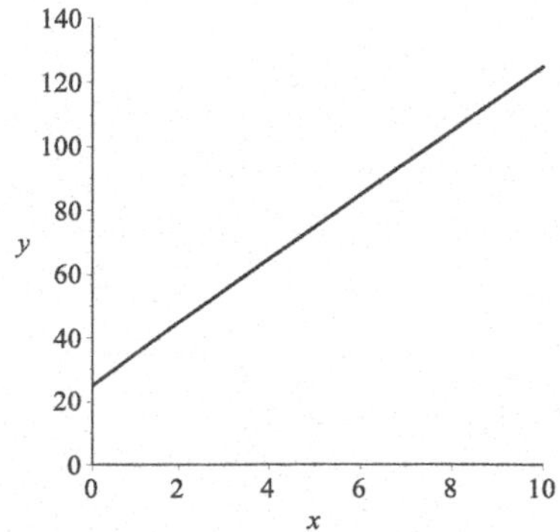


**35.** The side length is given by $P/4$, so the area is $A(P) = (P/4)^2 = P^2/16$.

**37.** The $x$ mg decreases to $0.33x$ during the first four hours; then $2 \cdot 325 = 650$ mg is added. After another four hours, the total $650+0.33x$ will decrease to $0.33(650+0.33x)$.

**39.** We assume that the maximum height is 2 (in), and after the cut the height is 0.5 (in).

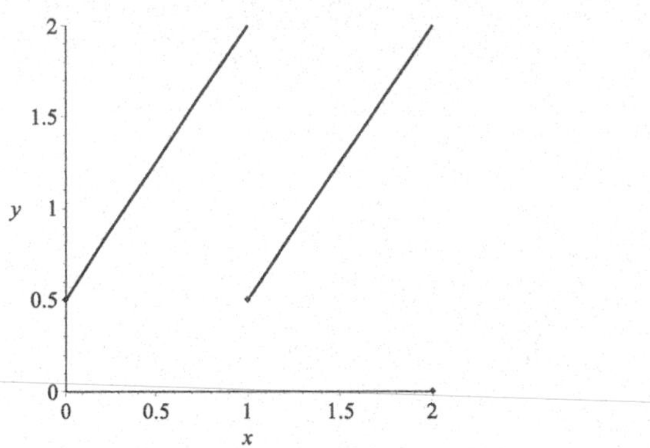


**41. a.** At the center, $r = 0$, so $S(0) = CR^2 = 1.76 \cdot 10^5 \cdot (1.2 \cdot 10^{-2})^2 \approx 25.3(\text{cm/sec})$.

**b.** Midway between the artery's wall and the central axis, $r = R/2 = 0.6 \cdot 10^{-2}$. Thus $S(R/2) = C(R^2 - r^2) = C(R^2 - (R/2)^2) = 1.76 \cdot 10^5 \cdot ((1.2 \cdot 10^{-2})^2 - (0.6 \cdot 10^{-2})^2) \approx 19(\text{cm/sec})$.

**c.** The domain is $[0, R] = [0, 1.2 \cdot 10^{-2}]$.

**d.**

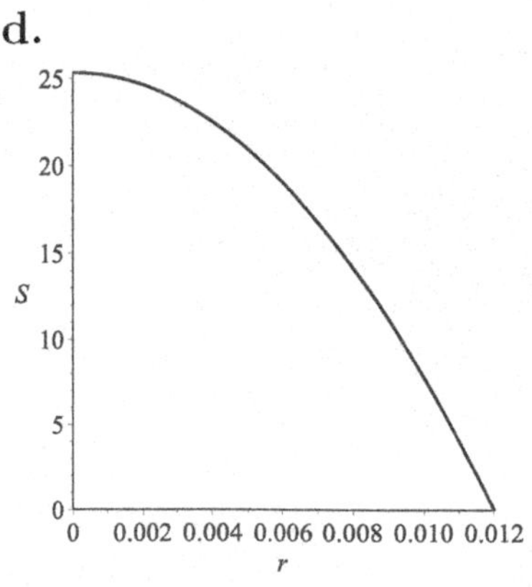


**43. a.** $D: (-\infty, 0) \cup (0, \infty)$.

**b.** $D: \{1, 2, 3, 4, \dots\}$

**c.**

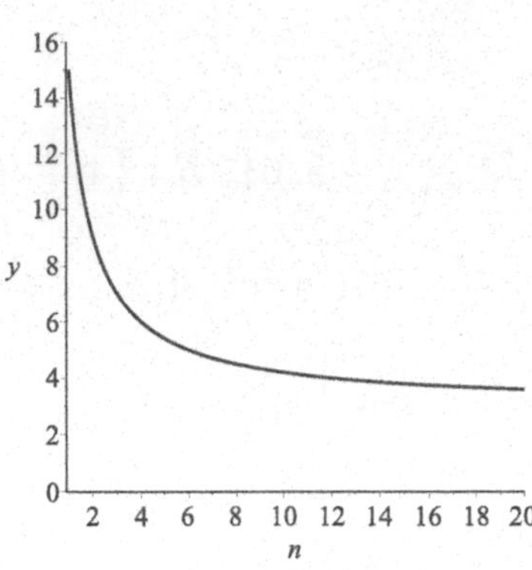


**d.** The time will decrease and approach 3 minutes. It will never be less than or equal to 3.

**45. a.** $D: 0 \le n \le 12.5$.

**b.**

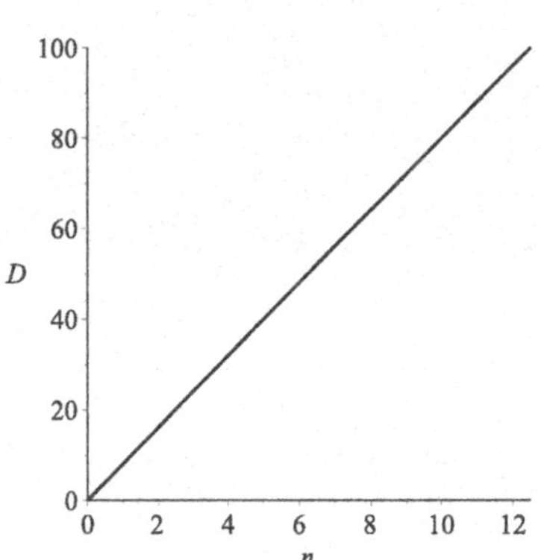


**c.** First, $D(3) = 2 \cdot 3A/25 = 100$, so $A = 2500/6$. Thus $D(5) = 2 \cdot 5 \cdot (2500/6)/25 = 500/3$ (mg).

**47. a.** $D: 0 \le w \le 150$, if we consider the meaning of the function (that a child can't get a larger amount of drug than an adult).

**b.**

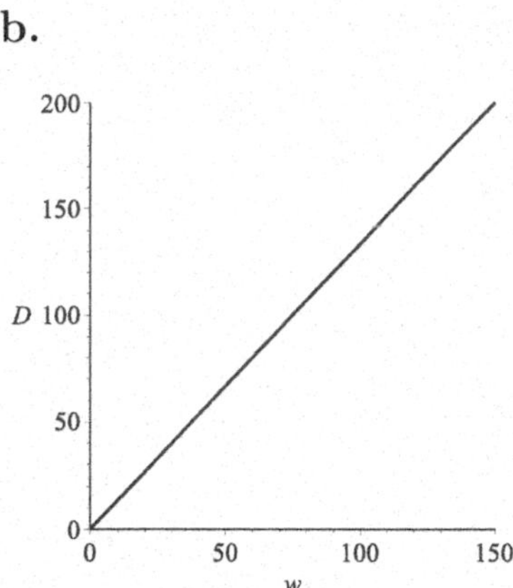


**c.** $D(70) = (70/150)A = 90$, thus $A = 1350/7$ (mg).

**49.** $2 = p^2/q^2$, thus $p^2 = 2q^2$. Thus $p^2$ is even, and then $p$ is even. So $p = 2k$; but then $2q^2 = p^2 = (2k)^2 = 4k^2$, so $q^2 = 2k^2$. Thus $q^2$ is even and then $q$ is even. We arrived to the contradiction, because we assumed that $p/q$ cannot be reduced.

## Problem Set 1.2 - Data Fitting with Linear and Periodic Functions

**1.** $y = \dfrac{5x}{4} - 2$

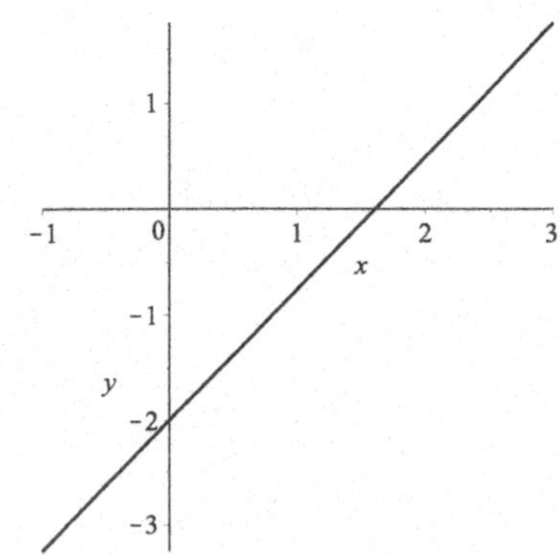


**3.** $y = \dfrac{2x}{5} + 2$

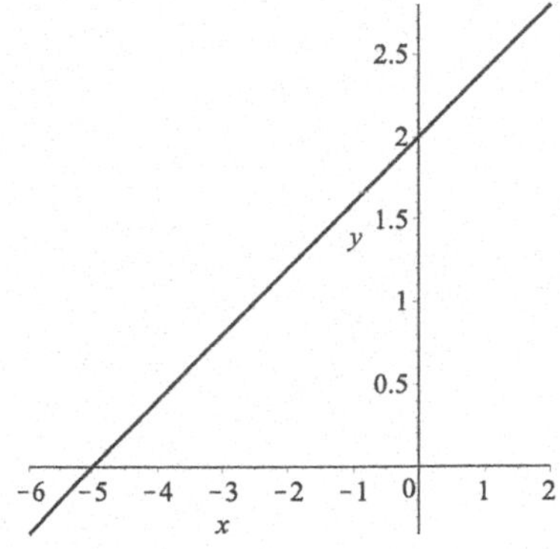


**5.** $y = -3x + 2$

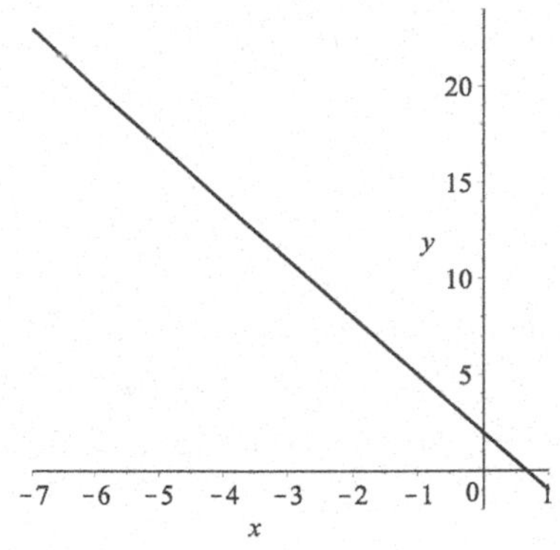


**7.**

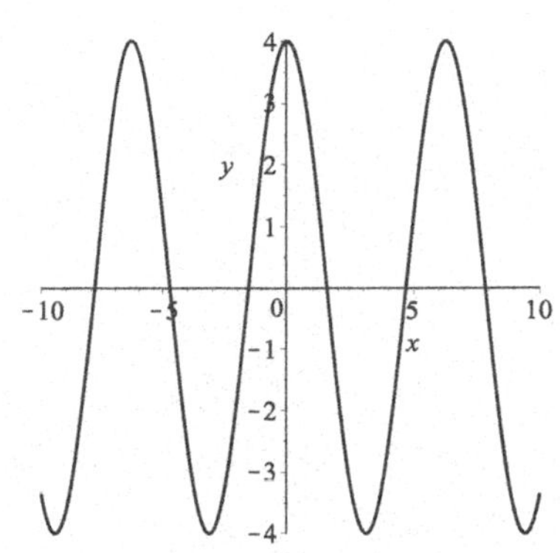

**9.**

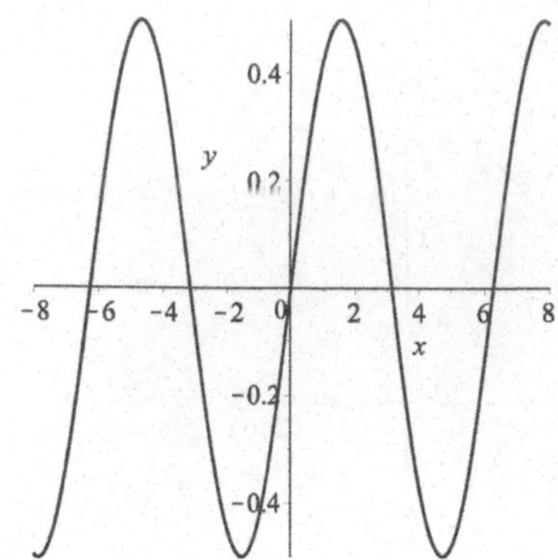


**11.** $y - 3 = 3(x - 1)$, after simplification $y = 3x$.

**13.** Slope is $(1 - 2)/(0 - (-1)) = -1$, so $y - 1 = 1 \cdot (x - 0)$, i.e. $y = -x + 1$.

**15.** Slope is 0, so $y = 4$.

**17.** The equation can also be written as $y = m(x - h) + k = mx + k - mh$. The slope is $m$, and the point $(h, k)$ satisfies the equation: $k = mh + k - mh$.

**19.** The slope is $4/2 = 2$, the $y$-intercept is $-4$. The equation is $y = 2x - 4$.

**21.** $y = 3\cos x$.

**23.** $y = -2\cos \pi x$.

**25.** D (the slope is $(8 - 2)/10$).

**27.** A (the slope is $(6 - 2)/10$).

**29.** C (the slope is $(-3 - 2)/10$).

**31.** Amplitude is $1/2$, period is $2\pi$.

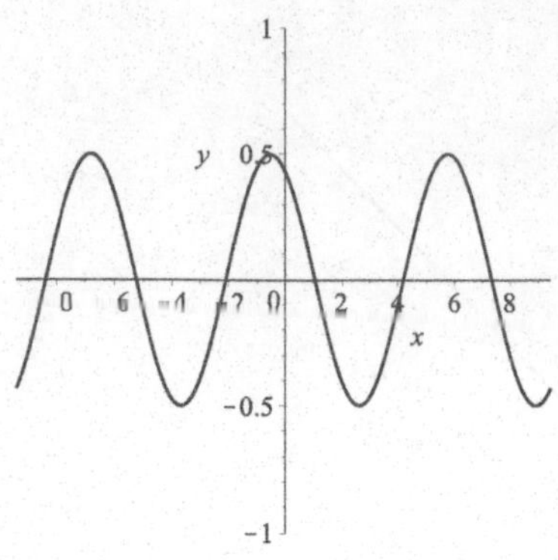


**33.** Amplitude is 2, period is $2\pi/2\pi = 1$.

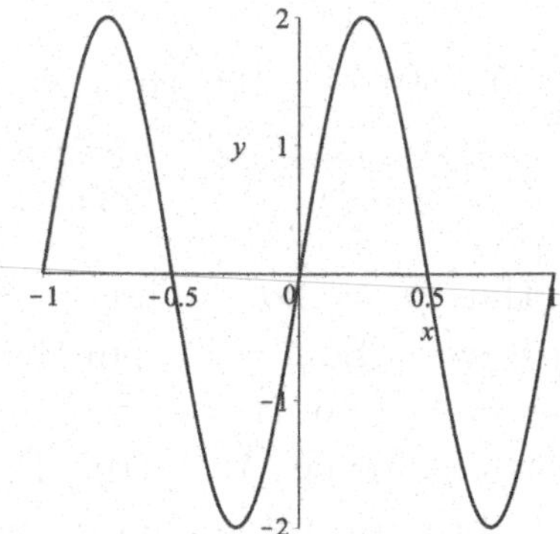


**35.** No amplitude, period is $\pi/2$.

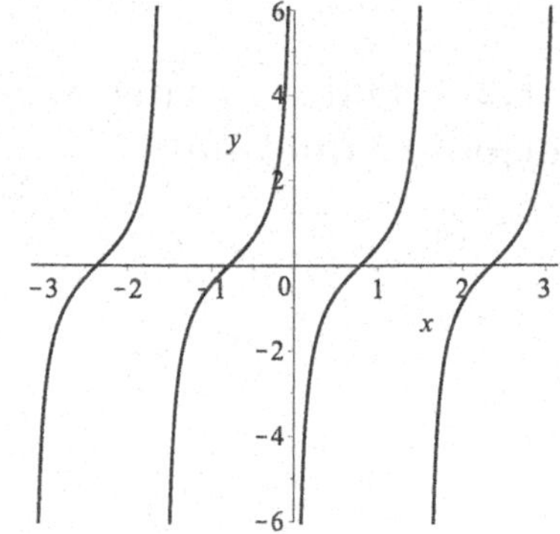


**37.** The slope is $(20 - 50)/(60 - 24) = -30/36 = -5/6$, and the equation is given by $E = -5A/6 + 70$.

**a.** We have to solve the equation $30 = -5A/6 + 70$. We obtain $A = 48$.

**b.** Her life expectancy is $E = 70$. $(A = 0)$

**c.** We have to solve $0 = -5A/6 + 70$; we

obtain 84.

**39.** B; the estimate is $0.231 \cdot 150 - 3 = 31.65$.

**41. a.** The data is close to linear.

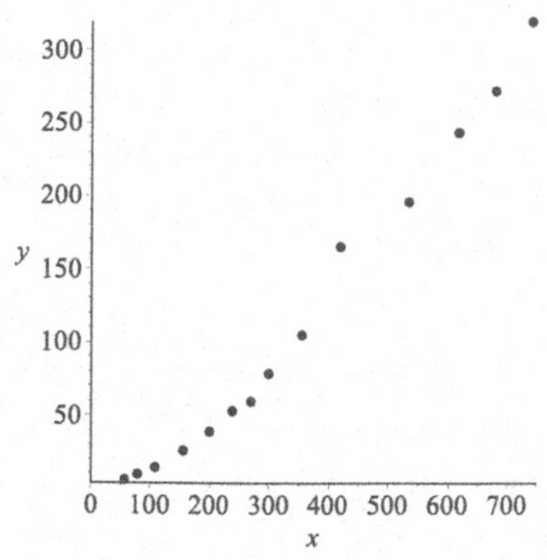

**b.**

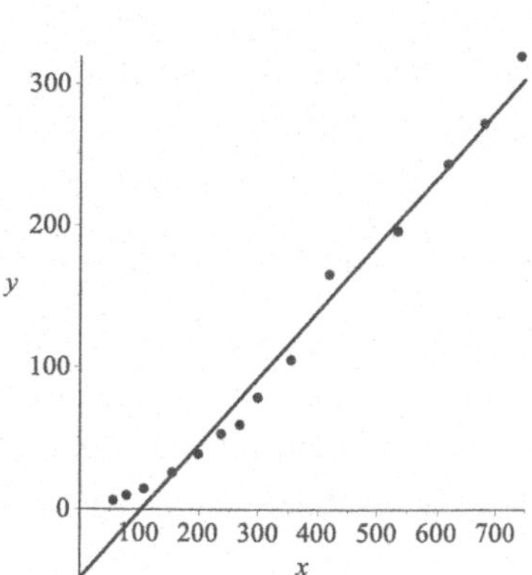

**43. a.** The data does not seem to be linear.

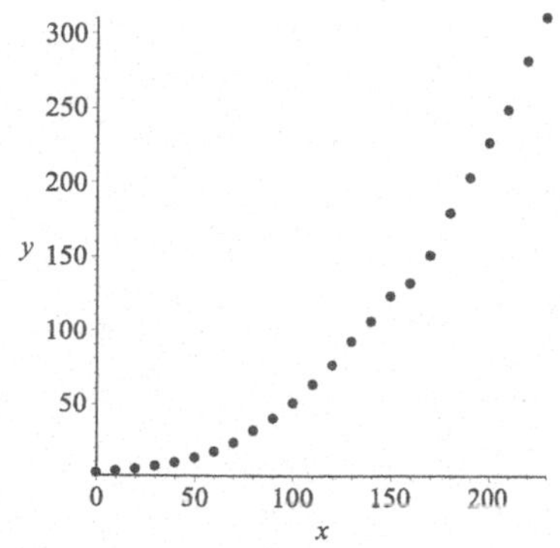

**b.**

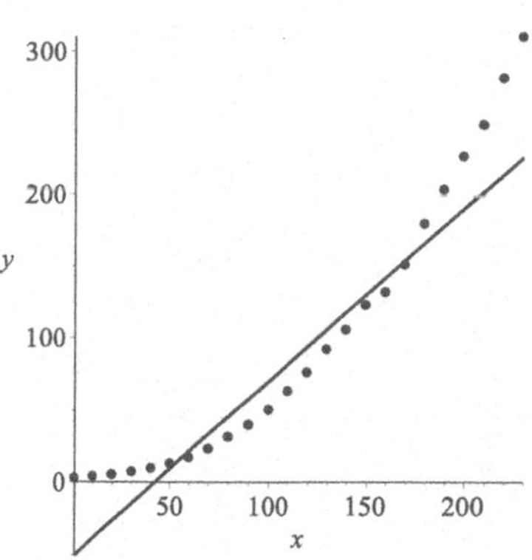

**45. a.**

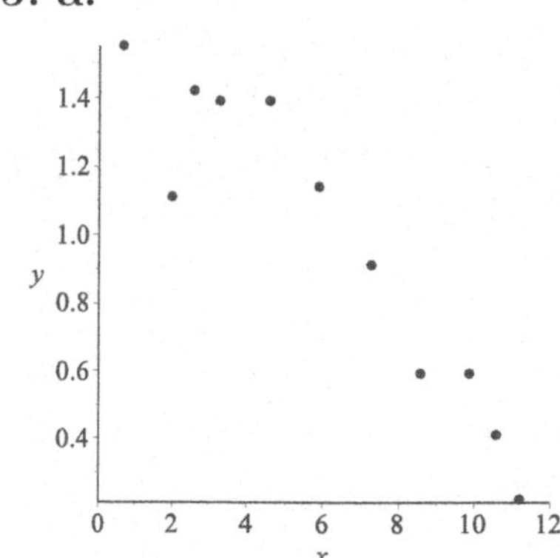

**b.** The first two data points give us the slope $(1.11 - 1.55)/(2.0 - 0.7) = -0.44/1.3$, equation is $y - 1.55 = -(0.44/1.3)(x - 0.7)$. The first and last data points give us the slope $(0.22 - 1.55)/(11.2 - 0.7) = -1.33/10.5$, equation is $y - 1.55 = -(1.33/10.5)(x-0.7)$. The second line seems to be a better fit.

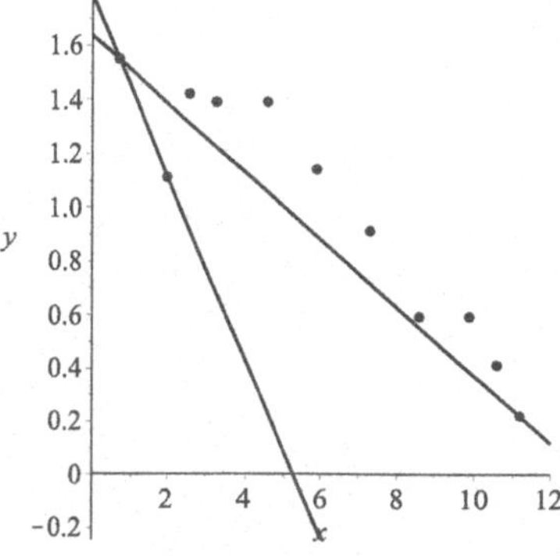

**c.** $y = -0.1165x + 1.681$.

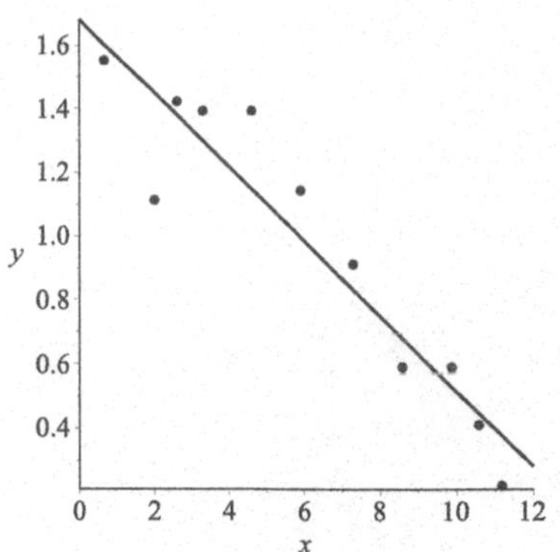


**d.** Using the line from part **c**, the estimate is 0.28 ft/sec at 12 feet, and −0.65 ft/sec at 20 feet. The second result is clearly out of the scope of the model.

**47. a.**

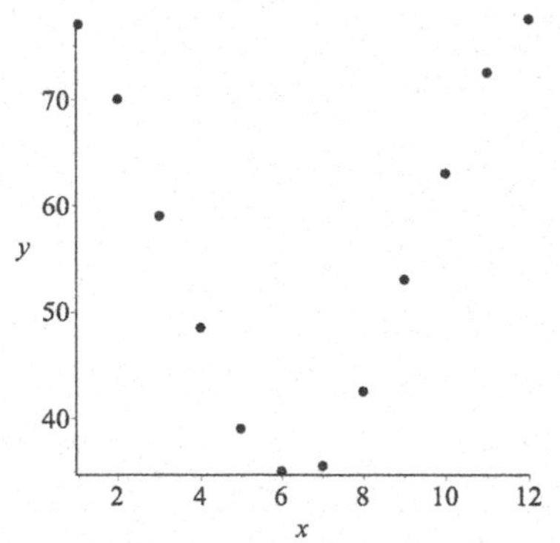


**b.** The amplitude is about $(77.5 - 35)/2 = 21.25$ (degrees), the period is 12 (months).

**c.** $a = 21.25$, $b = 2\pi/12 = \pi/6$.

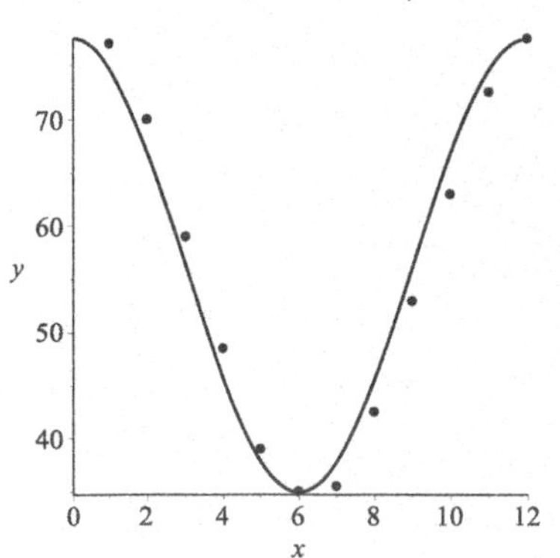


## Problem Set 1.3 - Power Functions and Scaling Laws

**1. a.** Yes, $y = (1/3)x^1$.

**b.** Yes, $y = (1/3)x^{-1}$.

**c.** Not a power function.

**3.** Not a power function.

**5.** Yes, $y = (1/4)x^{-3/2}$.

**7.** Not a power function.

**9.** Yes, $y = 6x^{-1/2}$.

**11.** We know that $y = ax^2$. Thus $10^{12} = 10^{15}/10^3 = y_2/y_1 = x_2^2/x_1^2 = (x_2/x_1)^2$, so $x$ increases $10^6$-fold.

**13.** This means $y = a\,10x^3 = b\,x^3$, so $y \propto x^3$ as well. Then $x \propto y^{1/3}$.

**15.** Using the general transitive property, $x \propto z^6$.

**17.** Decreasing on $(-\infty, 0]$, increasing on $[0, \infty)$.

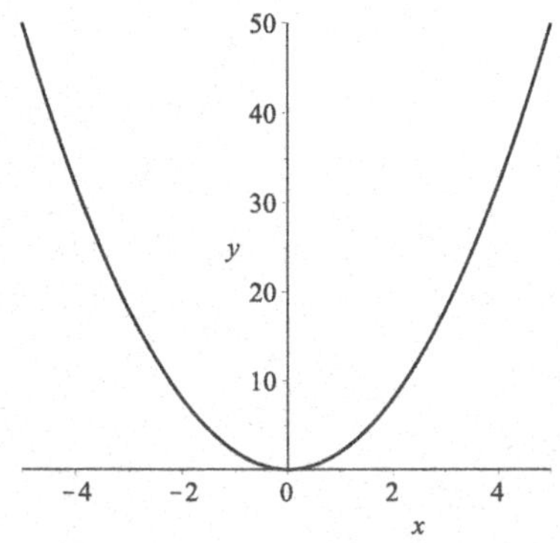


**19.** Decreasing on $(-\infty, \infty)$.

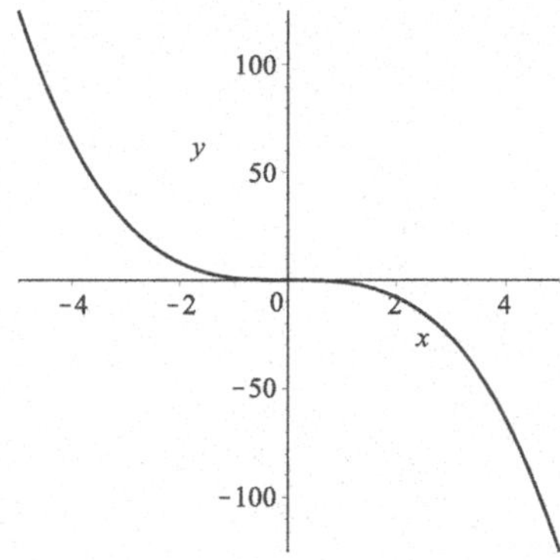


**21.** The domain is $[0, \infty)$, because $12x^{1/2} = 12\sqrt{x}$. Increasing on $[0, \infty)$.

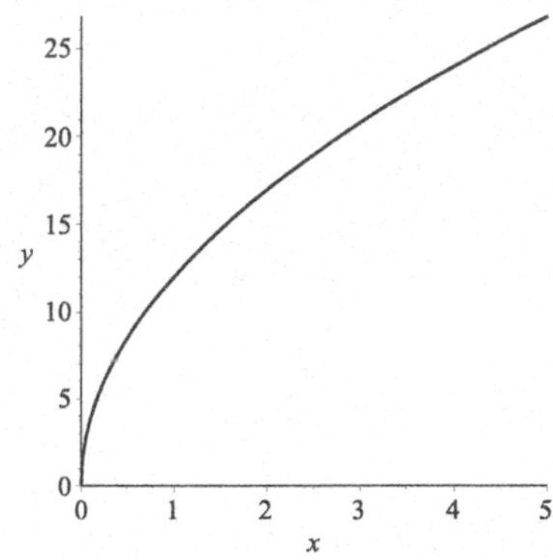


**23.** Power function only when $b = 0$.

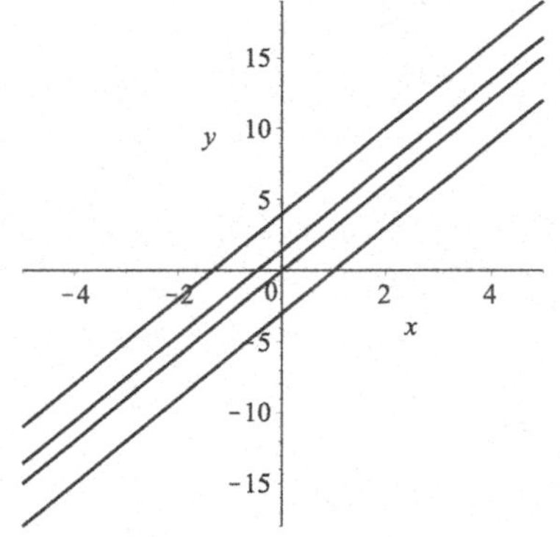


**25.** From $S = 4\pi r^2$, we get $r = \sqrt{S/(4\pi)}$. Thus $V = 4\pi r^3/3 = 4\pi(\sqrt{S/(4\pi)})^3/3 = S^{3/2}/(6\sqrt{\pi})$. If $S$ is quadrupled, $r$ is doubled.

**27.** The volume is $V = (1/3)(h/2)^2\pi h = (\pi/12)h^3$; the surface area is given by $S = (\sqrt{5}+1)\cdot\pi h^2/4$. If $h$ is doubled, $S$ is increased 4-fold.

**29.** $((10.4 \cdot (4.6)^{2/3}/100)/1.73)200 \approx 33$ mg.

**31.** $((10.4 \cdot (5.3)^{2/3}/100)/1.73)500 \approx 91$ mg.

**33.** $((10.4 \cdot (4.8)^{2/3}/100)/1.73)50 \approx 9$ mg.

**35.** The assumption is that strength on Krypton is proportional to weight; this gives $s \propto w$. This means that $s_2/s_1 = w_2/w_1 = 150 \cdot 16/(1/500) \approx 1.2 \cdot 10^6$, so a 150-lb man can lift a weight of $s_2 = 0.2 \cdot 1.2 \cdot 10^6 = 240000$ oz $= 15000$ lb (i.e. 100 times his weight).

**37.** The assumption is $A = ad^{2.99}$. We can find $a$ from the equation $78 = a(30)^{2.99}$; we obtain $a = 78/(30)^{2.99} \approx 0.00299$. Thus $A = 0.00299d^{2.99}$.

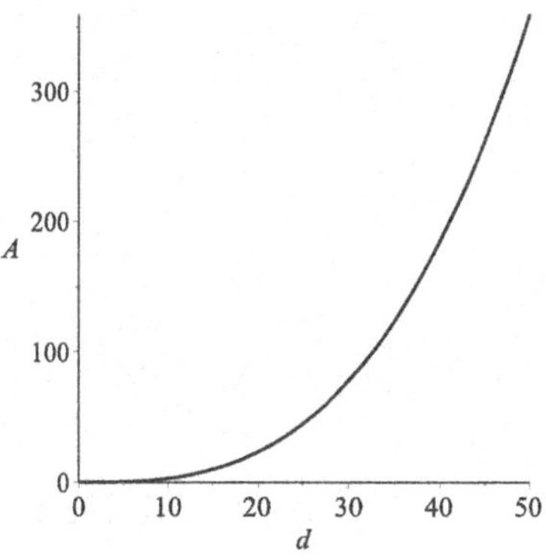


**39.** The assumption is $L = aW^{0.95}$ for pumpkins, $L = aW^{2.2}$ for snake gourds. We can find the $a$ values from the equations $10 = a(10)^{0.95}$ (pumpkins) and $10 = a(10)^{2.2}$ (snake gourds) ; we obtain $a = 10/(10)^{0.95} \approx 1.12$ (pumpkins) and $a = 10/(10)^{2.2} \approx 0.063$ (snake gourds). Thus $L = 1.12W^{0.95}$ (pumpkins) and $L = 0.063W^{2.2}$ (snake gourds). This also means $W = 0.888L^{1.05}$ and $W = 3.51L^{0.45}$.

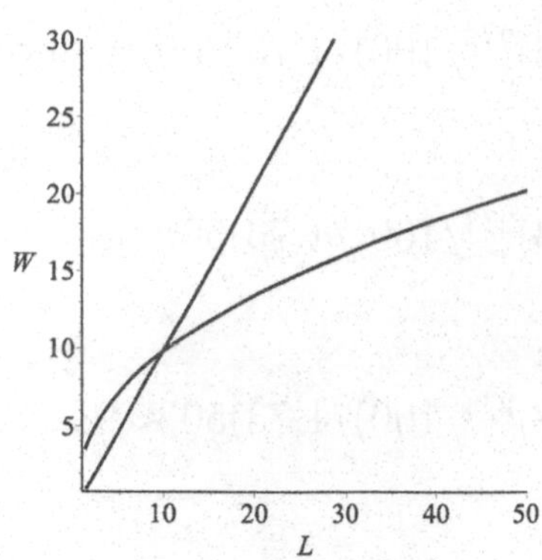


**41.** We have $1724 = F_2/F_1 = L_2^b/L_1^b = (L_2/L_1)^b = 12^b$, which means that $b \approx 3$. Weight is proportional to volume.

**43. a.** We obtain that $A \propto M^{2/3}$, which implies $M/A \propto M^{1/3}$.

**b.**

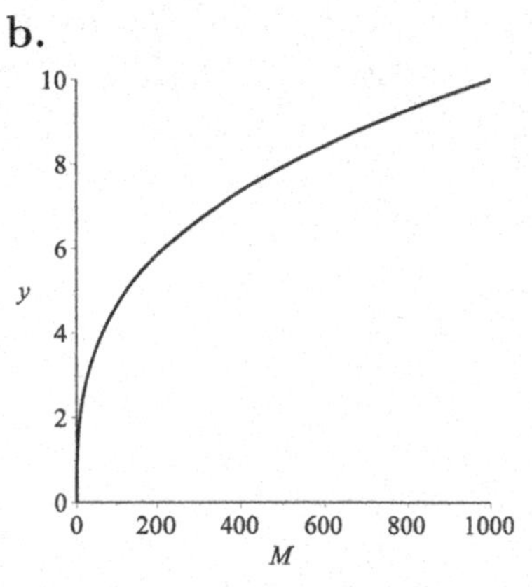


## Problem Set 1.4 - Exponential Growth

**1.**

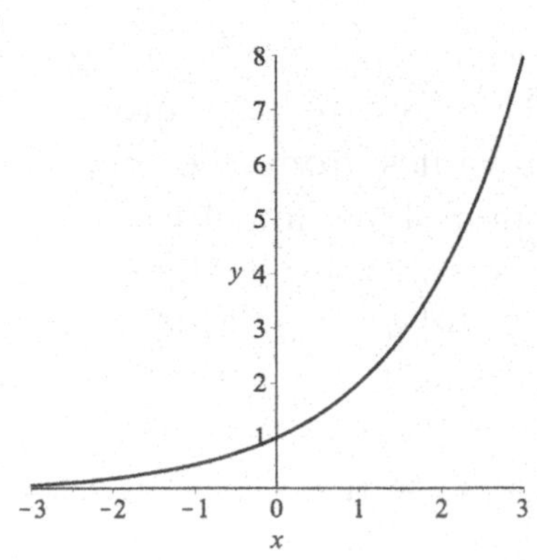


**3.**

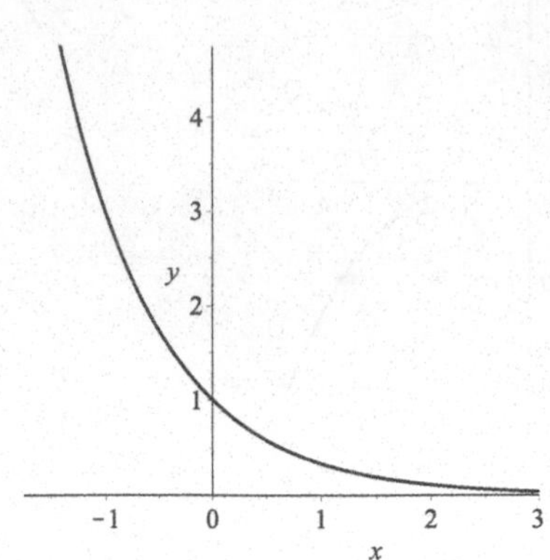


**5.**

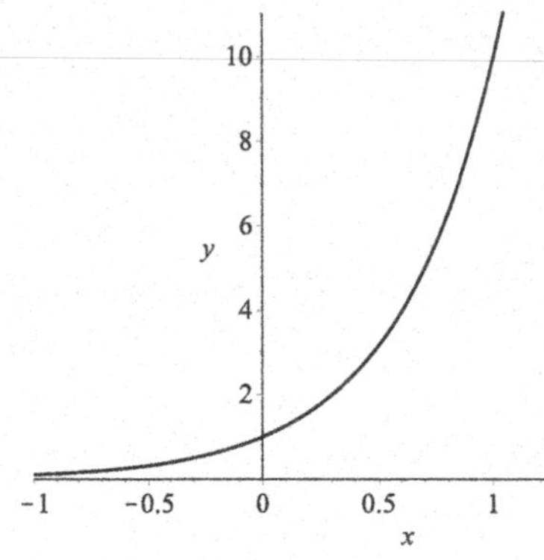


**7.** $2^x \geq 2x$ when $x \leq 1$ or $x \geq 2$.

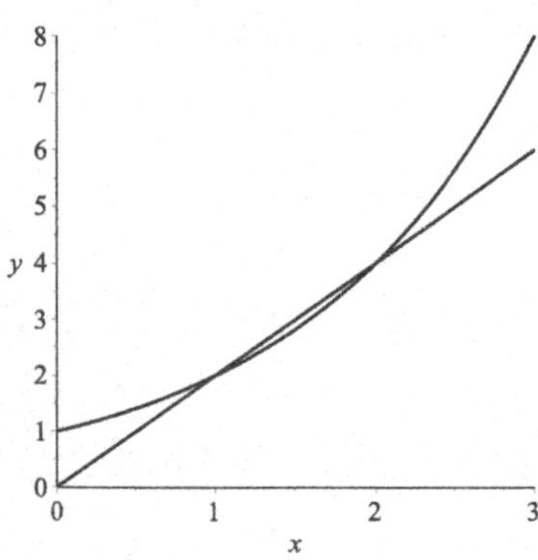


**9.** $\pi^x \geq x^4 - 4$ when $-1.43 \leq x \leq 1.89$ or $x \geq 6.58$.

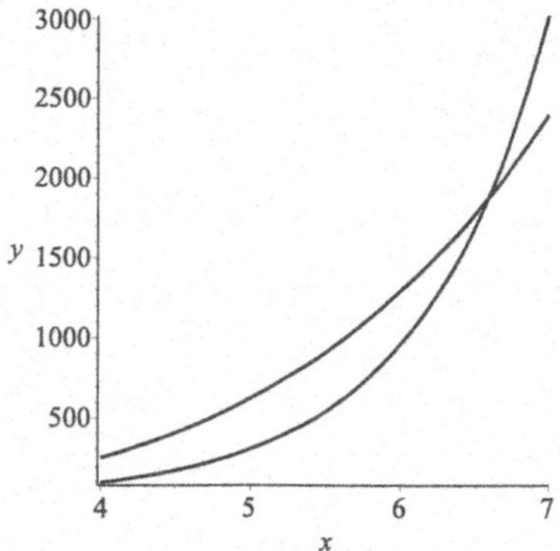


**11.** $x \approx 1.15$.

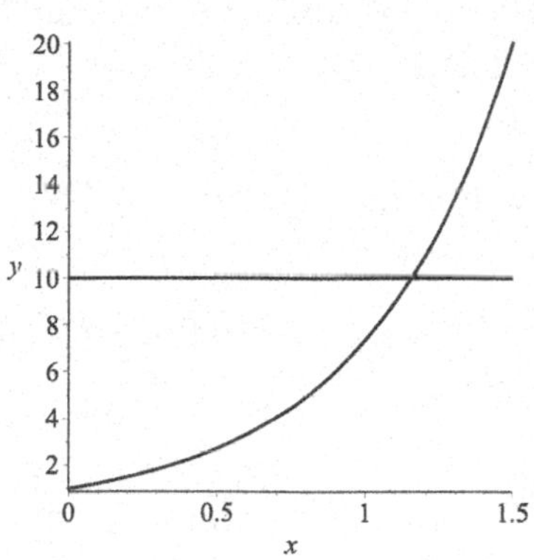


**13.** $x \approx 0.57$.

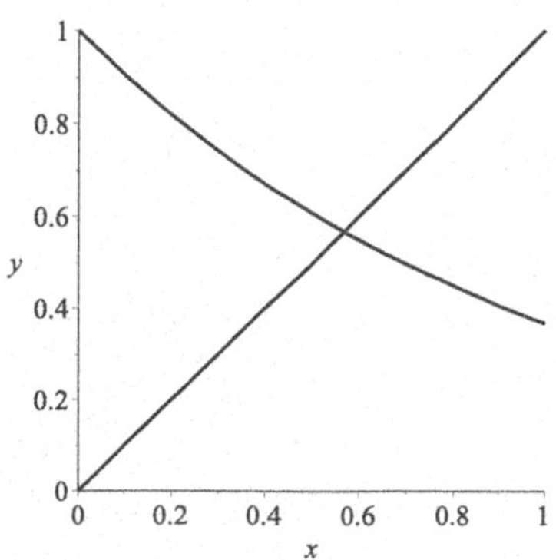


**15.** $x \approx 1.86$, $x \approx 4.54$.

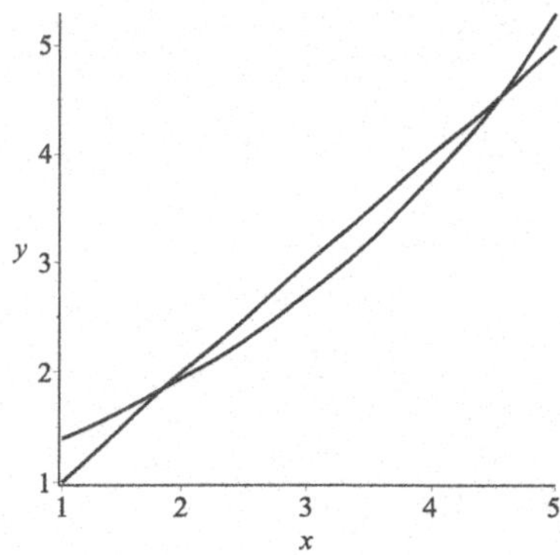


**17.** $x \approx -1.32$, $x \approx 0.54$.

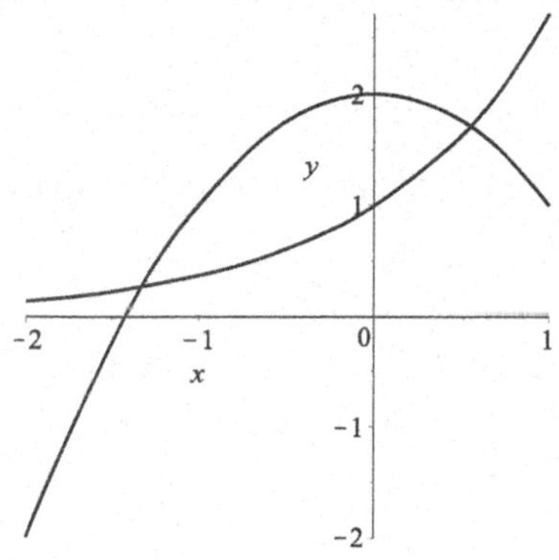


**19.** The future value is $1000 \cdot (1 + 0.07)^{25} \approx 5427.43$.

**21.** The future value is $1000 \cdot e^{0.01} \approx 1010.05$.

**23.** The future value is $1000 \cdot (1 + 0.16/12)^4 \approx 1054.41$.

**25.** $2 = ba^0 = b$, $5 = 2a^1$, so $f(x) = 2(5/2)^x$.

**27.** $3 = ba^{1/2}$, $1 = ba$, so (by dividing the two equations) $a^{1/2} = 1/3$, and then $a = 1/9$. The second equation gives $b = 9$. Thus $f(x) = 9(1/9)^x$.

**29.** $t \approx 168$ seconds.

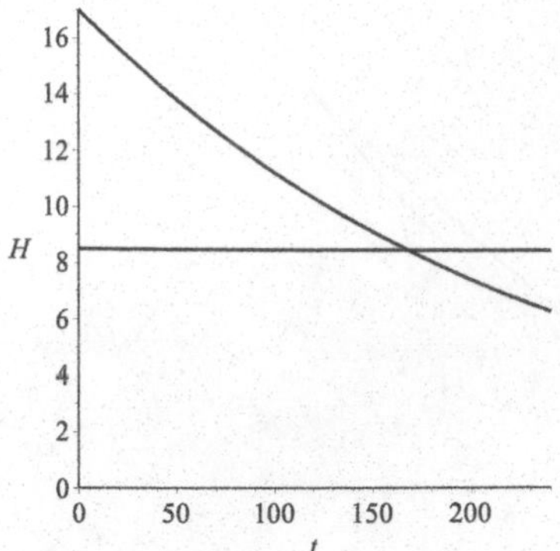


**31. a.** Compounded once a year, we obtain $100 \cdot (1 + 0.2) = 120$, twice a year: $100 \cdot (1 + 0.2/2)^2 = 121$, four times a year: $100 \cdot (1 + 0.2/4)^4 = 121.55$.

**b.** Compounded $n$ times a year, we obtain $100 \cdot (1 + 0.2/n)^n$.

**c.** For $n = 100$, 122.116, for $n = 1000$, 122.138, for $n = 10000$, 122.14. The values approach $100e^{1/5} \approx 122.14$.

**33. a.** Once a day: $N = (1 + 10)^{365}$, twice a day: $N = (1 + 10/2)^{730}$, four times a day: $N = (1 + 10/4)^{1460}$

**b.** $n$ times a day: $(1 + 10/n)^{365n}$.

**c.** The values approach $e^{3650}$.

**35. a.** At $t = 0$, Country #3 has the largest population size (20 million).

**b.** Country #1.

**c.** Country #3, losing 5% of its population every decade.

**37. a.** After 14 days, the amount of T4 in the body is $100(1/2)^{14/7} = 25$ mcg.

**b.** After $t$ days, the amount of T4 in the body is $100(1/2)^{t/7}$ mcg.

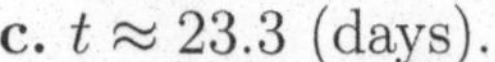
**c.** $t \approx 23.3$ (days).

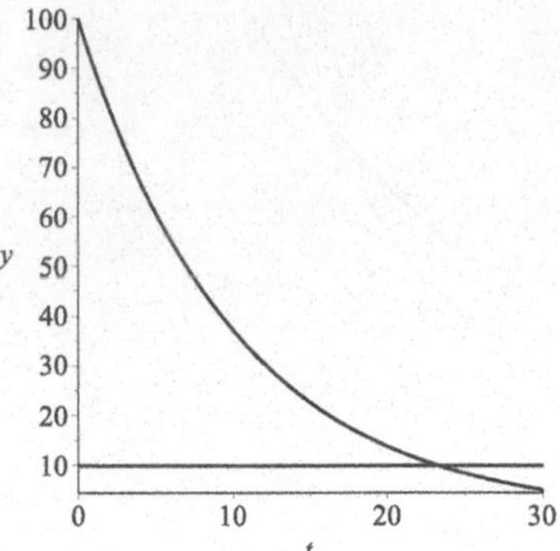


**39.** The remaining amount is $500 \cdot (1/2)^{t/5730}$.

**41. a.** Using the headline, the population size is given by $350(1.12)^t$.

**b.** The estimate is $350(1.12)^{29} \approx 9362$.

**43. a.** It is $1 - (1 - 0.1)^{10} \approx 0.65$, so about 65%.

**b.** It is $1 - (1 - 0.01)^{100} \approx 0.63$, so about 63%.

**c.** It is $1 - (1 - 1/N)^N$. For large values of $N$, these seem to approach $1 - 1/e \approx 0.632$.

**45. a.** The size is $0.5(1/2)^{t/5.7}$ (cm$^3$).

**b.** $t \approx 18.9$ (days).

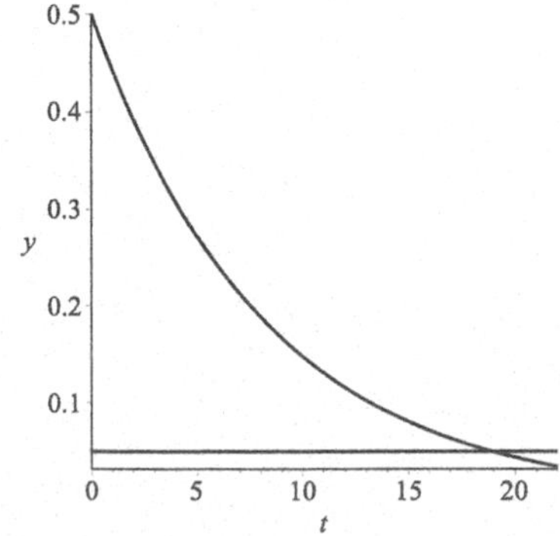

**47. a.**

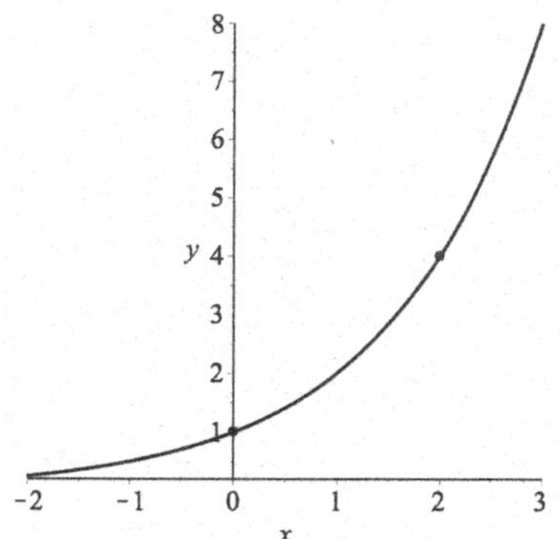


**b.**

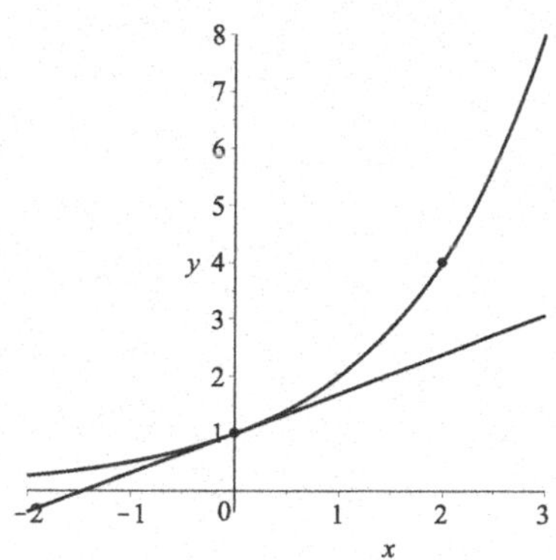


**c.** An estimate is 0.7.

**49. a-b-c.**

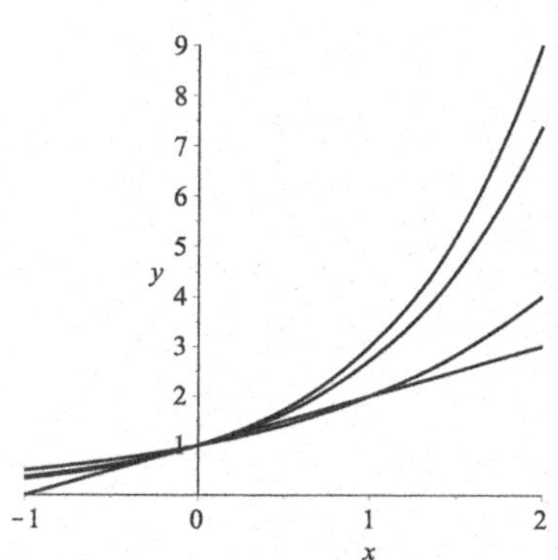


## Problem Set 1.5 - Function Building

**1.**

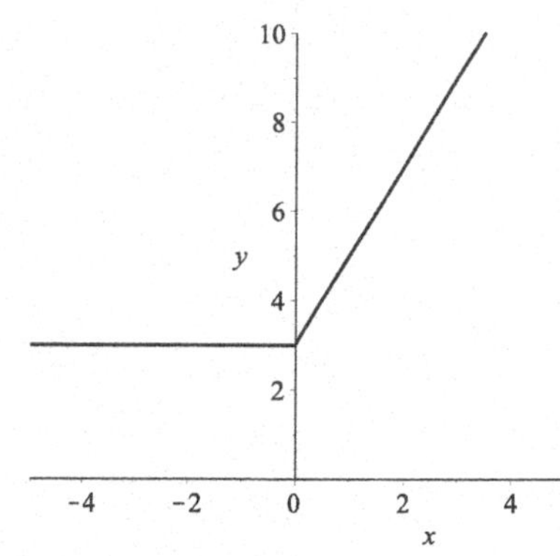


**3.**

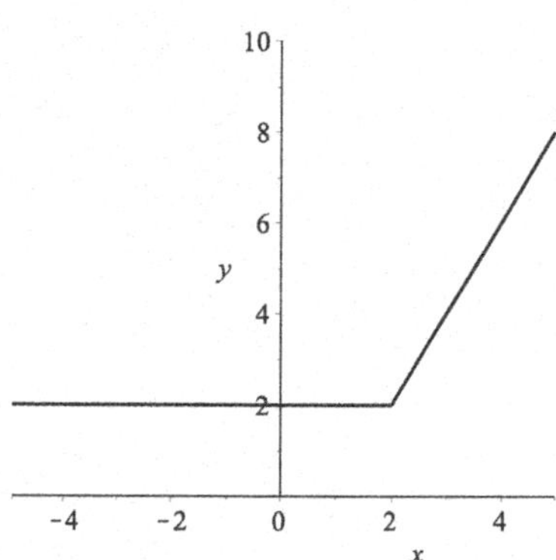


**5.**

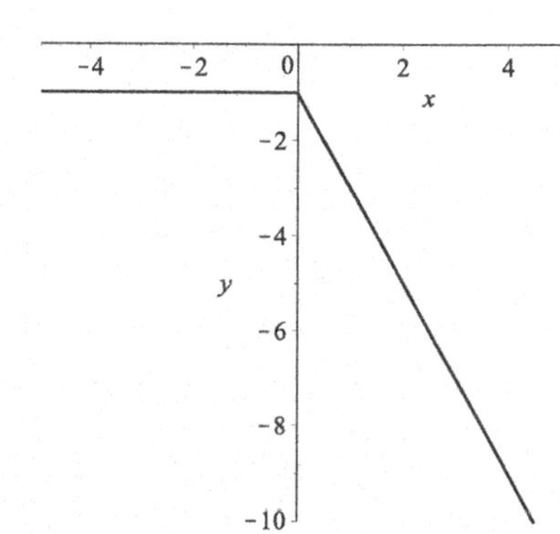

**7.**

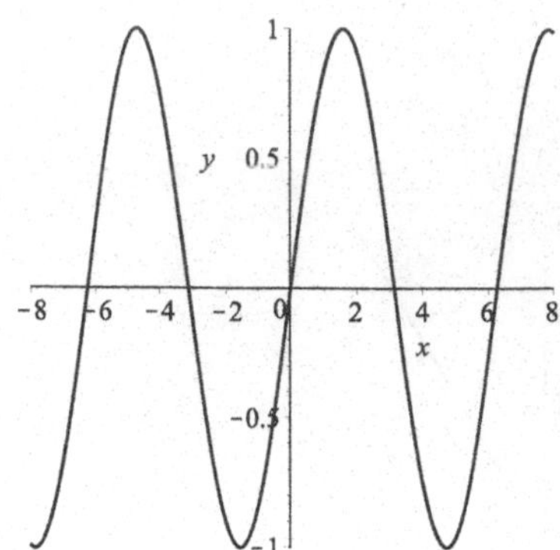


**9.**

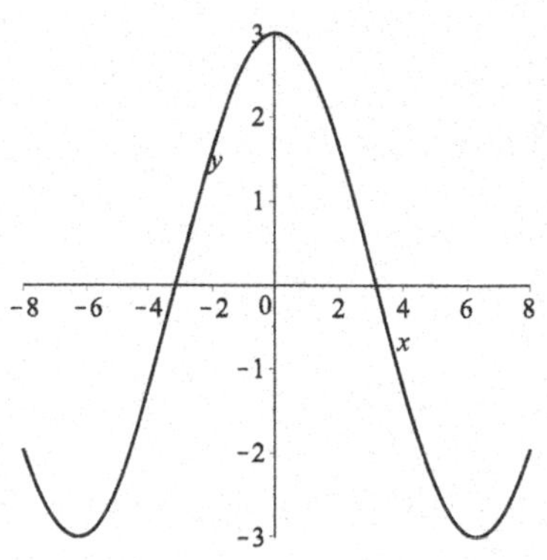


**11.**

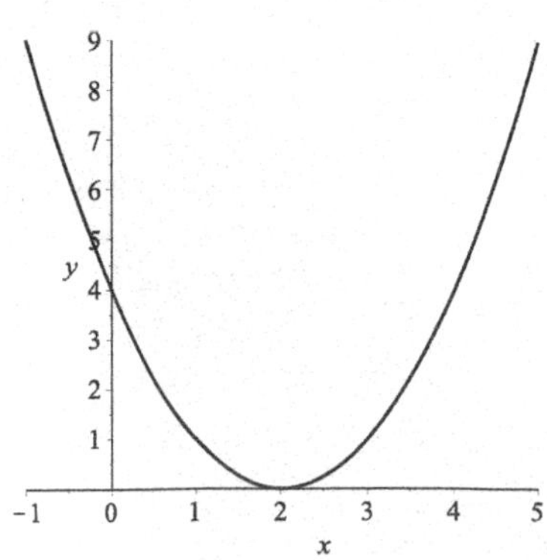


**13.**

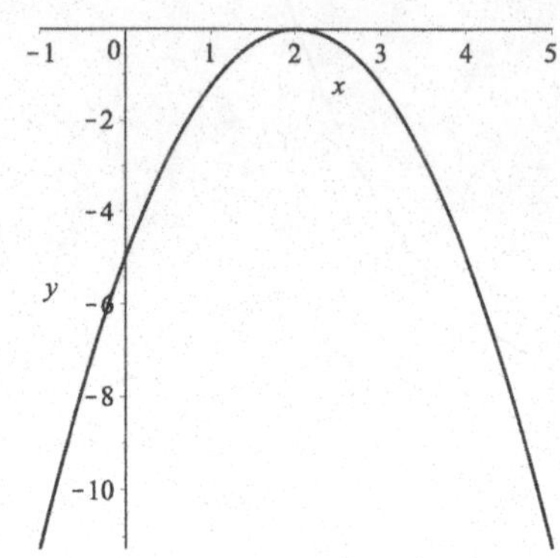


**15.**

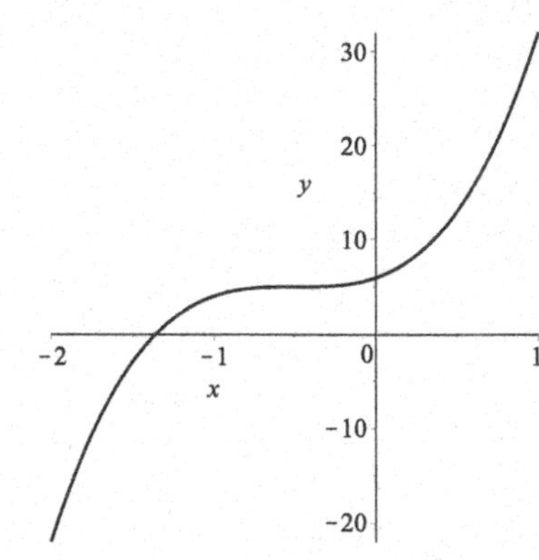


**17.**

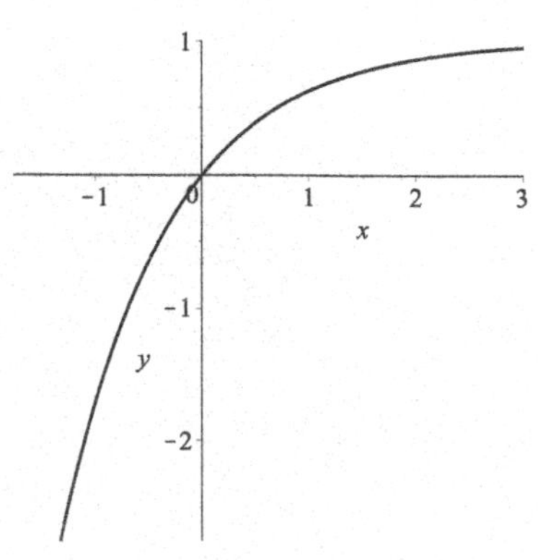

**19.**

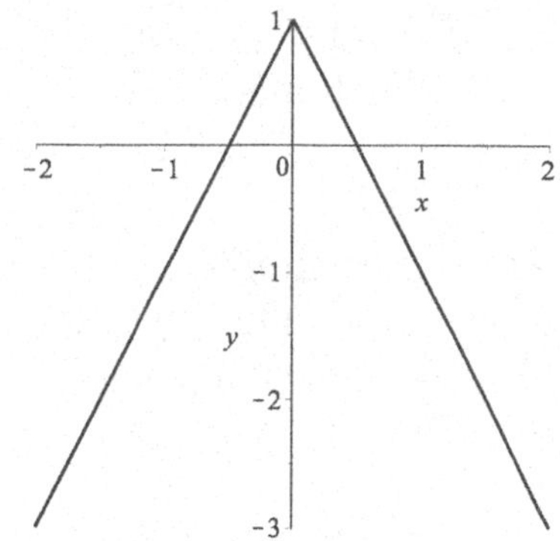


**21. a.** $(f+g)(-1) = f(-1)+g(-1) = -3+0 = -3$.

**b.** $f$ is not defined at 2.

**c.** $(fg)(9) = f(9)g(9) = 17 \cdot 70 = 1190$.

**d.** $(f/g)(99) = f(99)/g(99) = 197/9700$.

**e.** $(f \circ g)(0) = f(g(0)) = f(-2) = -5$.

**23. a.** $f(t+p) = f(t)$, so $f(t-1+p)+2 = f(t-1)+2$. The period is $p = T$, the amplitude is $a = A$.

**b.** The period is $p = T$ again with the same method, the amplitude is $4A$.

**c.** Using the periodicity of $f$ and some algebra, we get that $-2f(3t) = -2f(3t+T) = -2f(3(t+T/3))$, so the period is $T/3$; the amplitude is $a = 2A$.

**d.** The period is $p = T$ again with the same method, the amplitude is $a = 2A$.

**25.** One possibility is $f(x) = \sqrt{x}$ and $g(x) = 1 - \sin x$.

**27.** One possibility is $f(x) = e^x$ and $g(x) = 1 - x^2$.

**29.** One possibility is $f(x) = x^3 + \sqrt{x} + 5$ and $g(x) = x^2 - 1$.

**31.** After simplification, we obtain $f+g = \dfrac{2x^2-x-3}{x+1} + x^2 - x - 2 = \dfrac{(2x-3)(x+1)}{(x+1)} + x^2 - x - 2 = x^2 + x - 5$, domain is $x \neq -1$. $fg = \dfrac{(2x^2-x-3)(x^2-x-2)}{x+1} = (2x-3)(x^2-x-2)$, domain is $x \neq -1$. $f/g = \dfrac{2x^2-x-3}{(x+1)(x^2-x-2)} = \dfrac{2x-3}{x^2-x+2}$, domain is $x \neq -1$, $x \neq 2$.

**33.** $f+g = \sqrt{4-x^2} + \sin(\pi x)$, domain is $-2 \le x \le 2$, $fg = \sqrt{4-x^2}\sin(\pi x)$, domain is $-2 \le x \le 2$, $f/g = \sqrt{4-x^2}/\sin(\pi x)$, domain is $-2 < x < 2$, $x \neq -1$, $x \neq 0$, $x \neq 1$.

**35.** The data suggests that the period of $T$ is approximately 12 hours, which gives $B = 2\pi/12 = \pi/6$. The amplitude can be approximated by $A = (5.8 - (-0.4))/2 = 3.1$. The vertical shift is $D = (5.8 + (-0.4))/2 = 2.7$. The high tide occurs at around 11:00AM, so $C = -11$.

**37. a.** $y = bx/(1+ax)$ and $t = 1/x$, $z = 1/y$ gives us $1/z = b(1/t)/(1+a(1/t))$. Multiplication of both numerator and denominator by $t$ on the right side gives $1/z = b/(t+a)$; taking reciprocals of both sides gives $z = t/b + a/b$.

**b.** Technology finds that $z = 0.503 + 0.499t$; this means $b = 1/0.499 \approx 2$, and $a/b = 0.503$, so $a \approx 1$. The approximation is $y = 2x/(1+x)$.

**39. a.** When $t = 2$, $h(2) = 4$, and we obtain $V = \pi 4^3/12 = 16\pi/3$.

**b.** $V \circ h = V(h(t)) = \pi(2t)^3/12 = 2\pi t^3/3$.

**c.** We need that $0 \le 2t \le 6$, so $0 \le t \le 3$.

**41.** The value of $r$ is $4/365$; thus the composite per-capita growth rate function is $G(x) = (g \circ f)(x) = g(f(x)) = \frac{4}{365}(1 - \frac{40}{f(x)}) = \frac{4}{365}(1 - \frac{40(300^4 + x^4)}{110x^4}) = \frac{4}{365}\frac{70x^4 - 40 \cdot 300^4}{110x^4}$.

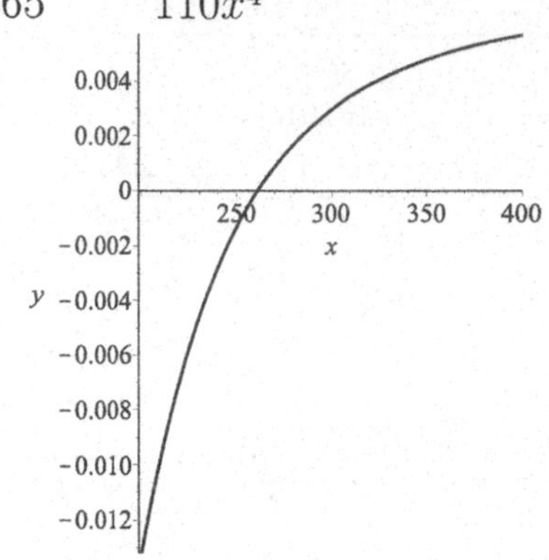


**43.** We have 13 hours of daylight when $n = 121$ and $n = 236$, i.e. on May 1 and August 24.

**45.** Let $x = 0$ correspond to the year 1844. The data suggests that the period of $y$ is approximately $(1997 - 1844)/14 \approx 11$ years, which gives $c = 2\pi/11$. The amplitude is about $b = (150 - 0)/2 = 75$. The vertical shift is $a = (150+0)/2 = 75$. The first maximum occurs at around $x = 11/2$, so $d = \pi$.

## Problem Set 1.6 - Inverse Functions and Logarithms

**1.** Not one-to-one, $f(2) = f(4)$.

**3.** One-to-one; $f^{-1}(0) = 11.9$, $f^{-1}(1) = 17$, $f^{-1}(4) = -2$, $f^{-1}(2) = 4$, and $f^{-1}(6) = 5$.

**5.** One-to-one.

**7.** Seems to be not one-to-one.

**9.** Set $y = x/(1 + x)$; then $y + yx = x$, so $y = x - yx = x(1 - y)$ and $x = y/(1 - y)$. Change the role of $x$ and $y$: we obtain $f^{-1}(x) = x/(1-x)$. The domain is $\mathbb{R}$, $x \neq 1$, the range is $\mathbb{R}$, $y \neq -1$.

**11.** Set $y = (x+1)^3-2$, then $y+2 = (x+1)^3$ and $\sqrt[3]{y+2} = x + 1$. Thus $f^{-1}(x) = \sqrt[3]{x+2} - 1$. The domain is $\mathbb{R}$, the range is $\mathbb{R}$.

**13.** Set $y = e^{x^2}$. Then $\ln y = x^2$, and $-\sqrt{\ln y} = x$ because $x \leq 0$ by assumption. Thus $f^{-1}(x) = -\sqrt{\ln x}$. The domain is $x \geq 1$, the range is $y \leq 0$.

**15. a.** $\log 10 = 1$, because $10^1 = 10$.

**b.** $\log 0.001 = -3$, because $10^{-3} = 0.001$.

**17. a.** $\log_5 125 = 3$, because $5^3 = 125$.

**b.** $\log_8 64 = 2$, because $8^2 = 64$.

**19. a.** $x = e^3$.

**b.** $x = 10^{4.5}$.

**21. a.** $\log_2 8^x = \log_2 2^{3x} = 3x$.

**b.** $\log_3 81^x = \log_3 3^{4x} = 4x$.

**c.** $\log_4 64^x = \log_4 4^{3x} = 3x$.

**d.** $\log_{1/2} 32^x = \log_{1/2}(1/2)^{-5x} = -5x$.

**e.** $\log_3 9^{-x} = \log_3 3^{-2x} = -2x$.

**23. a.** $5^x = e^{\ln 5^x} = e^{x \ln 5}$.

**b.** $(1/2)^x = e^{\ln(1/2)^x} = e^{x \ln(1/2)} = e^{-x \ln 2}$.

**c.** $5^{1/x} = e^{\ln 5^{1/x}} = e^{(\ln 5)/x}$.

**d.** $4^{x^2} = e^{\ln 4^{x^2}} = e^{x^2 \ln 4}$.

**e.** $3^{x^e} = e^{\ln 3^{x^e}} = e^{x^e \ln 3}$.

**25. a.** $\log(x+1) = \ln(x+1)/\ln 10$.

**b.** $\log(ex+e) = \ln(ex+e)/\ln 10 = (1+\ln(x+1))/\ln 10$.

**c.** $\log_2(x^2-2) = \ln(x^2-2)/\ln 2$.

**d.** $\log_7(2x-3) = \ln(2x-3)/\ln 7$.

**27.** $\ln e + \ln 1 + \ln e^{542} = 1 + 0 + 542 = 543$.

**29.** $10^{\log 0.5} = 0.5$.

**31.**

$10^{-2}$ $10^{-1}$ $10^{0}$ $10^{1}$ $10^{2}$ $10^{3}$ $10^{4}$

**33.**

$10^{-3}$ $10^{0}$ $10^{1}$ $10^{2}$ $10^{3}$ $10^{4}$ $10^{5}$

**35. a.** We have to solve the equation $1 = 1.2078C/(1+0.0506C)$. The result is $C = 0.864155$ (mg/l).

**b.** We solve $U = 1.2078C/(1+0.0506C)$ for $C$; the resulting function is $C = U/(1.2078 - 0.0506U)$ (mg/l).

**37.** The doubling time can be found by solving $700 = 350(1.12)^T$. This is the same as $2 = 1.12^T$; take the natural logarithm of both sides and divide to obtain $T = \ln 2/\ln 1.12 \approx 6.12$ years.

**39.** We solve the equation $0.5 = 0.1(2)^{t/2.9}$. Division by 0.1 gives $5 = 2^{t/2.9}$. Take the base two logarithm of both sides, then multiply by 2.9 to obtain $t = 2.9\log_2 5 \approx 6.73$ days.

**41.** The first equation gives $\ln 28 = \ln c + m\ln 0.4$, the second $\ln 100 = \ln c + m\ln 0.6$. Subtract the first equation from the second: $\ln 100 - \ln 28 = m(\ln 0.6 - \ln 0.4)$, so $m \approx 3.14$; then $c = 28/(0.4)^{3.14} \approx 497.4$. We obtain $W = 497L^{3.14}$.

**43. a.**

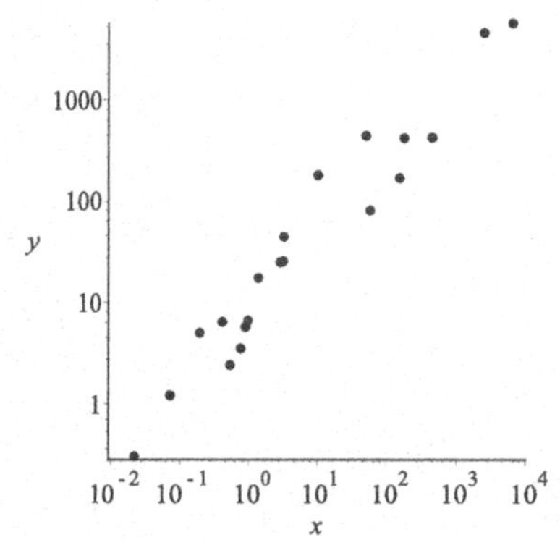


**b.** Technology gives that the best fitting line is $\ln y = 0.9366 + 0.7488\ln x$.

**45. a.**

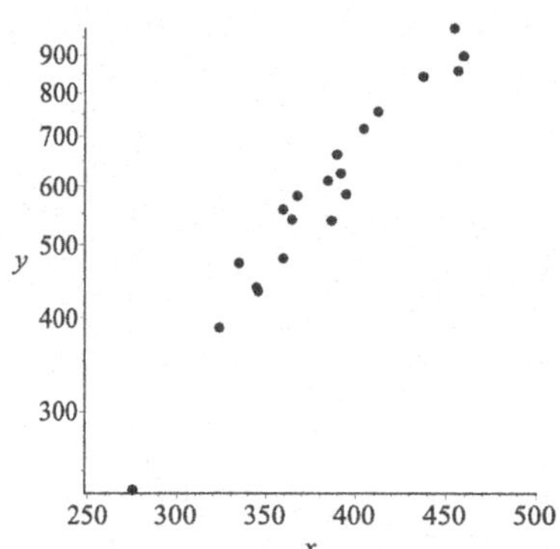


**b.** Technology gives that the best fitting line is $\ln y = 1.6314 + 0.00296x$.

**47.** The figure shows the plot of $(t, \ln x)$. The best fitting line is $-9.13 + 0.005t$. Because $x = ce^{rt}$, we obtain that $\ln x = \ln c + rt$, and then $r = 0.005$.

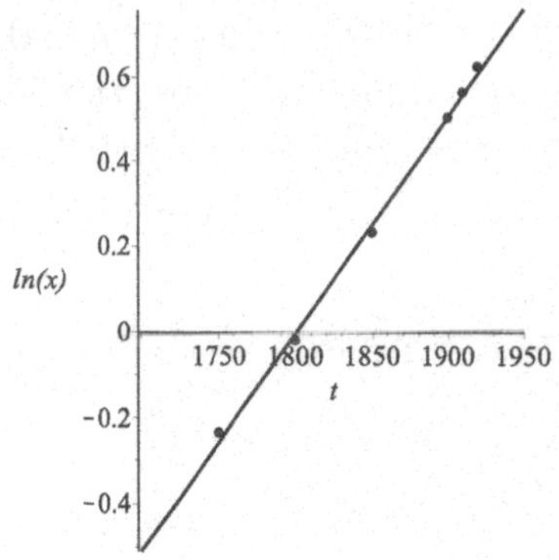


## Problem Set 1.7 - Sequences and Difference Equations

**1.** $a_1 = 1 - 1/1 = 0$, $a_2 = 1 - 1/2 = 1/2$, $a_3 = 1 - 1/3 = 2/3$, $a_4 = 1 - 1/4 = 3/4$, $a_5 = 1 - 1/5 = 4/5$.

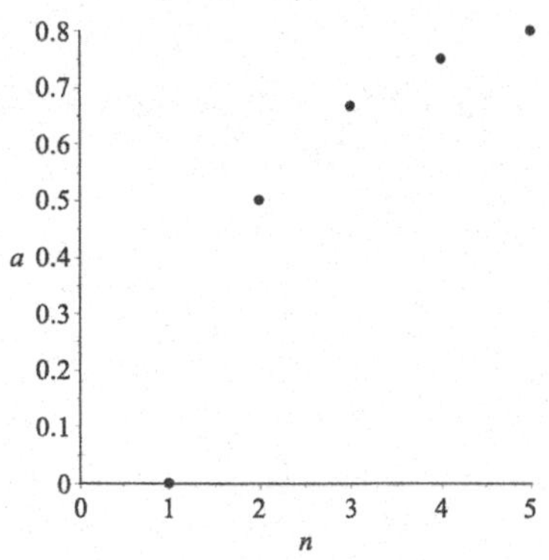

**3.** $a_1 = \cos(\pi/2) = 0$, $a_2 = \cos(\pi) = -1$, $a_3 = \cos(3\pi/2) = 0$, $a_4 = \cos(2\pi) = 1$, $a_5 = \cos(5\pi/2) = 0$.

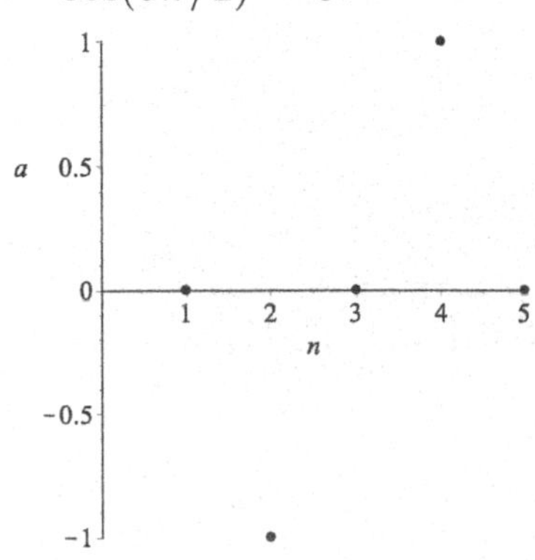

**5.** $a_1 = 1$, $a_2 = 4$, $a_3 = 2$, $a_4 = 8$, $a_5 = 5$.

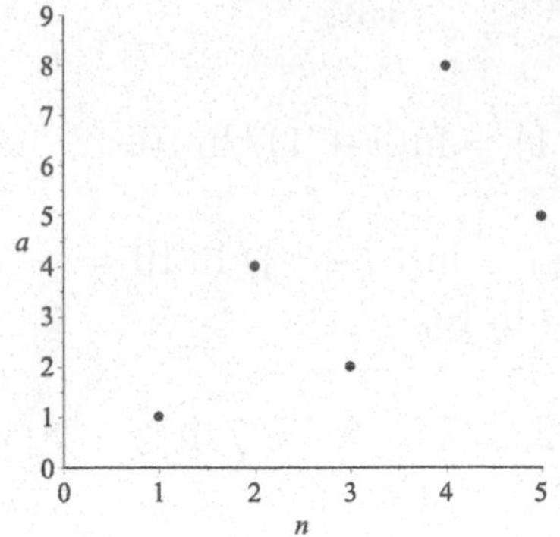

**7.** $a_1 = 256$, $a_2 = \sqrt{256} = 16$, $a_3 = \sqrt{16} = 4$, $a_4 = \sqrt{4} = 2$, $a_5 = \sqrt{2}$.

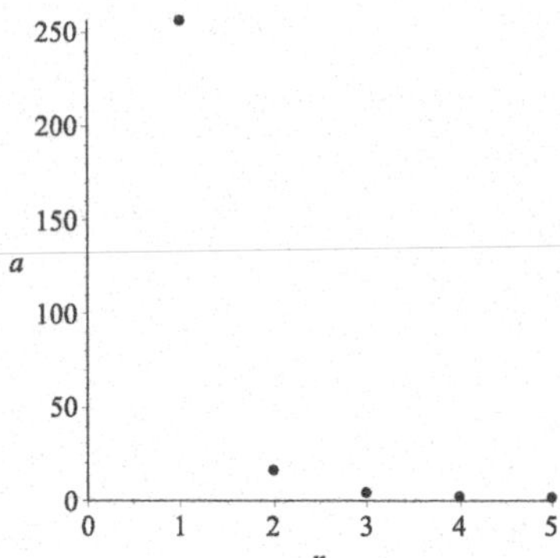

**9.** $a_1 = -4$, $a_2 = 6$, $a_3 = 6 - 4 = 2$, $a_4 = 2 + 6 = 8$, $a_5 = 8 + 2 = 10$.

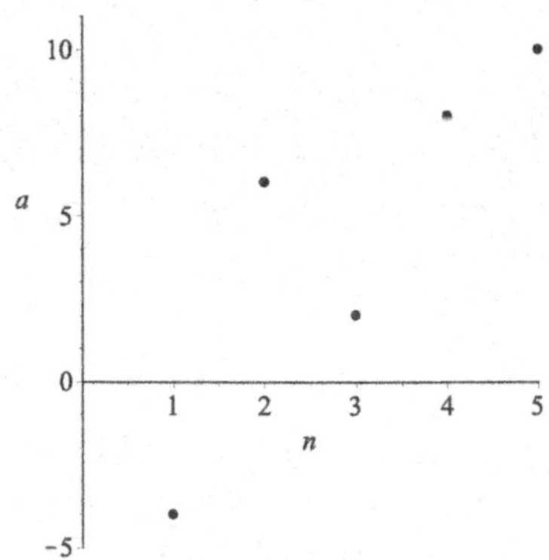

**11.** $a_1 = 0$, $a_2 = 0 + 8 = 8$, $a_3 = 8 + 8 = 16$, $a_4 = 16 + 8 = 24$, $a_5 = 24 + 8 = 32$.

**13.** $a_1 = 100$, $a_2 = 100/2 + 2 = 52$, $a_3 = 52/2 + 2 = 28$, $a_4 = 28/2 + 2 = 16$, $a_5 = 16/2 + 2 = 10$.

**15.** $a_1 = 0$, $a_2 = 5 \cdot 0 + 2 = 2$, $a_3 = 5 \cdot 2 + 2 =$

12, $a_4 = 5\cdot 12+2 = 62$, $a_5 = 5\cdot 62+2 = 312$.

**17.** $a_1 = 8$, $a_2 = 2\cdot 8+1 = 17$, $a_3 = 2\cdot 17+1 = 35$, $a_4 = 2\cdot 35+1 = 71$, $a_5 = 2\cdot 71+1 = 143$.

**19.** $a_1 = 1$, $a_2 = 2/2 = 1$, $a_3 = 2/2 = 1$, $a_4 = 2/2 = 1$, $a_5 = 2/2 = 1$.

**21.** The equilibria can be found by solving $x = 2x(1-x)$; subtracting $x$ from both sides and factoring gives $0 = x(1-2x)$. This product is zero when $x = 0$ or $x = 1/2$.

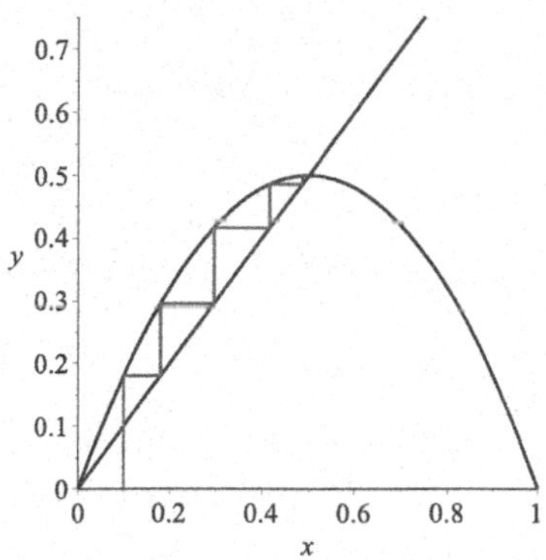


**23.** The equilibria can be found by solving $x = 3x/(1+x)$; subtracting $x$ from both sides and factoring gives $0 = x(3/(1+x)-1)$. This product is zero when $x = 0$ or $x = 2$.

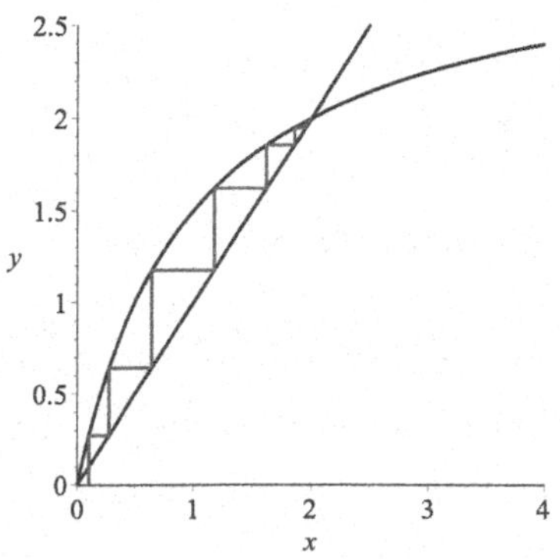


**25.** The equilibria can be found by solving $x = 1 + x/2$; we get $x = 2$.

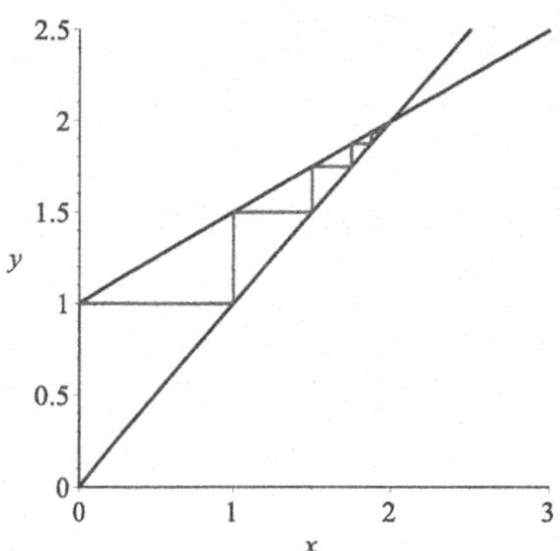


**27.** Equilibria are 0 and 5.

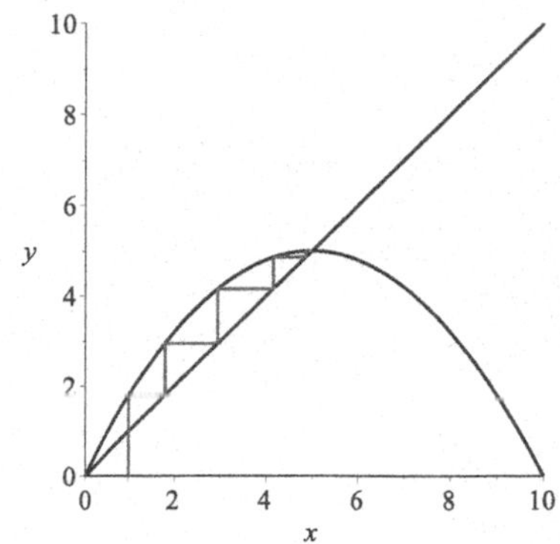


**29.** Equilibria are 0, $\approx 1$, and $\approx -1$.

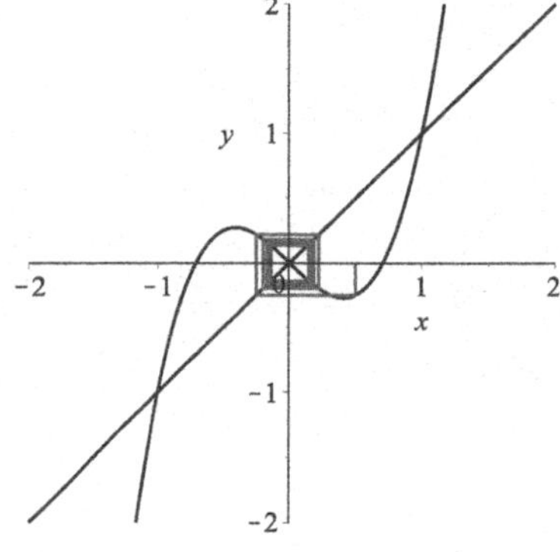


**31.** Let the original amount be $A$. At the end of the first hour, the amount is $A/2$. At the end of the second hour, the amount is $A/2/2 = A/4$. At the end of the third hour, the amount is $A/8$. At the end of 4 hours, we get $A/16$. This is 6.25%. At the end of $n$ hours, the drug present is $A/2^n$.

**33. a.** First, $a_1 = 0$. Then $a_2 = (1-c)A + (1-c)a_1 = (1-c)(A + a_1)$. Continuing,

$a_3 = (1-c)A + (1-c)a_2 = (1-c)(A+a_2)$. We obtain that $a_n = (1-c)(A+a_{n-1})$.

**b.** The equilibrium is the solution of $x = (1-c)(A+x)$, which is $(1-c)A/c$.

**c.** The equilibrium value is bigger than $A$ when $(1-c)/c > 1$; i.e. when $1-c > c$, which is $c < 1/2$.

**35. a.** We need that $x_{18} = a^{17}x_1$, i.e. $1249 = a^{17}263$, which gives $a = \sqrt[17]{1249/263} \approx 1.096$.

**b.** The model gives a very good fit.

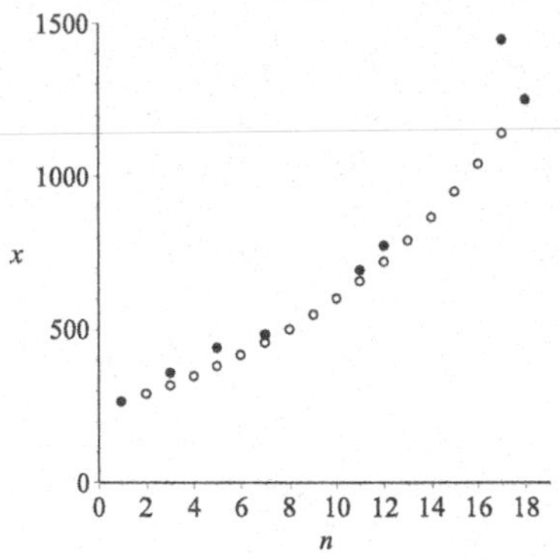

**c.** This means the difference equation is $x_{n+1} = ax_n - 150$. We choose $a_1 = 1249$ and iterate. We obtain the values 1219, 1186, 1150, 1110, 1067, 1019, 967, 910, 847, 779, 703, 621, 530, 431, 322, 203, 73, and then we get negative values. The model predicts extinction after 17 years.

**37. a.** $a_1 = 1$, $a_2 = 1 + 1/1 = 2$, $a_3 = 1 + 1/2 = 3/2$, $a_4 = 1 + 1/(3/2) = 5/3$, $a_5 = 1 + 1/(5/3) = 8/5$.

**b.** The equilibria are the solutions of $x = 1 + 1/x$; multiplication by $x$ gives the quadratic $x^2 - x - 1 = 0$ with solutions $(1 \pm \sqrt{5})/2$.

**c.** The iterates converge toward the positive equilibrium with the initial condition $a_1 = 1$.

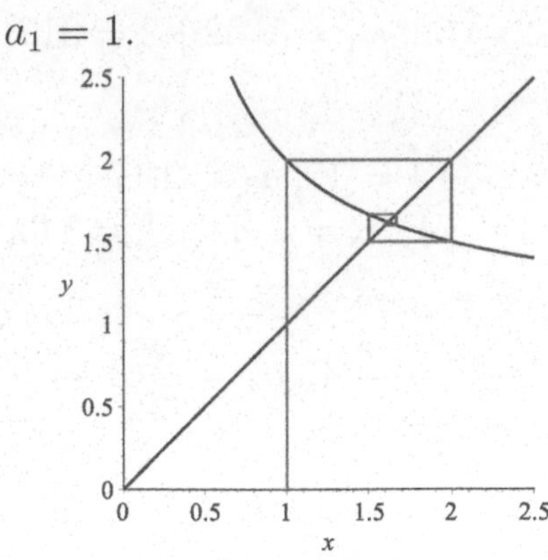

**39. a.** $x_2 = x_1/(1+x_1)$, $x_3 = x_2/(1+x_2) = x_1/((1+x_1)(1+x_1/(1+x_1)) = x_1/(1+2x_1)$, $x_4 = x_1/(1+3x_1)$, $x_5 = 1/(1+4x_1)$.

**b.** We guess $x_n = x_1/(1+(n-1)x_1)$.

**c.** We check that $x_{n+1} = x_n/(1+x_n) = (x_1/(1+(n-1)x_1))/(1+x_1/(1+(n-1)x_1)) = x_1/(1+nx_1)$.

**41. a.** With the assumption given, $x_{n+1}$ = number of $a$ alleles/total number of alleles $= (x_n(1-x_n)N/3)/(2x_n(1-x_n)N/3 + (1-x_n)^2N) = (x_n/3)/(2x_n/3 + (1-x_n)) = x_n/(3-x_n)$.

**b.** $x_1 = 1/2$, $x_2 = 1/5$, $x_3 = 1/14$, $x_4 = 1/41$, $x_5 = 1/122$, $x_6 = 1/365$, $x_7 = 1/1094$, $x_8 = 1/3281$, $x_9 = 1/9842$, $x_{10} = 1/29525$.

**c.** Equilibria are given by $x = x/(3-x)$, i.e. $x = 0$ and $x = 2$.

**d.** $a$ disappears faster in this case.

**43.** As expected, $a$ disappears more rapidly when it kills a bigger proportion of $Aa$ types.

## Review Questions

**1. a.** We need that $\log x + 1 \geq 0$, which

means $\log x \geq -1$, so $x \geq 1/10$ is the domain. The range is $[0, \infty)$.

**b.** For the inverse, we have to solve $y = \sqrt{\log x + 1}$ for $x$. We get $x = 10^{y^2-1}$. The domain is $[0, \infty)$, the range is $[1/10, \infty)$.

**3.** The equilibria are at $x = -1.4$ and $x = 1.4$.

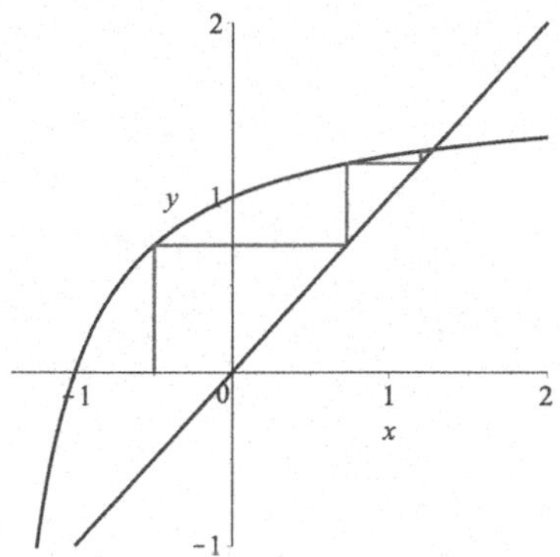


**5.** For $y = x$, the residuals are $e_1 = 0.7$, $e_2 = 0.1$, $e_3 = 0.2$, so the sum-of-squares is $0.49 + 0.01 + 0.04 = 0.54$; for $y = x/2 + 1$, the residuals are $e_1 = 0.3$, $e_2 = 0.4$, $e_3 = 0.2$, so the sum-of-squares is $0.09 + 0.16 + 0.04 = 0.29$. The second line is a better fit.

**7.** The best fitting line for the pairs $(\ln x, \ln y)$ is $2.35 - 0.373x$, thus the $m$ value is $-0.373$. If we need an integer, $m = -1$ is the best estimate.

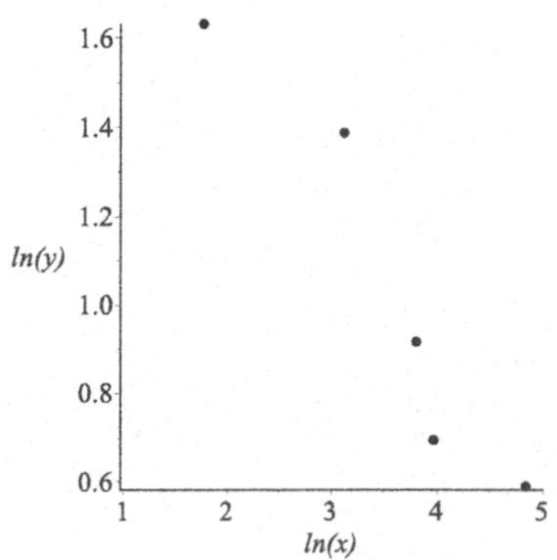


**9.** $y = \log a + x \log b$, thus $a = 100$ and $b = 10^{-1} = 0.1$.

**11. a.** With the assumption given, $x_{n+1} =$ number of $a$ alleles/total number of alleles $= (9x_n(1 - x_n)N/10)/(9x_n(1 - x_n)N/5 + (1 - x_n)^2N) = (9x_n/10)/(9x_n/5 + (1 - x_n))$ $= 9x_n/(10 + 8x_n)$.

**b.** $x_1 = 0.9$, $x_2 = 0.47$, $x_3 = 0.31$, $x_4 = 0.22$.

**13. a.** $f(x/2)$ is B - consider the $x$-axis.

**b.** $2f(x)$ is D - consider the $y$-axis.

**c.** $f(-x)$ is C - the graph is reflected on the $y$-axis.

**d.** $-f(x)$ is A - the graph is reflected on the $x$-axis.

**15. a.** $a_1 = 2-(1/2) = 3/2$, $a_2 = 2-(2/3) = 4/3$, $a_3 = 2 - (3/4) = 5/4$, $a_4 = 2 - (4/5) = 6/5$, $a_5 = 2 - (5/6) = 7/6$.

**b.** $a_1 = (1/2)^0 = 1$, $a_2 = (1/2)^1 = 1/2$, $a_3 = (1/2)^2 = 1/4$, $a_4 = (1/2)^3 = 1/8$, $a_5 = (1/2)^4 = 1/16$.

**c.** $a_1 = 2$, $a_2 = 3$, $a_3 = 5$, $a_4 = 7$, $a_5 = 11$.

**17.** $n(d)$ is not a linear function. $w(n)$ is a linear function.

**a.** We graph $w \circ n = w(n(d)) = 70 - 50d/(6 + d)$. The mathematical domain is $d \neq -6$; the physical meaning implies that $d > 0$.

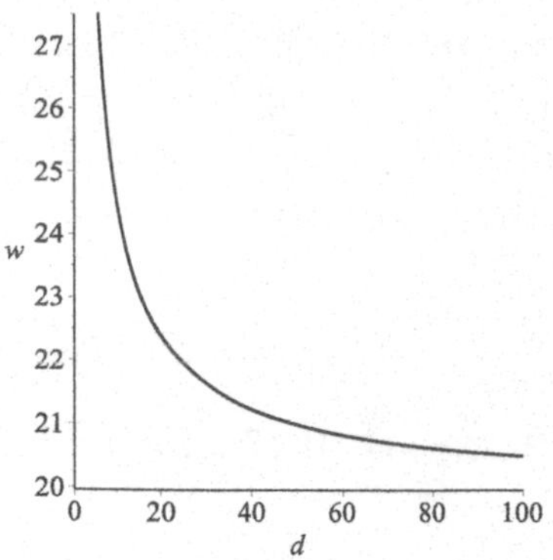


**b.** The weight decreases, toward the minimum weight of 20 (grams).

**19.** The data implies that one female produces $150 \cdot (1/3) \cdot 0.5$ female larvae. Thus after one cycle, we have 25 female larvae, after two, we have $25^2$, and so on. So the total number of descendants in one year is $25 + 25^2 + 25^3 + 25^4 + 25^5$, and these eat a total of 203.5 kg of wool.

## Problem Set 2.1 - Rates of Change and Tangent Lines

**1.** By definition, the average rate of change is $\dfrac{f(2) - f(-3)}{2 - (-3)} = \dfrac{-2 - 13}{5} = -3.$

**3.** By definition, the average rate of change is $\dfrac{f(3) - f(1)}{3 - 1} = \dfrac{27 - 3}{2} = 12.$

**5.** By definition, the average rate of change is $\dfrac{f(9) - f(4)}{9 - 4} = \dfrac{3 - 2}{5} = \dfrac{1}{5}.$

**7.** By definition, the instantaneous rate of change at this point is

$$\lim_{b \to -3} \frac{f(b) - f(-3)}{b - (-3)} = \lim_{b \to -3} \frac{4 - 3b - 13}{b - (-3)} = \lim_{b \to -3} \frac{-3(b + 3)}{b + 3} = -3.$$

**9.** By definition, the instantaneous rate of change at this point is

$$\lim_{b \to 1} \frac{f(b) - f(1)}{b - 1} = \lim_{b \to 1} \frac{3b^2 - 3}{b - 1} = \lim_{b \to 1} \frac{3(b - 1)(b + 1)}{b - 1} = \lim_{b \to 1} 3(b + 1) = 6.$$

**11.** By definition, the instantaneous rate of change at this point is

$$\lim_{b \to 4} \frac{f(b) - f(4)}{b - 4} = \lim_{b \to 4} \frac{\sqrt{b} - 2}{b - 4} = \lim_{b \to 4} \frac{\sqrt{b} - 2}{(\sqrt{b} - 2)(\sqrt{b} + 2)} = \lim_{b \to 4} \frac{1}{\sqrt{b} + 2} = \frac{1}{4}.$$

**13.** Similarly to Example 4, the instantaneous velocity at $t = 1$ is

$$\lim_{h \to 0} \frac{64 - 16(1 + h)^2 - (64 - 16 \cdot 1^2)}{1 + h - 1} = \lim_{h \to 0} \frac{-32h - 16h^2}{h} = \lim_{h \to 0} -32 - 16h = -32 \text{ (ft/s)}.$$

**15.** Similarly to Example 4, the instantaneous velocity at $t = 1/2$ is

$$\lim_{h\to 0} \frac{4 - 16(1/2 + h)^2 - (4 - 16\cdot(1/2)^2)}{1/2 + h - 1/2} = \lim_{h\to 0} \frac{-16h - 16h^2}{h} = \lim_{h\to 0} -16 - 16h = -16\,(\text{ft/s}).$$

Observe that this is the time instant when the falling cup hits the ground, so if we want to be precise, we have to choose $h$ negative, as the the function $f(t)$ can not describe the height of the cup when $t > 1/2$.

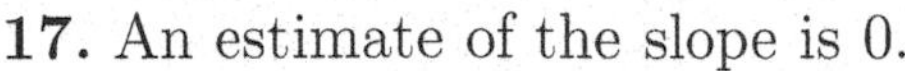

**17.** An estimate of the slope is 0.

**19.** An estimate of the slope is 2.

**21.** An estimate of the slope is 0.

**23.** An estimate of the slope is $-3$.

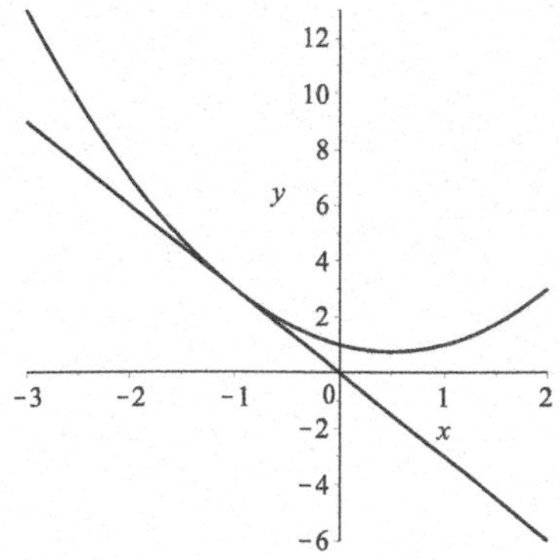


**25.** An estimate of the slope is 1.

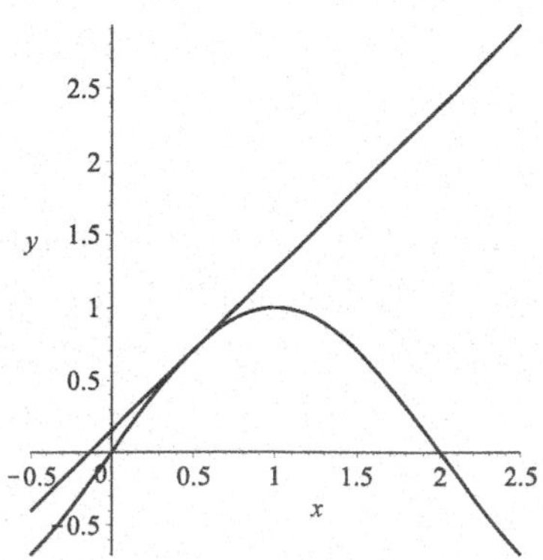


**27.** An estimate of the slope is 3.

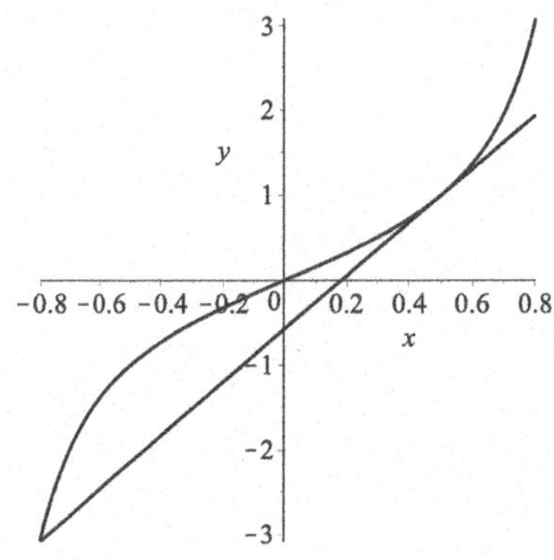


**29.** An estimate of the slope is $-0.03$.

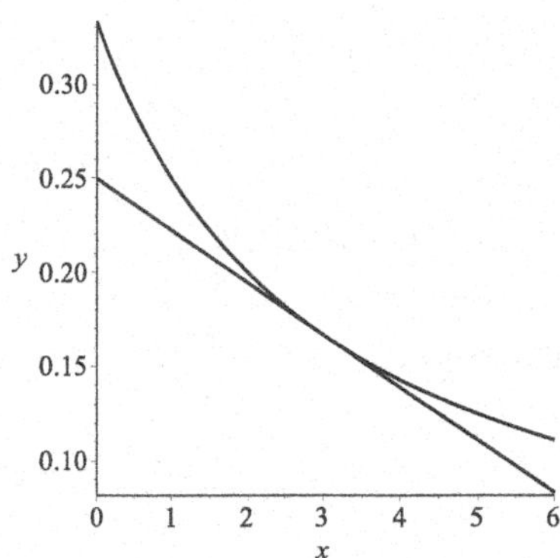


**31.** An estimate of the slope is 0.1.

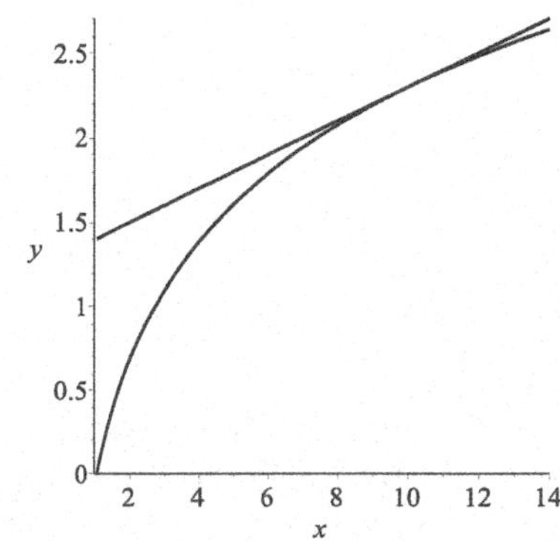


**33.** An estimate of the slope is 250.

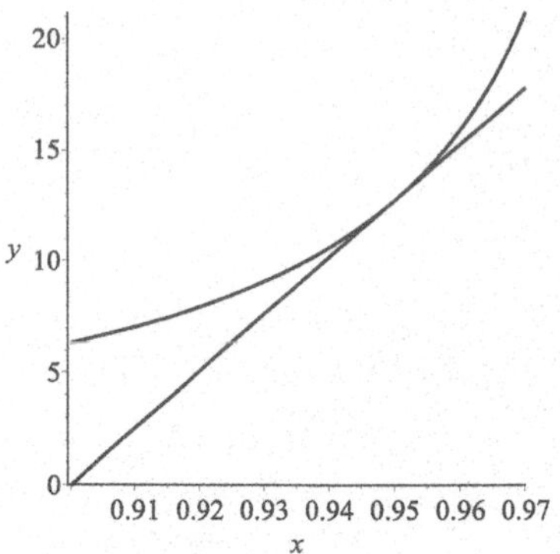


**35.** An estimate of the slope is $-10$.

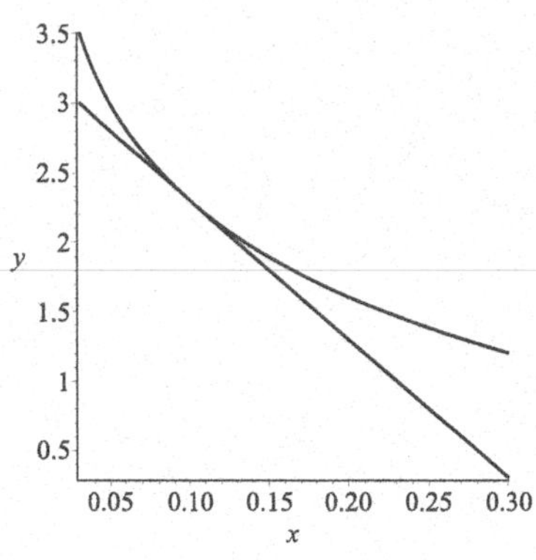


**37.** The slope of a secant line is given by $(3(3+h)-7-(3(3)-7))/h = 3$; thus this is the slope of the tangent line. The equation of the tangent line is $y = 3x - 7$.

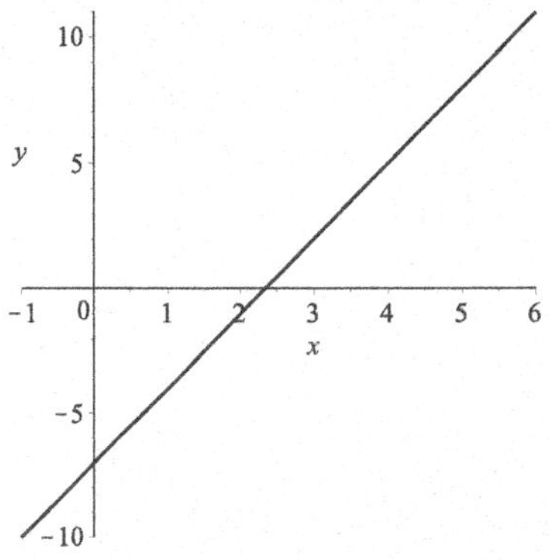


**39.** The slope of a secant line is given by $(3(-2+h)^2 - 3(-2)^2)/h = -12 + 3h$; thus the slope of the tangent line is $-12$. The equation of the tangent line is $y - 3(-2)^2 = -12(x+2)$, i.e. $y = -12x - 12$.

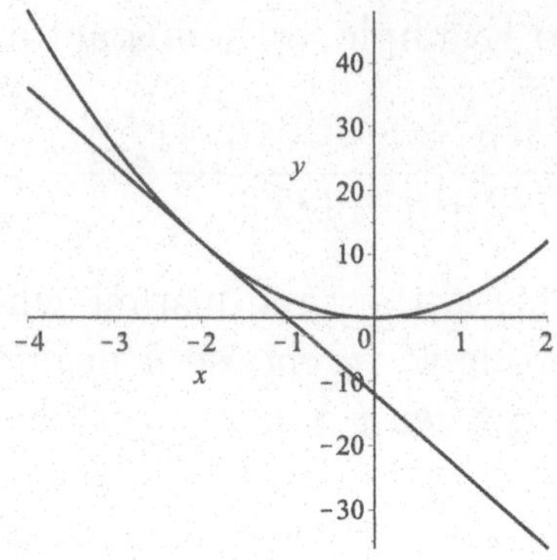


**41.** The slope of a secant line is given by $\dfrac{\sqrt{9+h}-3}{h} = \dfrac{\sqrt{9+h}-3}{h} \cdot \dfrac{\sqrt{9+h}+3}{\sqrt{9+h}+3} = \dfrac{1}{\sqrt{9+h}+3}$; thus the slope of the tangent line is $1/6$. The equation of the tangent line is $y - 3 = (1/6)(x-9)$, i.e. $y = x/6 + 3/2$.

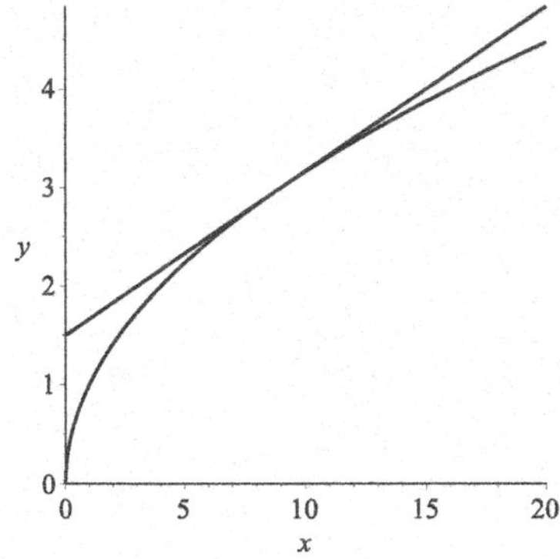


**43.** On $[0, 2]$, the average rate of change is $(8.3 \cdot 1.33^2 - 8.3)/2 \approx 3.19$ millions of people per decades, on $[2, 4]$, the average rate of change is $(8.3 \cdot 1.33^4 - 8.3 \cdot 1.33^2)/2 \approx 5.64$ millions of people per decades.

**45.** On $[0, 30]$, the average rate of change is $(14.9 - 17)/30 \approx -0.07$ cm of froth decrease per second, on $[60, 90]$, the average rate of change is $(11.9 - 13.2)/30 \approx -0.043$ cm of froth decrease per second.

**47.** At $t = 0$, 2.4 millions of people increase in a decade, at $t = 2$, 4.2 millions of people increase in a decade.

**49.** At $x = 0$, $-0.06$ cm of froth decrease per second, at $x = 60$, $-0.05$ cm of froth decrease per second.

**51.** The velocities (in m/s) are 5.9, 9.6, 10.8, 11.6, 11.9, 12.0, 11.9, 11.8, 11.5, and 11.1. He accelerates through the first 6 splits, then decelerates.

**53.** After simplification, the instantaneous rate of change at $t = 16$ can be found by $\lim_{h\to 0} \frac{B(t+h)-B(t)}{h} = \lim_{h\to 0} \frac{10}{\sqrt{16+h}+4} =$ 5/4 biomass units per year.

## Problem Set 2.2 - Limits

**1. a.** As $x$ gets closer to $-4$, the value of the function gets closer to 0, so $\lim_{x\to -4} f(x) = 0$.

**b.** As $x$ gets closer to 0, the value of the function gets closer to 4, so $\lim_{x\to 0} f(x) = 4$.

**3. a.** As $x$ gets closer to 2, the value of the function gets closer to 6, so $\lim_{x\to 2} f(x) = 6$.

**b.** As $x$ gets closer to $-4$, the value of the function gets closer to 3, so $\lim_{x\to -4} g(x) = 3$.

**5. a.** As $x$ gets closer to 1 from below, the value of the function gets closer to 0; $\lim_{x\to 1^-} F(x) = 0$.

**b.** As $x$ gets closer to 1 from above, the value of the function gets closer to 1, so $\lim_{x\to 1^+} F(x) = 1$.

**c.** According to parts **a** and **b**, $\lim_{x\to 1} F(x)$ does not exist because the left- and right-hand limits are not the same.

**7. a.** As $x$ gets closer to 0 from below, the value of the function gets closer to 0; $\lim_{x\to 0^-} G(x) = 0$.

**b.** As $x$ gets closer to 0 from above, the value of the function gets closer to 0, so $\lim_{x\to 0^+} G(x) = 0$.

**c.** According to parts **a** and **b** above, $\lim_{x\to 0} G(x) = 0$.

**9.** The figure shows that $\lim_{x\to 1^+} f(x) = 2$.

**11.** The figure shows that $\lim_{x\to 3^-} h(x) = 2$.

**13.**

| $x$ | 2 | 3 | 4 | 4.5 | 4.9 | 4.99 |
|---|---|---|---|---|---|---|
| $f(x)$ | 4 | 7 | 10 | 11.5 | 12.7 | 12.97 |

; we obtain $\lim_{x\to 5^-} 3x - 2 = 13$.

**15.**

| $x$ | 1 | 1.5 | 1.9 | 1.99 | 1.999 | 1.9999 | 3 | 2.5 | 2.1 | 2.01 | 2.001 | 2.0001 |
|---|---|---|---|---|---|---|---|---|---|---|---|---|
| $h(x)$ | 7 | 8.5 | 9.7 | 9.97 | 9.997 | 9.9997 | 13 | 11.5 | 10.3 | 10.03 | 10.003 | 10.0003 |

; we obtain $\lim_{x\to 2} \frac{3x^2-2x-8}{x-2} = 10$.

**17.** $\lim_{x\to -3^+} \frac{|x+3|}{x+3} = 1$; as $x$ gets closer to $-3$ from above, $|x+3| = x+3$, and the function is constant 1.

**19.** $\lim_{x\to -1} \cos(\pi x) = -1$; as $x$ gets closer to $-1$, $\cos(\pi x)$ gets closer to $\cos(-\pi) = -1$.

**21.** $\lim_{x\to 2} \frac{\sqrt{x+2}-2}{x-2} = \frac{1}{4}$; as $x$ gets closer to 2, the value of the function gets closer to 1/4.

**23.** $\lim\limits_{x \to 0} \dfrac{x^2}{|x|} = 0$; as $x$ gets closer to 0, the value of the function gets closer to 0.

**25.** Yes, $\lim\limits_{x \to 0^+} \sqrt{x} \cos(1/x) = 0$. As $x$ gets closer to 0 from above, the value of the function gets closer to 0.

**27. a.** We need that $4.9 \leq 4 + x \leq 5.1$, i.e. $0.9 \leq x \leq 1.1$. Thus $x$ need to be closer to 1 than 0.1.

**b.** We need that $4.99 \leq 4 + x \leq 5.01$, i.e. $0.99 \leq x \leq 1.01$. Thus $x$ need to be closer to 1 than 0.01.

**c.** We need that $4.999 \leq 4 + x \leq 5.001$, i.e. $0.999 \leq x \leq 1.001$. Thus $x$ need to be closer to 1 than 0.001.

**29. a.** We need that $\sqrt{x} \leq 0.1$, i.e. that $0 < x \leq 0.01$. Thus $x$ need to be closer to 0 than 0.01.

**b.** We need that $\sqrt{x} \leq 0.01$, i.e. that $0 < x \leq 0.0001$. Thus $x$ need to be closer to 0 than 0.0001.

**c.** We need that $\sqrt{x} \leq 0.001$, i.e. that $0 < x \leq 0.000001$. Thus $x$ need to be closer to 0 than 0.000001.

**31. a.**

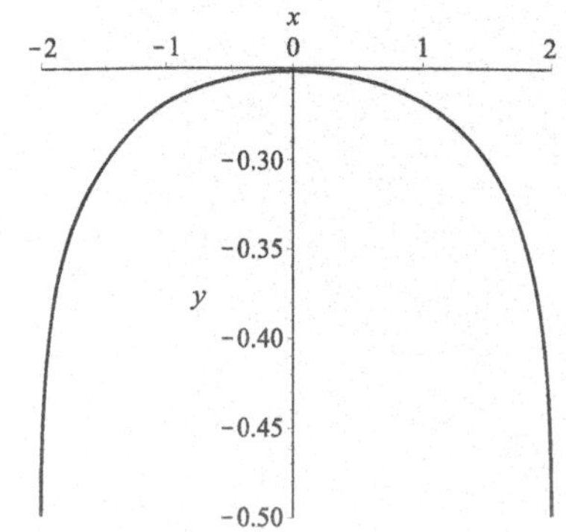


**b.**

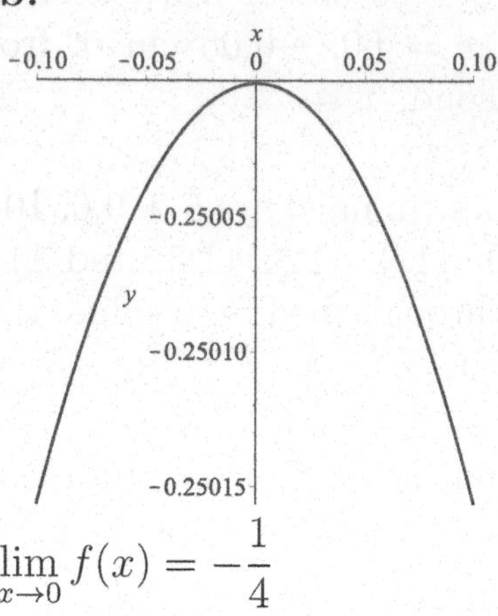


$\lim\limits_{x \to 0} f(x) = -\dfrac{1}{4}$

**c.**

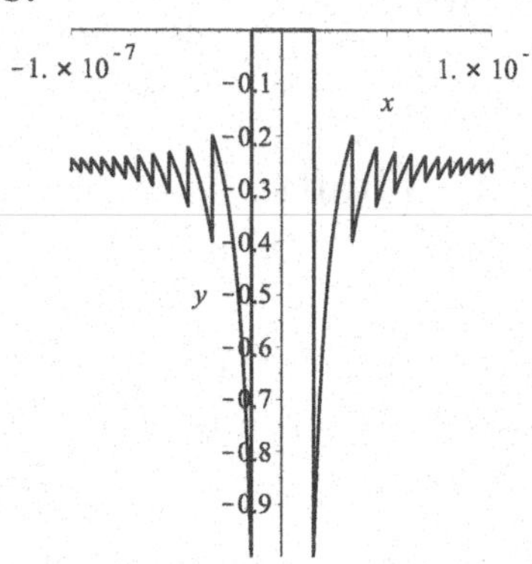


$\lim\limits_{x \to 0} f(x) = 0$

**d.** When $x$ becomes small, in the numerator $x^2$ becomes zero, and thus the numerator becomes 0. Thus the value of $f$ becomes 0 instead of the actual value.

**33.** Fix $\varepsilon > 0$, and let $\delta = \varepsilon$. Then whenever $|x - 5| < \delta$, $|f(x) - 6| = |(x + 1) - 6| = |x - 5| < \delta = \varepsilon$.

**35.** Fix $\varepsilon > 0$, and let $\delta = \min\{1, \varepsilon/2\}$. Then whenever $|x-1| < \delta$,

$$|f(x) - 0| = |x^2 - 3x + 2| = |(x-1)^2 - (x-1)| \le |x-1|^2 + |x-1| < \delta^2 + \delta < \varepsilon.$$

**37.** Fix $\varepsilon > 0$, and let $\delta = \min\{1, \varepsilon/8\}$. Then whenever $|x-2| < \delta$,

$$|f(x) - 6| = |(x^2+2) - 6| = |x^2 - 4| = |x-2||x+2| < \delta(4+\delta) < \varepsilon.$$

**39. a.**

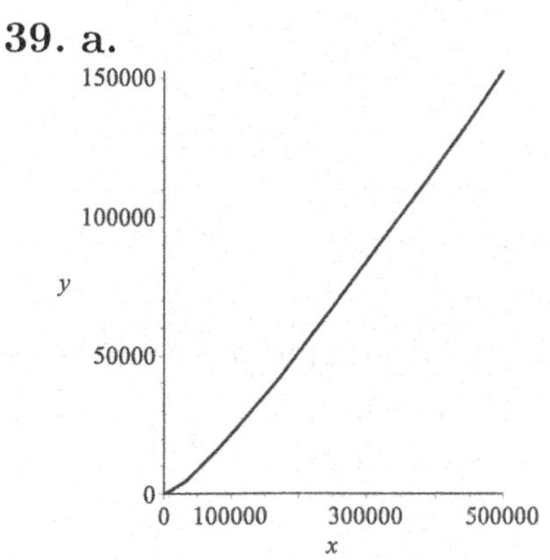


**b.** The limit exists for all values of $a$.

**41.** According to the description, the function will have the form

$$f(x) = \begin{cases} ax & \text{if } 0 \le x \le 3 \\ 1/25 & \text{if } x > 3 \end{cases}$$

We need that the limit exists at $x = 3$, so the left-hand and right-hand limits are the same. This gives us the equation $3a = 1/25$, so $a = 1/75$ above.

## Problem Set 2.3 - Limit Laws and Continuity

**1.** $f(x)$ is made up by two polynomials, so $\lim_{x\to 1^-} f(x) = 2(1) - 3 = -1$; $\lim_{x\to 1^+} f(x) = (1)^2 - 2 = -1$ and this gives $\lim_{x\to 1} f(x) = -1$.

**3.** $\lim_{x\to 0^-} x/|x| = \lim_{x\to 0^-} x/(-x) = -1$; while $\lim_{x\to 0^+} x/|x| = \lim_{x\to 0^-} x/x = 1$ and this implies that the limit at 0 does not exist.

**5.** $\lim_{x\to 1^-} f(x) = 0$; $\lim_{x\to 1^+} f(x) = 1$ and this implies that the limit at 1 does not exist.

**7.** We get $\lim_{x\to 3} (x^2+3x-10)/(3x^2+5x-7) = (9+9-10)/(27+15-7) = 8/35$, because $f(x)$ is a rational function and 3 is in its domain.

**9.** We obtain that $\lim_{t\to -3} (t^2+5t+6)/(t+3) = \lim_{t\to -3} (t+3)(t+2)/(t+3) = \lim_{t\to -3} (t+2) = -1$, because the last expression is a polynomial.

**11.** $\lim_{s\to 3} \dfrac{s-2}{s+2} + \sin s = 1/5 + \sin 3$, because of the composition limit law and continuity of elementary functions.

**13.** $\lim_{x\to \pi} \dfrac{1+\tan x}{2-\cos x} = \dfrac{1+\tan \pi}{2-\cos \pi} = 1/3$, because of the composition limit law and continuity of elementary functions.

**15.** $f$ is not continuous at $a = -1$, because the left and right hand side limits differ; $f$ cannot be made continuous here by redefining the value of the function. $f$ is not continuous at $a = 1$ because the value of the function is not equal to the limit at that point. Redefining $f$ to be 1 at this point makes the function continuous.

**17.** $f$ is not continuous at $a = 1$, because the left and right hand side limits differ; $f$ cannot be made continuous here by redefining the value of the function

**19.** $\lim_{x\to 2}\dfrac{x^2-x-2}{x-2} = \lim_{x\to 2}\dfrac{(x-2)(x+1)}{x-2} = \lim_{x\to 2}(x+1) = 3$, so if we define $f(2)=3$ the resulting function is continuous.

**21.** We get $\lim_{x\to 2^-} f(x) = 15-(2)^2 = 11$ and $\lim_{x\to 2^+} f(x) = 2(2)+5 = 9$; thus the function cannot be made continuous at 2.

**23.** Let $f(x) = -x^7+x^2+4$. This is a continuous function on $\mathbb{R}$; $f(0) = 4$ and $f(2) = -120$, so there is a $c$ between 0 and 2 such that $f(c) = 0$.

**25.** Let $f(x) = x^2+2x-1-\sqrt[3]{x}$. This is a continuous function on $\mathbb{R}$; $f(0) = -1$ and $f(1) = 1$, so there is a $c$ between 0 and 1 such that $f(c) = 0$, which gives a solution of the equation.

**27.** Let $f(x) = x\,2^x - \pi$. This is a continuous function on $\mathbb{R}$; $f(0) = -\pi$ and $f(2) = 8-\pi > 0$, so there is a $c$ between 0 and 2 such that $f(c) = 0$, which gives a solution of the equation.

**29.**

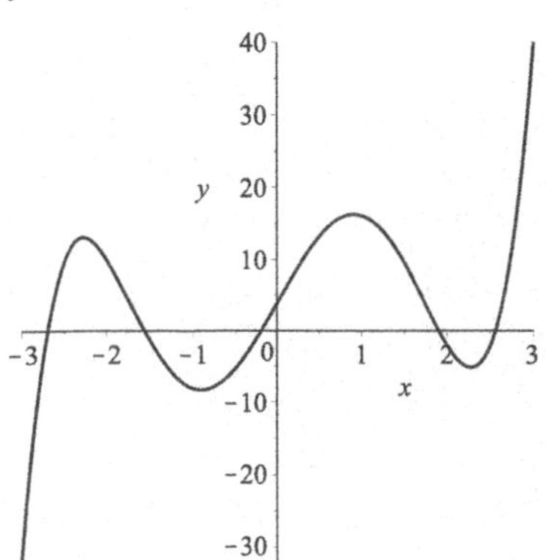


**a.** According to the graph, we can start with $a=-3$, $b=-2$. The result is $-2.709$.

**b.** According to the graph, we can start with $a=-2$, $b=-1$. The result is $-1.561$.

**c.** According to the graph, we can start with $a=-1$, $b=0$. The result is $-0.193$.

**d.** According to the graph, we can start with $a=1$, $b=2$. The result is $1.903$.

**31.** For $x\neq 0$, $-1\le \cos(1/x)\le 1$, thus (for $x$ close to 0), we get $(5+x^2-x)/(2+x) \le (5+x^2-x\cos(1/x))/(2+x) \le (5+x^2+x)/(2+x)$. The limit of the upper and lower bounds is $5/2$, thus the original limit is $5/2$ as well.

**33.** The continuity of elementary functions and the composition limit law imply that $\lim_{x\to\pi/2} x/(8+\cos x\sin(\cos x)) = \pi/16$.

**35.** $f$ is continuous at $a$: $\lim_{x\to a} f(x) = f(a)$, $g$ is continuous at $a$: $\lim_{x\to a} g(x) = g(a)$. Using the limit law for sums gives $\lim_{x\to a}(f+g)(x) = \lim_{x\to a} f(x)+g(x) = \lim_{x\to a} f(x) + \lim_{x\to a} g(x) = f(a)+g(a) = (f+g)(a)$. The two ends of the identities prove the statement.

**37.** $f$ is continuous at $a$: $\lim_{x\to a} f(x) = f(a)$, $g$ is continuous at $a$: $\lim_{x\to a} g(x) = g(a)$. Using the limit law for quotients (and using $g(a)\neq 0$) gives $\lim_{x\to a}(f/g)(x) = \lim_{x\to a} f(x)/g(x) = \lim_{x\to a} f(x)/\lim_{x\to a} g(x) = f(a)/g(a) = (f/g)(a)$. The two ends of the identities prove the statement.

**39.** $f(x) = x^3+ax^2+bx+c$ is a polynomial, so it is a continuous function oh $\mathbb{R}$. Let $x > M = \max\{2, 2|a|, 2|b|, 2|c|\} \ge 2$. Then $f(M) = M^3+aM^2+bM+c \ge M^3-(M/2)M^2-(M/2)M-M/2 = (M/2)(M^2-M-1) > 0$ because $M\ge 2$. Also, $f(-M) = -M^3+aM^2-bM+c \le -M^3+(M/2)M^2+(M/2)M+M/2 = (M/2)(-M^2+M+1) < 0$

because $M \geq 2$. The continuity implies that somewhere between $-M$ and $M$ there is a root.

**41. a.** The function is

$$g(N) = \begin{cases} \dfrac{100}{1+N}, & N < c \\ 5, & N \geq c \end{cases}$$

**b.** The function will be continuous when $100/(1+c) = \lim_{N \to c^-} g(N) = \lim_{N \to c^+} g(N) = 5$, i.e. when $c = 19$.

**43.**

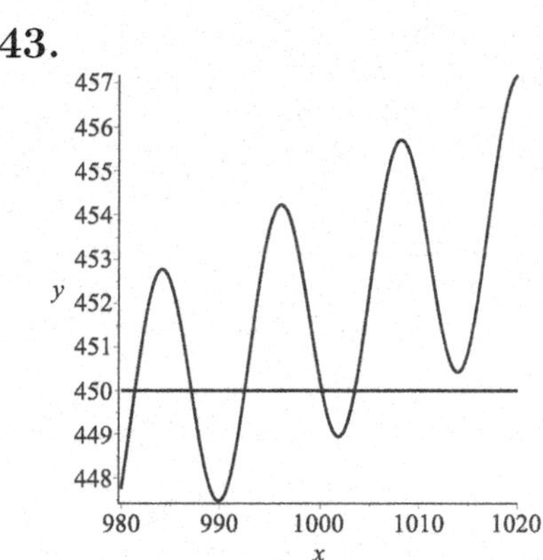


According to the graph, we can start with $a = 1003$, $b = 1005$. The result is 1003.48.

**45.** The graph of the function is the following:

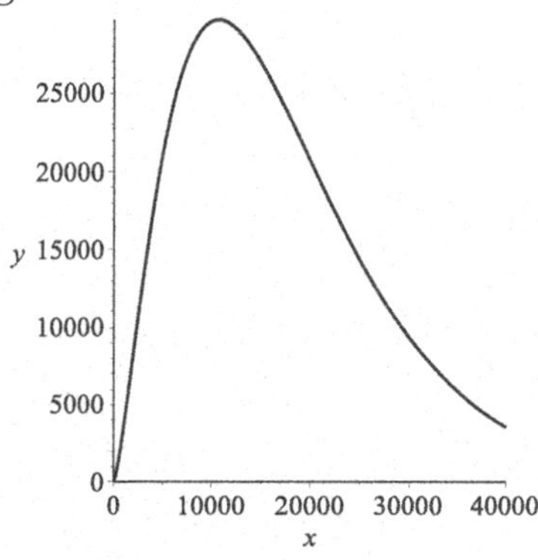


According to the graph, we will have two solutions. For the first one, we can start with $a = 2380$, $b = 2400$. The result is 2383. For the second one, try $a = 29000$, $b = 30000$. The result is 29248.

**47.** The graph of the function is the following:

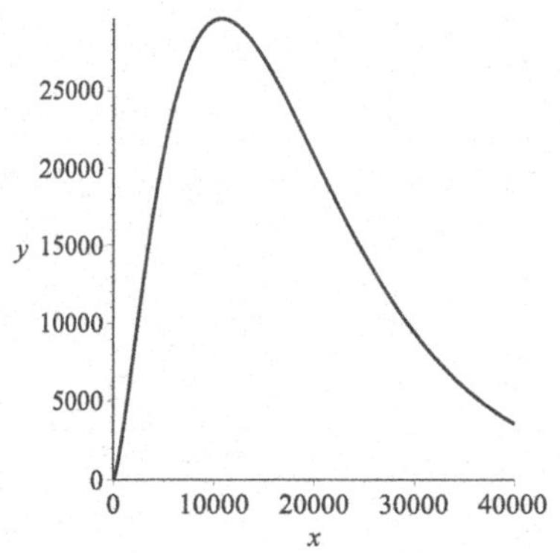


According to the graph, we will have two solutions. For the first one, we can start with $a = 1300$, $b = 1400$. The result is 1365. For the second one, try $a = 36000$, $b = 37000$. The result is 36602.

## Problem Set 2.4 - Asymptotes and Infinity

**1.** We get $\lim_{x \to -\infty} e^x = \lim_{x \to \infty} e^{-x} = 0$.

**3.** We get $\lim_{x \to 2^+} 1/(x-2) = \infty$.

**5.** We get $\lim_{x \to 3^-} (3 + 2x/(x-3)) = -\infty$.

**7.** We get $\lim_{x \to 3^+} (3 + 2x/(x-3)) = \infty$.

**9.** $\lim_{x \to 3^+} \dfrac{x^2 - 4x + 3}{x^2 - 6x + 9} = \lim_{x \to 3^+} \dfrac{(x-3)(x-1)}{(x-3)(x-3)} = \lim_{x \to 3^+} \dfrac{x-1}{x-3} = \infty$.

**11.** $\lim_{x \to \infty} \cos x^2$ does not exist, the function takes every value between $-1$ and 1 for arbitrarily large numbers.

**13.** Division gives $\lim_{x \to \infty} \dfrac{(2x+5)(x-2)}{(7x-2)(3x+1)} =$

$$\lim_{x\to\infty} \frac{(2+5/x)(1-2/x)}{(7-2/x)(3+1/x)} = 2/21.$$

**15.** We know that $-1 \leq \sin x \leq 1$, thus $-1/(1+x) \leq \sin x/(x+1) \leq 1/(1+x)$. The left and right side limit is 0, so we obtain that $\lim_{x\to\infty} \frac{\sin x}{x+1} = 0$.

**17.** $\lim_{x\to\infty} \frac{Ae^x+3}{Be^{2x}+4} = \lim_{x\to\infty} \frac{Ae^{-x}+3e^{-2x}}{B+4e^{-2x}} = 0/B = 0$.

**19.** $\lim_{x\to\infty} \frac{1+5e^{ax}}{7+2e^{ax}} = \lim_{x\to\infty} \frac{e^{-ax}+5}{7e^{-ax}+2} = 5/2$ (we used that if $a > 0$, then $\lim_{x\to\infty} e^{-ax} = 0$).

**21.** We compute: $\lim_{x\to\infty} \sqrt{x+1} - \sqrt{x} = \lim_{x\to\infty} \frac{(\sqrt{x+1}-\sqrt{x})(\sqrt{x+1}+\sqrt{x})}{\sqrt{x+1}+\sqrt{x}} = \lim_{x\to\infty} \frac{1}{\sqrt{x+1}+\sqrt{x}} = 0$.

**23.** Multiplication by the conjugate gives that $\lim_{x\to\infty} \sqrt{x^4+ax^2+x+1} - x^2 =$

$$= \lim_{x\to\infty} \frac{(\sqrt{x^4+ax^2+x+1+1} - x^2)(\sqrt{x^4+ax^2+x+1}+x^2)}{\sqrt{x^4+ax^2+x+1}+x^2} =$$

$$= \lim_{x\to\infty} \frac{ax^2+x+1}{\sqrt{x^4+ax^2+x+1}+x^2} = \lim_{x\to\infty} \frac{a+1/x+1/x^2}{\sqrt{1+a/x^2+1/x^3+1/x^4}+1} = a/2.$$

**25.** If $x > 2$, then $x-2 > 0$, so $1/(x-2) \geq 10^6$ means that $10^{-6} \geq x-2$, and we need that $2 < x \leq 2+10^{-6}$.

**27.** If $x < 1$, then $1/(1-x) \geq 10^6$ means $10^{-6} \geq 1-x$, so we need $1-10^{-6} \leq x < 1$.

**29.** If $x < 0$, then $-1/\sin x \geq 10^6$ means $-\sin x \leq 10^{-6}$, so $\sin x \geq -10^{-6}$ and we need $-\arcsin(10^{-6}) \leq x < 0$.

**31.** We need $|1/x^2 - 0| = 1/x^2 \leq 0.05$; this means $x^2 \geq 20$, so we get $x \leq -\sqrt{20}$.

**33.** We need $\left|\frac{x}{1+x} - 1\right| = \left|\frac{x-(1+x)}{1+x}\right| = \left|\frac{1}{1+x}\right| \leq 0.05$. This is the same as $|1+x| \geq 20$, and because $x$ is negative, we need that $-1-x \geq 20$, i.e. $x \leq -21$.

**35.** We need that $x^2 > 10^6$, this happens when $x > 10^3 = 1000$.

**37.** We need that $x^2/(1+x) > 10^6$, which is the same as $x^2 > 10^6 + 10^6 x$ (when $x > 1$). Thus we want that $x^2 - 10^6 x - 10^6 > 0$; this happens when $x$ is greater than the bigger root of this quadratic equation, i.e. we need that $x > (10^6 + \sqrt{10^{12} + 4 \cdot 10^6})/2$.

**39. a.** We want that $8.3(1.33)^t > 500$, which happens when $(1.33)^t > 500/8.3$, i.e. when $t > \ln(500/8.3)/\ln 1.33 \approx 14.37$ decades.

**b.** We want that $8.3(1.33)^t > 1000$, which happens when $(1.33)^t > 1000/8.3$, i.e. when $t > \ln(1000/8.3)/\ln 1.33 \approx 16.8$ decades.

**41. a.** $\lim_{x\to\pm\infty} 1.2708x/(1+0.0506x) = \lim_{x\to\pm\infty} 1.2708/(1/x+0.0506) = 1.2708/0.0506 \approx 25.11$, so the only horizontal asymptote is $y = 25.11$. This corresponds to the maximum uptake rate. The vertical asymptote is at $x = -1/0.0506 \approx -19.76$, which is a physically meaningless value. The function should be considered only for $x \geq 0$.

**b.**

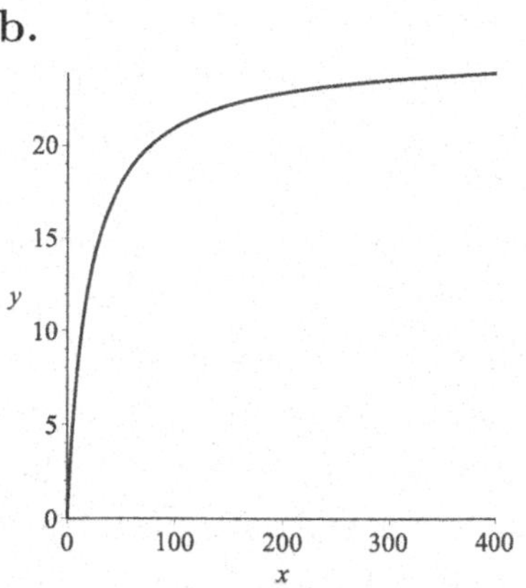


**43. a.** To be within 0.1 of $L$, we need that $58.7(x-0.03)/(0.76+x) \geq 58.6$. This is the same as $58.7(x-0.03) \geq 58.6(0.76+x)$, i.e. $0.1x \geq 58.6 \cdot 0.76 + 58.7 \cdot 0.03$. Thus we need $x \geq 462.97$.

**b.** To be within 0.01 of $L$, we need that $58.7(x-0.03)/(0.76+x) \geq 58.69$. This is the same as $58.7(x-0.03) \geq 58.69(0.76+x)$, i.e. $0.01x \geq 58.69 \cdot 0.76 + 58.7 \cdot 0.03$. Thus we need $x \geq 4636.54$.

**45.** After division, we obtain that $\lim\limits_{t\to\infty} 9.7417e^{0.53t}/(1 + 0.01476e^{0.53t}) = \lim\limits_{t\to\infty} 9.7417/(1e^{-0.53t} + 0.01476) = 9.7417/0.01476 \approx 660$; this is the maximum population density.

**47.** After division, we get that the horizontal asymptote is $\lim\limits_{x\to\infty} 16e^{5x}/(3.2 + e^{5x}) = \lim\limits_{x\to\infty} 16/(3.2e^{-5x} + 1) = 16$. We need that $16e^{5x}/(3.2+e^{5x}) = 16/(3.2e^{-5x}+1) \geq 0.995 \cdot 16$. Thus we need $1 \geq 0.995(3.2e^{-5x} + 1)$, which is the same as $0.005/(3.2 \cdot 0.995) \geq e^{-5x}$. Taking the logarithm of both sides, we obtain $x \geq 1.291$.

## Problem Set 2.5 - Sequential Limits

**1.** $\lim\limits_{n\to\infty} \dfrac{n^2-n}{1+3n^2} = \lim\limits_{n\to\infty} \dfrac{1-1/n}{1/n^2+3} = 1/3.$

**3.** $\lim\limits_{n\to\infty} \dfrac{e^n}{1+e^n} = \lim\limits_{n\to\infty} \dfrac{1}{e^{-n}+1} = 1.$

**5.** The sequence is $2, -2, 2, -2, \ldots$, thus it does not have a limit.

**7.** The sequence $\cos n$ does not have a limit. To see this, consider the value $y = 1/2$. $\cos x = 1/2$ when $x = \pi/3 + 2k\pi$ and $x = 5\pi/3 + 2k\pi$. This means that $\cos x \geq 1/2$ on the intervals $[5\pi/3, 7\pi/3]$, $[11\pi/3, 13\pi/3]$, and so on. Similarly, $\cos x \leq 1/2$ on the intervals $[2\pi/3, 4\pi/3]$, $[8\pi/3, 10\pi/3]$, and so on. The length of these intervals is bigger than 2, thus there are infinitely many integers $n$ such that $\cos n \geq 1/2$ and infinitely many integers $n$ such that $\cos n \leq -1/2$. This shows the nonexistence of the limit.

**9. a.** $\lim\limits_{n\to\infty} \dfrac{n}{3+n} = \lim\limits_{n\to\infty} \dfrac{1}{3/n+1} = 1.$

**b.** We need $\left|\dfrac{n}{3+n} - 1\right| = \left|\dfrac{n-3-n}{3+n}\right| = \left|\dfrac{-3}{3+n}\right| = \dfrac{3}{3+n} < 0.001$. This is the same as $3/0.001 < 3+n$, i.e. we need $n > 2997$.

**11. a.** $\lim\limits_{n\to\infty} 1000/n = 0.$

**b.** We need $|1000/n - 0| = 1000/n < 0.001$. This is the same as $10^6 < n$.

**13. a.** $\lim\limits_{n\to\infty} \dfrac{n^2+1}{n^3} = \lim\limits_{n\to\infty} \dfrac{1/n+1/n^3}{1} = 0.$

**b.** We want $\left|\dfrac{n^2+1}{n^3} - 0\right| = \left|\dfrac{1}{n} + \dfrac{1}{n^3}\right| < \dfrac{2}{n} < 0.001$. This is the same as $2/0.001 < n$, i.e. $n > 2000$.

**15.** $a_n = 2n \geq 10^6$ when $n \geq 5 \cdot 10^5$.

**17.** $a_n = 2^n - 10^4 \geq 10^6$ when $2^n \geq 10^6 + 10^4$, which is the same as $n \ln 2 \geq \ln(10^6 + 10^4)$, so we need that $n > \ln(10^6 + 10^4)/\ln 2$.

**19.** $x_1 = 0$, $x_2 = 2$, $x_3 = 2 + 2 = 4$, $x_4 = 6$, and we can see $x_n = 2(n-1)$. Thus $\lim_{n\to\infty} x_n = \infty$.

**21.** $x_1 = 100$, $x_2 = \sqrt{100} = 10 = 100^{1/2}$, $x_3 = 100^{1/4}$, $x_4 = 100^{1/8}$, and we can see $x_n = 100^{1/2^{n-1}}$. Thus $\lim_{n\to\infty} x_n = 100^0 = 1$.

**23.** We can use the monotone convergence theorem with $f(x) = x^2$, $I = [0,1]$ here. The sequence $x_1 = 0.99999$, $x_2 = 0.99999^2$, $x_3 = 0.99999^4$, ... is monotone decreasing, bounded from below by 0, so it is convergent. The limit satisfies $f(a) = a$, so it is 0.

**25.** The equilibrium must satisfy the equation $a = 3/(2+a)$, thus $a^2 + 2a - 3 = 0$, and then $a = -3$ or $a = 1$. In this case, $a_1 = 1$ and so $a_2 = 1$, $a_3 = 1$ ... and we get $\lim_{n\to\infty} a_n = 1$.

**27.** The equilibrium satisfies $a = 3a(1-a)$, thus $a = 0$ or $a = 2/3$. In this case, $a_1 = 0.1$ and iteration with a computer suggests that $\lim_{n\to\infty} a_n = 2/3$.

**29.** We can use the monotone convergence theorem with $f(x) = 2x/(1+x)$, $I = [0,1]$. The equilibrium must satisfy $a = 2a/(1+a)$, thus $a = 0$ or $a = 1$. In this case, $a_1 = 0.5$ and so $a_2 = 2(1/2)/(3/2) = 2/3$, $a_3 = 2(2/3)/(1+2/3) = 4/5$ and we can show that the sequence is increasing and bounded above by 1. Thus using the possible values for $a$, $\lim_{n\to\infty} a_n = 1$.

**31.** The sequence terminates after the first iteration; $a_1 = 1$, $a_2 = 2\ln 1 = 0$ and $f$ is not defined at 0.

**33.** We can use the monotone convergence theorem with $f(x) = \sqrt{5+x}$, $I = [0,3]$. The equilibrium must satisfy $a = \sqrt{5+a}$, thus $a = (1+\sqrt{21})/2$. (Solving the equation by squaring both sides gives another $a$ value, but it is negative, thus it does not satisfy the original equation.) In this case, $a_1 = 0$ and so $a_2 = \sqrt{5}$, ... and we can show that the sequence is increasing and bounded above. Thus using the possible value for $a$, $\lim_{n\to\infty} a_n = (1+\sqrt{21})/2$.

**35. a.** We will use mathematical induction. $x_1 = 1/4 = 0.25$. Suppose the statement is true for $n$: $x_n = 1/(n+3)$. Then for $n+1$ we obtain $x_{n+1} = x_n/(1+x_n) = 1/(n+3)/(1+1/(n+3)) = 1/(n+3)/((n+4)/(n+3)) = 1/(n+4) = 1/((n+1)+3)$, so the statement is also true for $n+1$. Hence $x_n = 1/(n+3)$ for all $n$.

**b.** Clearly $\lim_{n\to\infty} x_n = 0$.

**c.** We need $1/(n+3) \leq 0.1$, which means $0.7 \leq 0.1n$, thus $n \geq 7$.

**d.** We need $1/(n+3) \leq 0.001$, which means $0.997 \leq 0.001n$, thus $n \geq 997$.

**37. a.** With the initial value $x_1 = 0.91$, the sequence is increasing and approaches 1.

**b.** With the initial value $x_1 = 0.89$, the sequence is decreasing and approaches 0.

**39.** The new model is $R_{n+1} = R_n + pR_{n-1}$, because the current population survives and $p$ is the proportion which becomes pregnant and gives birth one cycle later. Thus if $a_n = R_n/R_{n-1}$, then division of the original equation by $R_n$ gives $a_{n+1} = 1 + p/a_n$. Thus the limit satisfies $a = 1 + p/a$, meaning $a^2 - a - p = 0$. We get that $a = (1 + \sqrt{1+4p})/2$. If $p = 3/4$, then $a = 3/2$, thus the rate of increase is 50% in one cycle.

**41.** The new model is $R_{n+1} = R_n + pR_{n-1}$, because the current population survives and $p$ is the proportion which becomes pregnant and gives birth one cycle later. Thus if $a_n = R_n/R_{n-1}$, then division of the original equation by $R_n$ gives $a_{n+1} = 1 + p/a_n$. Thus the limit satisfies $a = 1 + p/a$, meaning $a^2 - a - p = 0$. We get that $a = (1+\sqrt{1+4p})/2$. If $p = 1/2$, then $a \approx 1.366$, thus the rate of increase is 36.6% in one cycle.

**43.** The new model is $R_{n+1} = R_n + pR_{n-1}$, because the current population survives and $p$ is the proportion which becomes pregnant and gives birth one cycle later. Thus if $a_n = R_n/R_{n-1}$, then division of the original equation by $R_n$ gives $a_{n+1} = 1 + p/a_n$. Thus the limit satisfies $a = 1 + p/a$, meaning $a^2 - a - p = 0$. We get that $a = (1+\sqrt{1+4p})/2$. If $p = 1/4$, then $a \approx 1.207$, thus the rate of increase is 20.7% in one cycle.

**45.** For $r = 0.9$ we obtain $N_1 = 0.5$, $N_2 = 0.725$, $N_3 = 0.9044375$, $N_4 = 0.9822247777$, $N_5 = 0.9979381151$, $N_6 = 0.9997899853$, $N_7 = 0.9999789588$, $N_8 = 0.9999978955$, $N_9 = 0.9999997895$ and $N_{10} = 0.9999999789$. It seems these converge to 1. For $r = 1.5$, we obtain $N_1 = 0.5$, $N_2 = 0.875$, $N_3 = 1.0390625$, $N_4 = 0.9781799316$, $N_5 = 1.010195861$, $N_6 = 0.9947461361$, $N_7 = 1.002585527$, $N_8 = 0.9986972091$, $N_9 = 1.000648850$ and $N_{10} = 0.9996749435$. It seems these converge to 1. For $r = 2.1$, we obtain $N_1 = 0.5$, $N_2 = 1.025$, $N_3 = 0.9711875$, $N_4 = 1.029950414$, $N_5 = 0.9651707873$, $N_6 = 1.035764678$, $N_7 = 0.9579727186$, $N_8 = 1.042520796$, $N_9 = 0.9494302864$ and $N_{10} = 1.050256364$. It seems these do not converge (oscillation), but the even and odd indexed terms (as two separate sequences) do.

**Problem Set 2.6 - Derivative at a Point**

**1.** We compute: $f'(-2) = \lim_{h\to 0} \frac{f(-2+h) - f(-2)}{h} = \lim_{h\to 0} \frac{3(-2+h) - 2 - (3(-2) - 2)}{h} =$
$= \lim_{h\to 0} \frac{3h}{h} = 3.$

**3.** We compute: $f'(1) = \lim_{h\to 0} \frac{f(1+h) - f(1)}{h} = \lim_{h\to 0} \frac{-(1+h)^2 - (-1^2)}{h} = \lim_{h\to 0} \frac{-2h - h^2}{h} =$
$= \lim_{h\to 0} (-2 - h) = -2.$

**5.** We compute: $f'(-4) = \lim_{h\to 0} \frac{f(-4+h) - f(-4)}{h} = \lim_{h\to 0} \frac{1/(2(-4+h)) - (1/2(-4))}{h} =$
$= \lim_{h\to 0} \frac{1/(2h-8) + 1/8}{h} = \lim_{h\to 0} \frac{2h/(8(2h-8))}{h} = \lim_{h\to 0} \frac{2}{8(2h-8)} = -\frac{1}{32}.$

**7.** We compute: $f'(-1) = \lim_{h\to 0} \dfrac{f(-1+h) - f(-1)}{h} = \lim_{h\to 0} \dfrac{(-1+h)^3 - (-1)^3}{h} =$

$= \lim_{h\to 0} \dfrac{3h - 3h^2 + h^3}{h} = \lim_{h\to 0} (3 - 3h + h^2) = 3.$

**9.** We compute: $f'(9) = \lim_{h\to 0} \dfrac{f(9+h) - f(9)}{h} = \lim_{h\to 0} \dfrac{\sqrt{9+h} - \sqrt{9}}{h} =$

$= \lim_{h\to 0} \dfrac{(\sqrt{9+h} - 3)(\sqrt{9+h} + 3)}{h(\sqrt{9+h} + 3)} = \lim_{h\to 0} \dfrac{9 + h - 9}{h(\sqrt{9+h} + 3)} = \lim_{h\to 0} \dfrac{1}{(\sqrt{9+h} + 3)} = \dfrac{1}{6}.$

**11.** The tangent line is given by $y - f(a) = f'(a)(x - a)$, thus using Problem 1, the answer is $y + 8 = 3(x - (-2))$, i.e. $y = 3x - 2$.

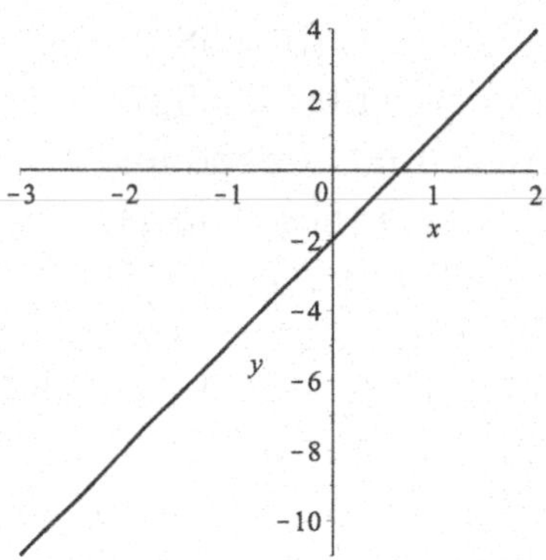


**13.** The tangent line is given by $y - f(a) = f'(a)(x - a)$, thus using Problem 3, the answer is $y + 1 = -2(x - 1)$, i.e. $y = -2x + 1$.

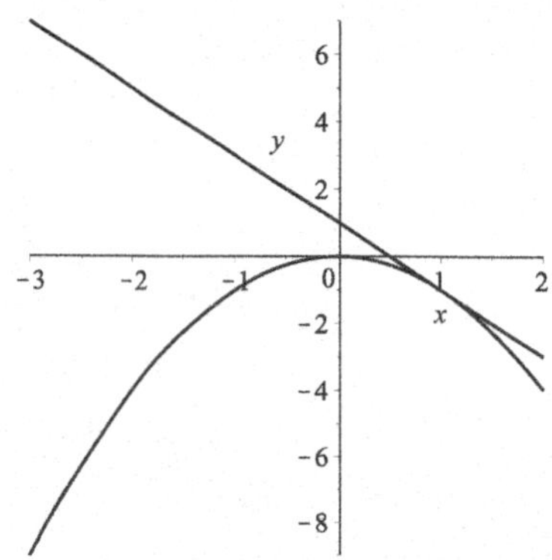


**15.** The tangent line is given by $y - f(a) = f'(a)(x - a)$, thus using Problem 5, the answer is $y - (-1/8) = -(1/32)(x - (-4))$, i.e. $y = -x/32 - 1/4$.

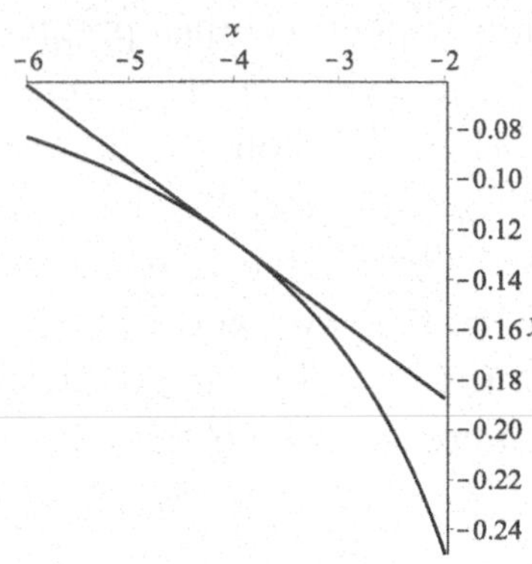


**17.** The tangent line is given by $y - f(a) = f'(a)(x - a)$, thus using Problem 7, the answer is $y - (-1) = 3(x - (-1))$, i.e. $y = 3x + 2$.

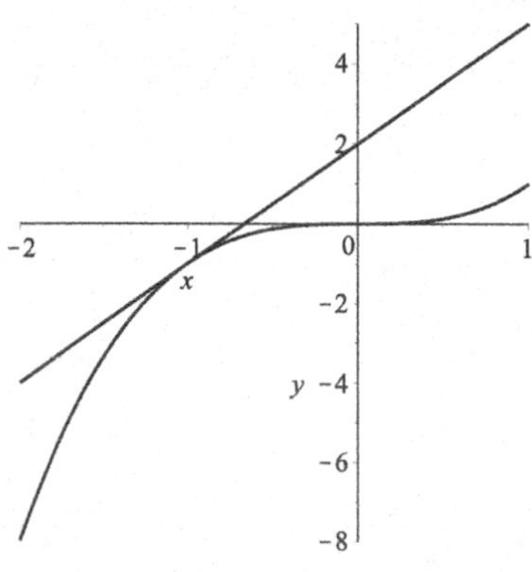


**19.** The tangent line is given by $y - f(a) = f'(a)(x - a)$, thus using Problem 9, the answer is $y - 3 = (1/6)(x - 9)$, i.e. $y = x/6 + 3/2$.

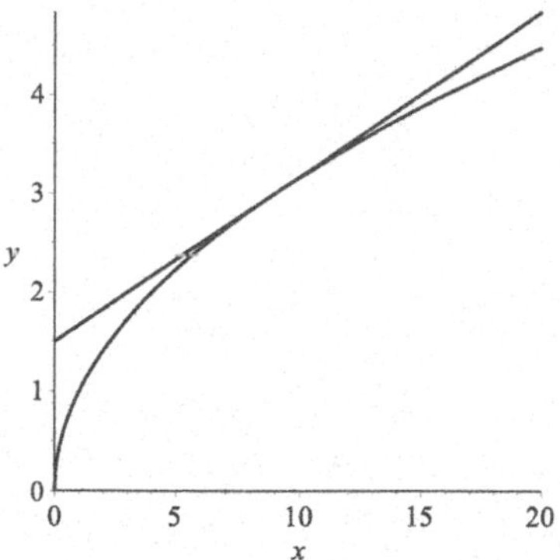

**21.** At $x = 0$; the limit of the difference quotients is $-1$ from the left side and 1 from the right side.

**23.** At $x = -1$, $x = 0$, and $x = 1$; the limit of the difference quotients is different from the left side and from the right side.

**25.** At $x = 2$; the limit of the difference quotients is $-1$ from the left side and 1 from the right side.

**27. a.**

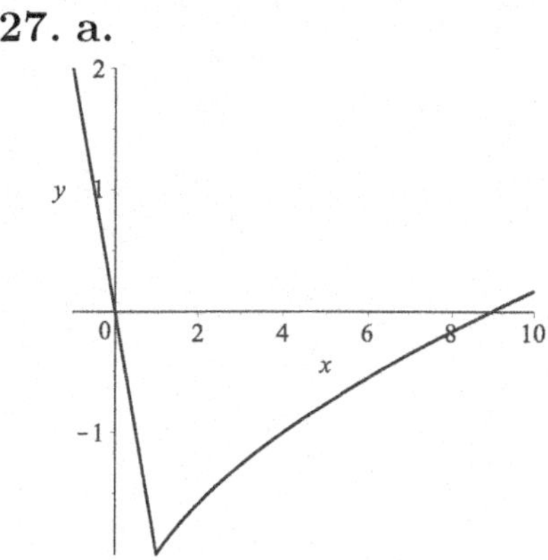

**b.** $\lim_{x\to 1+} f(x) = \sqrt{1} - 3 = -2$ and $\lim_{x\to 1^-} f(x) = (-2)\cdot 1 = -2$, so the function is continuous. $\lim_{h\to 0^+} (f(1+h) - f(1))/h = 1/2$ and $\lim_{h\to 0^-} (f(1+h) - f(1))/h = -2$, so the function is not differentiable.

**29.** At $x = 0$, the function is continuous, because the left and right hand side limits are both 1. It is not differentiable there, because the left and right hand side limits of the difference quotients are not the same (1 and $-1$).

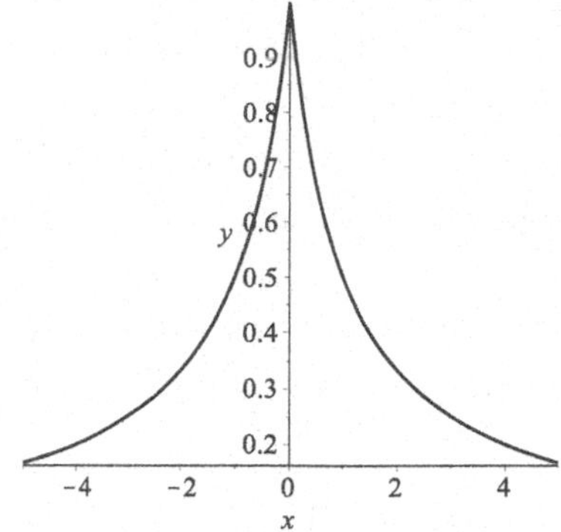

**31. a.** We get $H'(2) = \lim_{h\to 0} \dfrac{H(2+h) - H(2)}{h} = \lim_{h\to 0} \dfrac{10(2+h) - 16(2+h)^2 - (10\cdot 2 - 16\cdot 2^2)}{h}$ $= \lim_{h\to 0} \dfrac{-54h - 16h^2}{h} = -54$ (ft/s).

**b.** It hits the ground when $10t - 16t^2 = 0$; this happens when $t = 5/8 = 0.625$ s. This also means that in part **a**, the computation is only theoretical.

**c.** After simplification, we obtain that $H'(5/8) = \lim_{h\to 0} \dfrac{H(5/8+h) - H(5/8)}{h} = \lim_{h\to 0} \dfrac{-10h - 16h^2}{h} = -10$ (ft/s).

**33.** $A'(50) = \lim_{h\to 0} \dfrac{A(50+h) - A(50)}{h} = \lim_{h\to 0} \dfrac{-0.9h - 0.2h^2}{h} = -0.9$. This means that the enzyme activity is decreasing when the temperature is rising through 50 °C.

**35. a.** $P'(40) = \lim_{h\to 0} \dfrac{P(40+h) - P(40)}{h} = \lim_{h\to 0} \dfrac{9.2h + 0.5h^2}{h} = 9.2$.

**b.** $P'(45) = \lim_{h\to 0} \dfrac{P(45+h) - P(45)}{h} =$

$\lim_{h\to 0} \frac{14.2h + 0.5h^2}{h} = 14.2.$

**c.** The increase in the prevalence of MS is speeding up as the latitude is increasing.

**37. a.** An estimate is $-29$ from the difference quotient using technology.

**b.** The units are particles per mL per day; it is the change in the value of viral load.

**39. a.** We obtain $g'(0) = \lim_{h\to 0} \frac{g(h) - g(0)}{h} = \lim_{h\to 0} \frac{-0.0611147}{1 + 0.0506h} = -0.0611$. Similarly, we get $g'(20) = \lim_{h\to 0} \frac{g(20+h) - g(20)}{h} = \lim_{h\to 0} \frac{-0.0303751}{2.012 + 0.0506h} = -0.0151$.

**b.** They are the rate of change of the rate of glucose consumption per concentration of glucose in the environment. As the glucose concentration in the environment increases, the rate decreases, and the rate at which this rate is changing is also decreasing.

**Problem Set 2.7 - Derivatives as Functions**

**1.** We compute: $f'(x) = \lim_{h\to 0} \frac{f(x+h) - f(x)}{h} = \lim_{h\to 0} \frac{8-8}{h} = \lim_{h\to 0} 0 = 0.$

**3.** We compute: $f'(x) = \lim_{h\to 0} \frac{f(x+h) - f(x)}{h} = \lim_{h\to 0} \frac{-(x+h)^2 - (-x^2)}{h} = \lim_{h\to 0} \frac{-2xh - h^2}{h} =$
$= \lim_{h\to 0} (-2x - h) = -2x.$

**5.** We compute: $f'(x) = \lim_{h\to 0} \frac{f(x+h) - f(x)}{h} = \lim_{h\to 0} \frac{(x+h)^4 - (x^4)}{h} =$
$= \lim_{h\to 0} \frac{4x^3h + 6x^2h^2 + 4xh^3 + h^4}{h} = \lim_{h\to 0} (4x^3 + 6x^2h + 4xh^2 + h^3) = 4x^3.$

**7.** We compute: $f'(x) = \lim_{h\to 0} \frac{f(x+h) - f(x)}{h} = \lim_{h\to 0} \frac{1/(x+h) - 1/x}{h} =$
$= \lim_{h\to 0} \frac{(x - (x+h))/x(x+h)}{h} = \lim_{h\to 0} \frac{-1}{x(x+h)} = -\frac{1}{x^2}.$

**9.** $\left.\frac{dy}{dx}\right|_{x=-2} = 0$ from Problem 1.

**11.** $\left.\frac{dy}{dx}\right|_{x=4} = -2(4) = -8$ from Problem 3.

**13.** $\left.\frac{dy}{dx}\right|_{x=2} = 4(2)^3 = 32$ from Problem 5.

**15.** $\left.\frac{dy}{dx}\right|_{x=10} = -1/10^2 = -1/100$ from Problem 7.

**17.** The average rate of change on the interval is 0, and the slope of the instantaneous rate of change is 0 at all points of the interval as well, according to Problem 1.

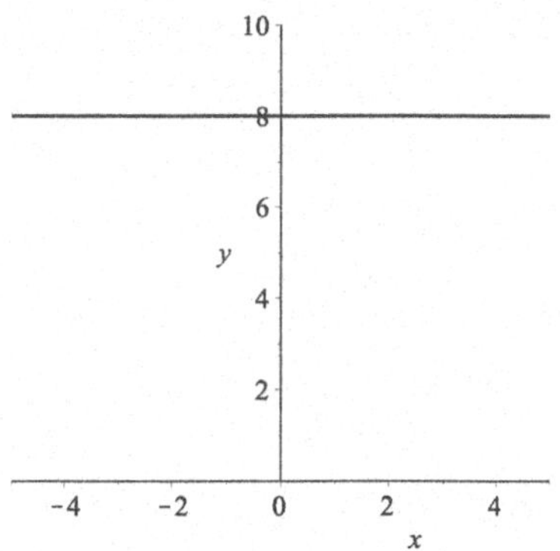


**19.** The average rate of change on the interval is $(-(1)^2-(-(-1)^2))/(1-(-1)) = 0$ and the slope of the instantaneous rate of change is $-2x$ according to Problem 3; they are the same when $x = 0$.

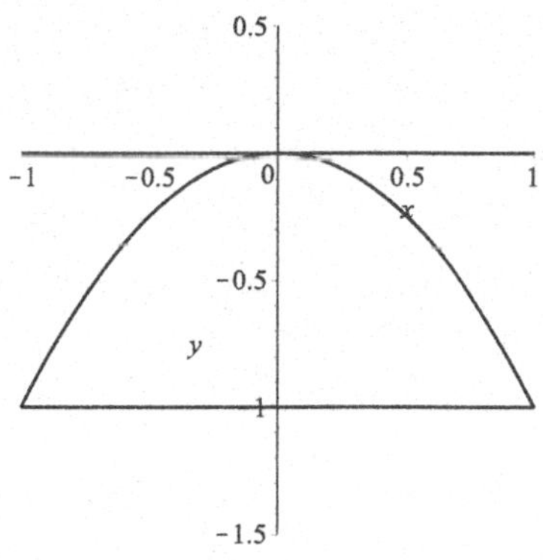


**21.** The average rate of change on the interval is $(1/2 - 1/1)/(2 - 1) = -1/2$ and the slope of the instantaneous rate of change is $-1/x^2$ according to Problem 7; they are the same when $x = \sqrt{2}$.

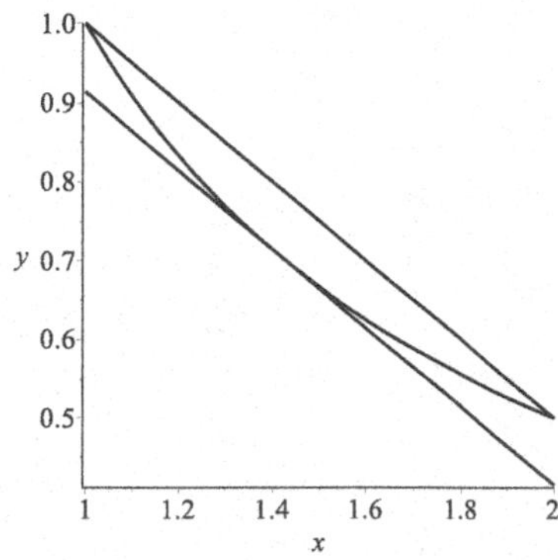


**23.** E - the slope is positive for negative values of $x$, then 0 at 0, finally negative for positive values of $x$.

**25.** B - the slope is positive, then negative, then positive, then negative again; the changes are at around $-1/2$, $1/2$ and 2, where the slope is 0.

**27.** D - the slope is positive for all $x$, and increasing.

**29.** $f'(x) = 2x - 1$; this is negative when $x < 1/2$ and positive when $x > 1/2$, thus $f$ is decreasing on $(-\infty, 1/2)$ and increasing on $(1/2, \infty)$.

**31.** $f'(x) = 3x^2 + 1$; this positive for all $x$, thus $f$ is increasing on $(-\infty, \infty)$.

**33.** $f'$ is negative when $x < -3.5$ and positive for all $x > -3.5$ except at $x = 0$. Thus $f$ is decreasing on $(-\infty, -3.5)$ (if we assume $f'$ is negative for all $x < -3.5$) and $f$ is increasing on $(-3.5, \infty)$ (if we assume $f'$ is positive for all $x > 3.5$).

**35.** $c \approx -2.2$, $c \approx -0.8$.

**37.**

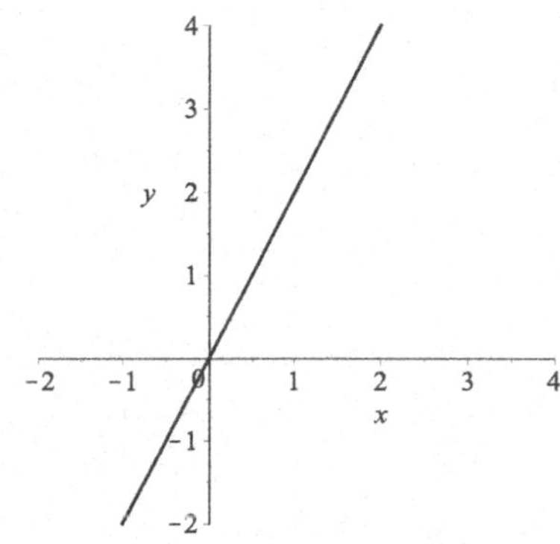

**39.**

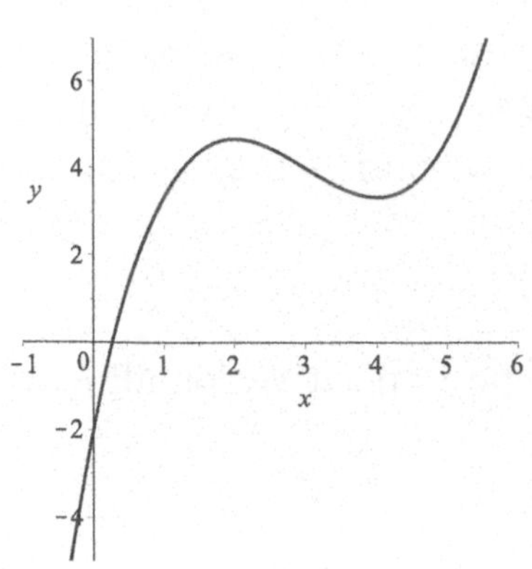


**41.** Let $a < x_1 < x_2 < b$. Then according to the Mean Value Theorem, $f(x_2) - f(x_1) = f'(c)(x_2 - x_1)$ for some $x_1 \leq c \leq x_2$. Because $f' < 0$ on $[a, b]$, $f'(c) < 0$ and thus $f(x_2) - f(x_1) < 0$, meaning $f(x_2) < f(x_1)$. This statement is true for any choice of $x_1$ and $x_2$ on $(a, b)$, so the function is decreasing.

**43.** Consider the function $g(x) = f(x) - f(a) - (f(b) - f(a))/(b - a)(x - a)$. This function is continuous and differentiable on the required intervals as well, and $g(a) = f(a) - f(a) - 0 = 0$ and $g(b) = f(b) - f(a) - (f(b) - f(a)) = 0$. Thus $g$ satisfies the assumptions of Rolle's theorem and there is a $c$ between $a$ and $b$ such that $g'(c) = 0$. But $g'(c) = f'(c) - (f(b) - f(a))/(b - a)$ and the Mean Value Theorem follows.

**45. a.** The function is increasing when $0 < t < 3$, then decreasing when $t > 3$. (The derivatives are $S'(t) = 1 > 0$ when $0 < t < 3$ and $S'(t) = -9/t^2 < 0$ when $t > 3$.)

**b.**

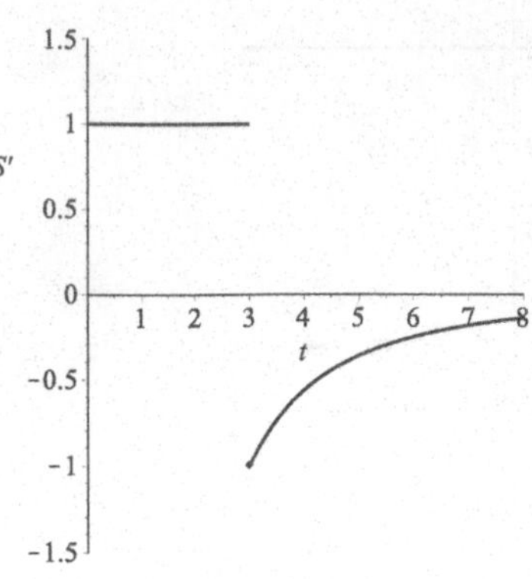


**47. a.** For $0 < t < 1$, $N'(t)$ can be approximated by $(18.3 - 9.6)/1 = 8.7$. Similarly, $N'$ is approximately 10.7, 18.2, 23.9, 48, 55.5, 82.7, 93.4, 90.3, 72.3, 46.4, 35.1, 34.6, 11.4, 10.3, 4.8 and 3.7 on the next 16 intervals.

**b.** The derivative is positive, the culture is growing all the time. It grows faster and faster for the first 8 hours, then the growth slows down.

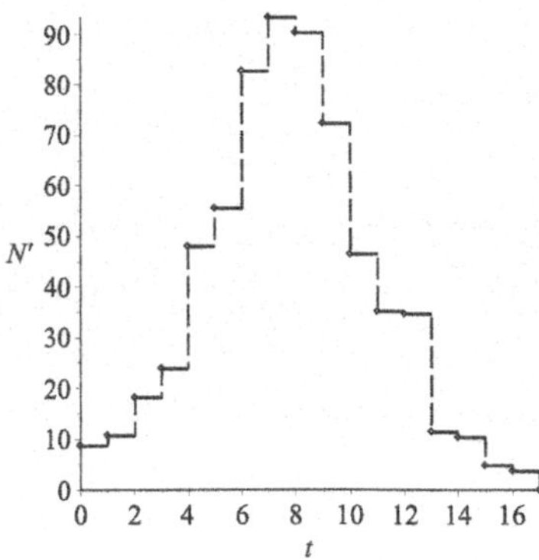


**49.** Let $d(t)$ be the distance of the sports car from the first patrol car. We can suppose that $d(t)$ is a differentiable function. By the Mean Value Theorem, there is some time instance $c$ between $t = 0$ and $t = 1/12$ (hours) such that its speed $d'(c) = (d(1/12) - d(0))/(1/12) = 6/(1/12) = 72$ miles per hour.

**Review Questions**

**1.** The average rate of change is given by $(f(2) - f(-1))/(2 - (-1)) = (6 - 0)/3 = 2$; the instantaneous rate of change is given by $3x^2 - 1$, thus we need the solution of the equation $3x^2 - 1 = 2$. We find two values: $a = -1$ and $a = 1$.

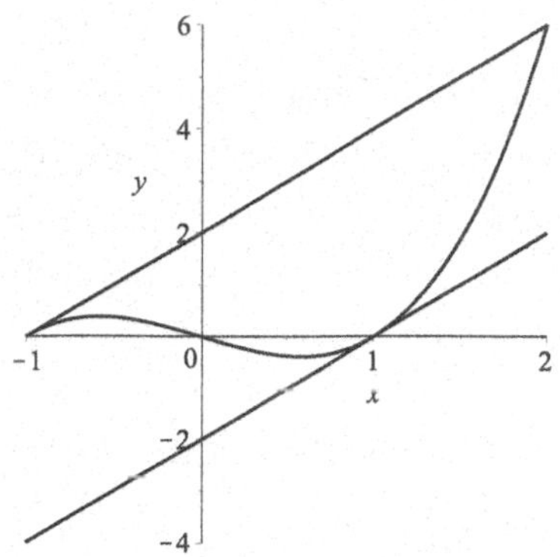

**3.** Simplification gives that $f'(x) = \lim_{h \to 0} \dfrac{1/(x+h)^2 - 1/x^2}{h} = \lim_{h \to 0} \dfrac{-2hx - h^2}{x^2(x+h)^2 h} = \lim_{h \to 0} \dfrac{-2x - h}{x^2(x+h)^2} = -\dfrac{2}{x^3}$.

**5. a.**

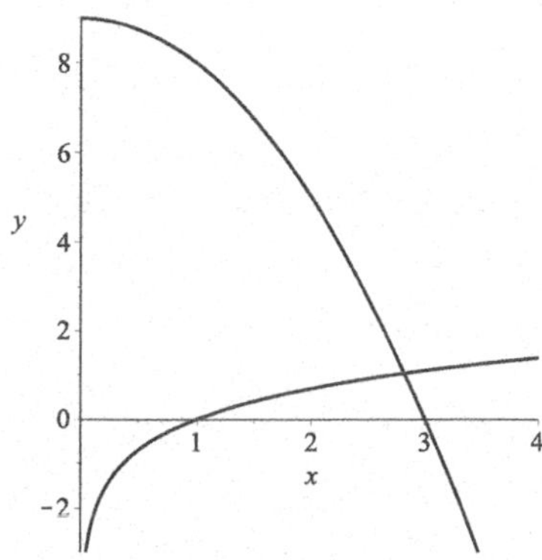

**b.**

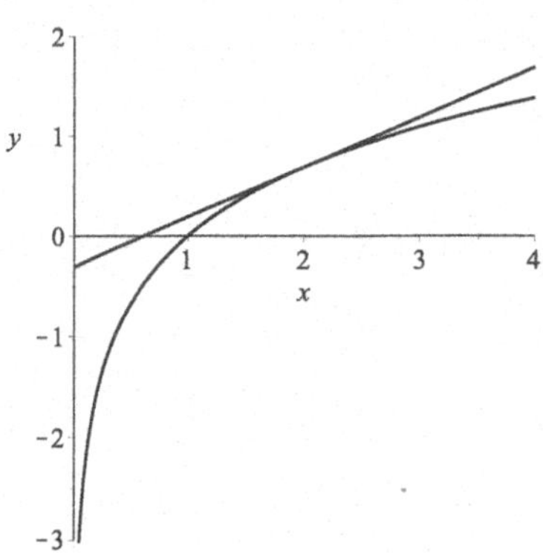

**c.** The derivative of $f$ is $f' = -2x$. Thus the equation of the tangent line at the point $(2, 5)$ is $y - 5 = -4(x - 2)$, i.e. $y = -4x + 13$.

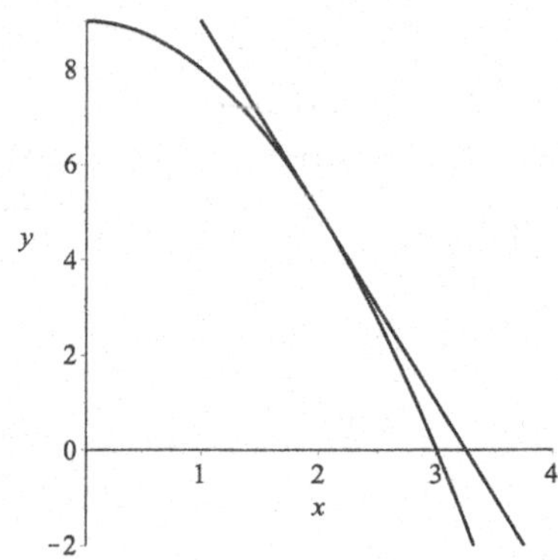

**7. a.** $\lim_{x \to 1} f(x) = \lim_{x \to 1} (2 - 2x) = 0$.

**b.** $\lim_{x \to 2^-} f(x) = \lim_{x \to 2^-} (2 - 2x) = -2$.

**c.** $\lim_{x \to 2^+} f(x) = \lim_{x \to 2^+} -1/(x - 4) = 1/2$.

**d.** $\lim_{x \to 7} f(x) = \lim_{x \to 7} -1/(x - 4) = -1/3$.

**e.** The function is not continuous at $x = 7$. This is reparable, if we define $f$ to be $-1/3$ here.

**9.**

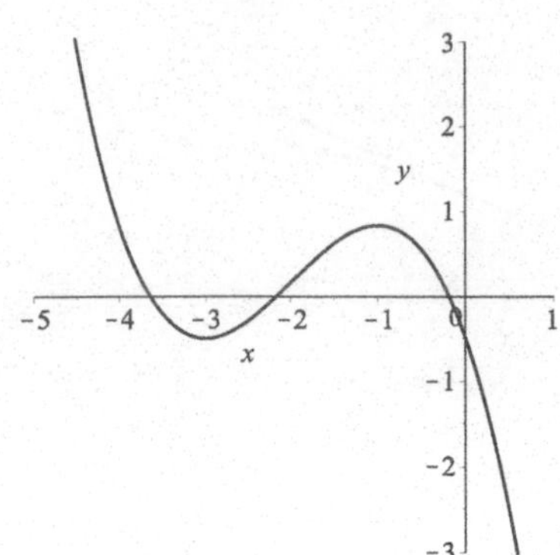


**11. a.** The horizontal and vertical asymptotes are given by $y = \lim_{x\to\pm\infty} \dfrac{2x^2+1}{x^2-2x} = \lim_{x\to\pm\infty} \dfrac{2+1/x^2}{1-2/x^2} = 2$.

**b.** The vertical asymptotes are at $x = 0$ and at $x = 2$. The numerator is always positive, thus $\lim_{x\to 0^-} y = \infty$, $\lim_{x\to 0+} y = -\infty$, $\lim_{x\to 2^-} y = -\infty$, and $\lim_{x\to 2+} y = \infty$.

**13. a.** For $0 < t < 0.1$, $C'(t)$ can be approximated by $(0.2-0)/(0.1-0) = 2$. Similarly, $C'$ is approximately 2, 2, 2, 1, 1, $-1$, 1, $-1$, $-2$ on the next 9 intervals.

**b.** The derivative is positive first, then negative, then positive and then returns to negative. The concentration increases first, then decreases, then increases again and then starts to decrease.

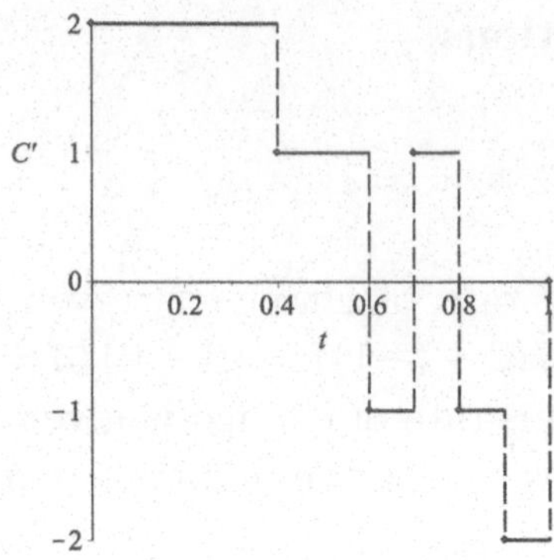


**15.** $\lim_{x\to\infty} e^x/(2+2e^x) = \lim_{x\to\infty} 1/(2e^{-x}+2) = 1/2$. We want to find $x$ so that $|e^x/(2+2e^x) - 1/2| < 10^{-6}/2$. This means we need $|e^x/(2+2e^x) - 1/2| = |(e^x - 1 - e^x)/(2+2e^x)| = (1/2)(1/(1+e^x)) < 10^{-6}/2$. This is the same as $10^6 < 1 + e^x$, so we need that $x > \ln(10^6 - 1) \approx 13.8155$.

**17.** $f$ is a continuous function, $f(0) = -0.1 < 0$ and $f(1) = 1/e - 0.1 > 0$, thus somewhere between 0 and 1 we have a root for $f(x)$.

**19.** Let $N(t)$ be measured in millions. The Mean Value Theorem implies that there is some time instant $a$ between 1981 and 1983, such that $N'(a) = 1.82$, and that there is some time instant $b$ between 1983 and 1985, such that $N'(b) = 1.915$. We also assumed that $N'(t)$ is continuous, so by the Intermediate Value Theorem, there is some $c$ value between $a$ and $b$ such that $N'(c) = 1.85$. This is exactly what we wanted to show.

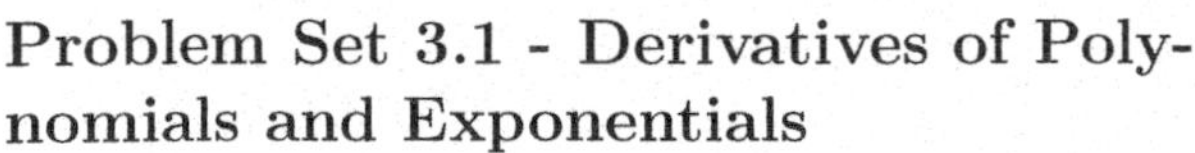
## Problem Set 3.1 - Derivatives of Polynomials and Exponentials

**1. a.** Using the power rule, $f'(x) = 7x^6$.

**b.** Using the derivative rule of exponential functions, $g'(x) = (\ln 7)\, 7^x$.

**3. a.** Using the power rule, $f'(x) = 15x^4$.

**b.** This is a constant, so $g'(x) = 0$.

**5.** Using the power rule, $f'(x) = 2x$ ($3\pi + C$ is a constant so its derivative is 0.)

**b.** Using the power rule, $g'(x) = -2$

($\pi^2 - C$ is a constant so its derivative is 0.)

**7.** Using the sum and power rules, $f'(x) = 5x^4 - 6x$.

**9.** Using the sum, power and exponential rules, $s'(t) = 4e^t - 5$.

**11.** Using the exponential rule, $f'(t) = 5.9(\ln 2.25)2.25^t$.

**13.** Using the sum, power and exponential rules $g'(x) = 2Cx + 5 - 2e^{-2x}$.

**15.** $f'(x) = 3x^2 - 2x = x(3x-2)$; so $f' > 0$ when $x < 0$ and $x > 2/3$ and $f' < 0$ when $0 < x < 2/3$. Thus $f$ is increasing on $(-\infty, 0)$ and $(2/3, \infty)$ and decreasing on $(0, 2/3)$.

**17.** $f'(x) = 5x^4 + 20x^3 - 1650x^2 - 4000x + 60000$. Numerical analysis gives the roots $x \approx -17.86$, $x \approx -7.79$, $x \approx 5.26$ and $x \approx 16.39$. This implies that $f$ is increasing on $(-\infty, -17.86)$, $(-7.79, 5.26)$ and $(16.39, \infty)$, and decreasing on $(-17.86, -7.79)$ and $(5.26, 16.39)$.

**19.** $H'(w) = 2 - e^w$, so $H' > 0$ when $w < \ln 2$ and $H'(w) < 0$ when $w > \ln 2$. Thus $H$ is increasing on $(-\infty, \ln 2)$ and decreasing on $(\ln 2, \infty)$.

**21. a.** Using the definition of the derivative,

$$(x^{3/2})' = (x\sqrt{x})' = \lim_{h\to 0} \frac{(x+h)\sqrt{x+h} - x\sqrt{x}}{h} = \lim_{h\to 0} \frac{x(\sqrt{x+h} - \sqrt{x})}{h} + \frac{h\sqrt{x+h}}{h} =$$

$$= \lim_{h\to 0} \frac{x}{\sqrt{x+h} + \sqrt{x}} + \sqrt{x+h} = \frac{x}{2\sqrt{x}} + \sqrt{x} = \frac{3}{2}\sqrt{x}.$$

**b.** $(x^{3/2})' = (3/2)x^{1/2}$.

**23.** $f(x) = x^{-5/3}$, so $f'(x) = -(5/3)x^{-8/3}$ when $x \neq 0$.

**25.** $h(t) = (3/2)^t + (1/6)^t$, so we obtain $h'(t) = \ln(3/2)(3/2)^t + \ln(1/6)(1/6)^t$.

**27.** $b^x = e^{\ln b^x} = e^{x \ln b}$, so $(b^x)' = (e^{x\ln b})' = (\ln b)e^{x \ln b} = (\ln b)b^x$.

**29.** We compute: $((cf)(x+h) - (cf)(x))/h = c(f(x+h) - f(x))/h$; then using the limit law $\lim_{h\to 0} cg(x) = c\lim_{h\to 0} g(x)$ we obtain that $(cf)' = cf'$.

**31. a.** $dW/dD = -0.41 + 0.34D$ dkg per unit dose.

**b.** $dW/dD > 0$ when $D > 0.41/0.34 \approx 1.21$; so $W$ is increasing when $1.21 < D < 8$.

**33. a.** $10 = N(0) = 50(1 - C)$, so $C = 4/5$.

**b.** $N'(t) = -50(-0.1)(4/5)e^{-0.1t} = 4e^{-t/10}$ and then $N'(5) \approx 2.43$ individuals per day.

**35.** $L'(W) = 1.12(0.95)W^{-0.05}$, so $L'(5) \approx 0.982$ and $L'(50) \approx 0.875$.

**37.** $dL/dM = 20.15(2/3)M^{-1/3}$, thus $L'(100) \approx 2.89$ kg additional weight lifted when going from 100kg to 101kg in competitor's weight.

**39. a.** $N'(t) = -0.63t^2 + 6.08t + 44.05$, thus $N'(10) \approx 41.85$ thousand per year.

**b.** The incidence starts to decline when $N'(t)$ turns negative, i.e. about $t \approx 14.48$ years, so in the middle of year 2014.

## Problem Set 3.2 - Product and Quotient Rules

**1.** Using the product rule, we get $p'(x) = 6x(7+2x^3) + (3x^2-1)(6x^2) = 30x^4 - 6x^2 + 42x$.

**3.** Using the quotient rule, we obtain that $q'(x) = \dfrac{4(3-x^2)-(-2x)(4x-7)}{(3-x^2)^2} = \dfrac{4x^2-14x+12}{(3-x^2)^2}$.

**5.** Using the product rule, we obtain $f'(x) = 1 \cdot 2^x + x(\ln 2)2^x = 2^x(1 + x\ln 2)$.

**7.** Using the product rule, we get $f'(x) = (1+2x)e^x + (1+x+x^2)e^x = e^x(x^2+3x+2)$.

**9.** Using the product rule, we get $F'(L) = (1+3L^2+4L^3)(L-L^2)+(1+L+L^3+L^4)(1-2L) = -6L^5 + 4L^3 - 3L^2 + 1$.

**11.** Using the product rule, we get $f'(x) = 4(4x+3)+(4x+3)4 = 8(4x+3) = 32x+24$.

**13.** Using the quotient rule, we get $f'(x) = \dfrac{e^x(1+e^x)-(e^x)e^x}{(1+e^x)^2} = \dfrac{e^x}{(1+e^x)^2}$.

**15.** Using the quotient rule we obtain that $f'(p) = \dfrac{a(1+2^p)-(\ln 2)2^p(ap)}{(1+2^p)^2} = \dfrac{a(2^p - 2^p p\ln 2 + 1)}{(1+2^p)^2}$.

**17.** Using the power and quotient rules, we obtain that $F'(x) = (-2)2/(3x^3) - 1/3 + (1 \cdot x - 1(x+1))/x^2 = -4/(3x^3) - 1/3 - 1/x^2$.

**19.** $f'(x) = (3x^2-4x)(x+2) + (x^3-2x^2)$, so $f'(1) = -4$ and then the tangent line is given by $y - f(1) = f'(1)(x-1)$, i.e. $y-(-3) = -4(x-1)$, which is $y = -4x+1$.

**21.** $F'(x) = (1(x-1) - 1(x+1))/(x-1)^2$, so $F'(0) = -2$ and then the tangent line is given by $y - F(0) = F'(0)(x-0)$, i.e. $y-(-1) = -2(x-0)$, which is $y = -2x-1$.

**23.** We obtain that $F'(x) = (6x(2x^2+x-3) - (3x^2+5)(4x+1))/(2x^2+x-3)^2$, so $F'(-1) = 9$ and then the tangent line is given by $y - F(-1) = F'(-1)(x-(-1))$, i.e. $y-(-4) = 9(x+1)$, which is $y = 9x+5$.

**25. a.** Using the power rule, $f'(x) = 4x - 5$.

**b.** We find the roots of $2x^2 - 5x - 3 = 0$ with the aid of the quadratic formula and obtain $x = (5 \pm \sqrt{25+24})/4 = 3, -1/2$. Thus the factorization is $f(x) = 2(x-3)(x+1/2) = (x-3)(2x+1)$. Then $f'(x) = 1(2x+1) + 2(x-3) = 4x-5$. This is the same as part **a**.

**27. a.** $B = 703w/63^2$, thus $dB/dw = 703/63^2 \approx 0.177$; the BMI increase is constant 0.177 per pound increase in weight.

**b.** $B = 703 \cdot 60/h^2$, thus we obtain $dB/dh = -2(42180)h^{-3} = -84360/h^3$; so $B'(54) \approx -0.536$. The BMI decreases by 0.536 units per inches for individuals that weigh 60 pounds at height 54 inches.

**29.** We obtain that the probability of survival is $p(4) = f(4)g(4) = 0.984 \cdot 0.996 = 0.9801$.

**b.** Using the product rule, we get $p'(4) =$

$f'(4)g(4) + f(4)g'(4) = (0.004) \cdot 0.996 + (0.001) \cdot 0.984 = 0.0050$.

**c.** The approximation is $0.9801 + 0.0050 = 0.9851$.

**31. a.** The graph shows the case $a = 1$, $b = 2$, and $c = 1$. $a$ gives the horizontal asymptote. The value at $x = c$ is $(a+b)/2$, so $b$ and $c$ determines the steepness of the decline.

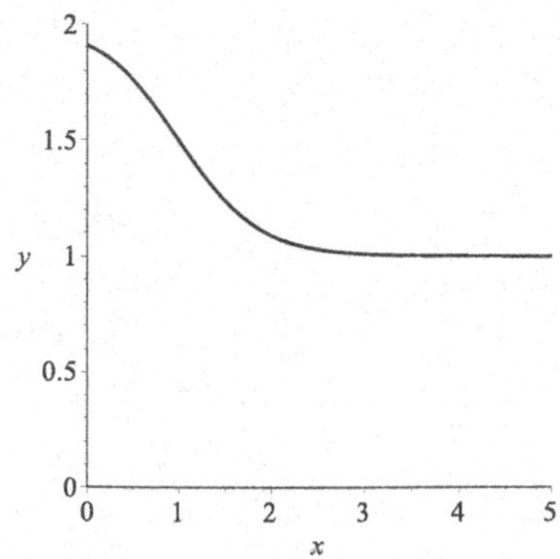


**b.** For the derivative, we obtain $dT/dx = -(b-a)(\ln 10)10^{x-c}/(1+10^{x-c})^2 < 0$ for all $x > 0$, because $b > a$; thus $T(x)$ is decreasing.

**33.** We obtain that $f'(x) = (3.36(0.46+x) - 3.36x)/(0.46+x)^2 = 1.5456/(0.46+x)^2$, thus $f'(1/2) \approx 1.677$ and $f'(2) \approx 0.255$ increase in moose killed per wolf per 100 days per number of moose per km$^2$.

**35.** We obtain that $f'(x) = (58.7(0.76+x) - 58.7(x-0.03))/(0.76+x)^2 = 46.373/(0.76+x)^2 > 0$, thus $f$ is increasing for all $x \geq 0$.

**37.** First, $\operatorname{sech}^2 x = 4/(e^x + e^{-x})^2 = 4/(e^{2x} + 2 + e^{-2x})$. Thus $f(t) = 890 \operatorname{sech}^2(0.2t - 3.4) = 890 \cdot 4/(e^{-6.8}e^{0.4t} + 2 + e^{6.8}e^{-0.4t})$ and then we get $f'(t) = 3560(0.4e^{-6.8}e^{0.4t} - 0.4e^{6.8}e^{-0.4t})/(e^{-6.8}e^{0.4t} + 2 + e^{6.8}e^{-0.4t})^2$. The function is increasing when $f' > 0$, i.e. when $0 \leq t < 17$ and decreasing when $f' < 0$, i.e. when $t > 17$ (weeks).

## Problem Set 3.3 - Chain Rule and Implicit Differentiation

**1.** $dy/dx = 2u(3) = 6u = 6(3x-2)$.

**3.** $dy/dx = (-4/u^3)2x = -8x/(x^2-9)^3$.

**5. a.** $dg/du = 5u^4$.

**b.** $du/dx = 3$.

**c.** $df/dx = 5(3x-1)^4 \cdot 3 = 15(3x-1)^4$.

**7. a.** $dg/du = 15u^{14}$.

**b.** $du/dx = 6x + 5$.

**c.** $df/dx = 15(3x^2 + 5x - 7)^{14}(6x+5)$.

**9.** Using the chain rule, we get $dy/dx = 9(5 - x + x^4)^8(-1 + 4x^3)$.

**11.** Using the chain rule, we get $dy/dx = (-12)(1 + x - x^5)^{-13}(1 - 5x^4)$.

**13.** Using the chain rule, we get $dy/dx = (1/x^2)2x = 2/x$.

**15.** Using the chain rule, $dy/dx = 2/(2x+5)$.

**17.** Using the product rule and chain rule, we get $dy/dx = 10(x^4-1)^9 4x^3(2x^4+3)^7 + (x^4-1)^{10}7(2x^4+3)^6 8x^3 = 8x^3(x^4-1)^9(2x^4+3)^6(17x^4+8)$.

**19.** Differentiate both sides with respect to $x$, using the chain rule: $2x + dy/dx = 3x^2 + 3y^2 dy/dx$; solve for $dy/dx$ to obtain $dy/dx = (3x^2 - 2x)/(1 - 3y^2)$.

**21.** Differentiate both sides with respect to $x$, using the product rule: $y(2x+3y) + x dy/dx(2x+3y) + xy(2 + 3dy/dx) = 0$;

solve for $dy/dx$ to obtain that $dy/dx = (-4xy - 3y^2)/(2x^2 + 6yx)$.

**23.** Differentiate both sides with respect to $x$, using the chain rule. We obtain that $2(2x+3y)(2+3dy/dx) = 0$; solve for $dy/dx$ to obtain $dy/dx = -2/3$.

**25.** Differentiate both sides with respect to $x$, using the chain rule: $e^{xy}(y + xdy/dx) + (1/y^2)2ydy/dx = 1$; solve for $dy/dx$ to obtain $dy/dx = (1 - ye^{xy})/(xe^{xy} + (2/y))$.

**27. a.** $u(2) = 5$, slope of the tangent line is about $1/2$.

**b.** $g(2) = 5$, thus $y = f(g(2)) = f(5) = 3$; slope of the tangent line is about $3/2$.

**c.** We get $(f \circ g)'(2) = f'(g(2))g'(2) = f'(5)g'(2) = (3/2) \cdot (1/2) = 3/4$.

**29. a.** Using the chain rule, we get $g'(x) = f'(3x-1)3 = 3/((3x-1)^2 + 1)$.

**b.** Using the chain rule, $h'(x) = f'(1/x)(-1/x^2) = (-1/x^2)(1/(1/x^2+1)) = -1/(1+x^2)$.

**31.** Using implicit differentiation, $3x^2 + 3y^2dy/dx - (9/2)y - (9/2)xdy/dx = 0$, thus $dy/dx = ((9/2)y - 3x^2)/(3y^2 - (9/2)x)$. The point $(2, 1)$ is clearly on the curve, and $dy/dx = (9/2 - 12)/(3 - 9) = 5/4$ there. Thus the tangent line is given by $y - 1 = (5/4)(x - 2)$, which is $y = 5x/4 - 3/2$.

**33.** We know that the cross-sectional area covered by plaque is $A = \pi R^2 - \pi(R - p(t))^2$. $dA/dt = -2\pi(R - p(t))p'(t) = -2\pi(R - p(t))(-1/2)R(0.009)(12350 - t^2)^{-1/2}(-2t)$. When $t = 60$, we obtain $dA/dt \approx 0.0305R^2$.

**35.** We obtain $df/dt = df/dx \cdot dx/dt = ((3.36(0.46+x) - 3.36x)/(0.46+x)^2) \cdot dx/dt = (1.5456/(0.46+x)^2) \cdot dx/dt$. Thus $df/dt = (1.5456/(0.46+0.5)^2) \cdot 0.1 \approx 0.168$ moose per wolf per 100 days per year.

**37.** We obtain $df/dt = df/dx \cdot dx/dt = -0.02e^{-0.02x}dx/dt = -0.02e^{-0.02 \cdot 10} \cdot 20 \approx -0.3275$.

**39.** When $x = 1/2$, the $y$ values are given by the solutions of $y^2(1-1/4) = (1/4+2y-1)^2$, which is the same as $13y^2/4 - 3y + 9/16 = 0$. Thus the $y$ values are $y_1 = (12 + 3\sqrt{3})/26$ and $y_2 = (12 - 3\sqrt{3})/26$. Differentiate both sides of the equation by $x$: $2ydy/dx(1-x^2) + y^2(-2x) = 2(x^2 + 2y - 1)(2x + 2dy/dx)$. From here, $dy/dx = (4x(x^2 + 2y - 1) + 2xy^2)/(2y(1 - x^2) - 4(x^2 + 2y - 1))$. Plugging in the corresponding values and simplifying, we get the two lines $y - y_1 = ((-128 - 45\sqrt{3})/169)(x - /12)$ and $y - y_2 = ((-128 + 45\sqrt{3})/169)(x - /12)$.

**41.** When $b \neq 1$, we know that $\log_b x = \ln x / \ln b$. Thus $(\log_b x)' = (\ln x/\ln b)' = (1/x)(1/\ln b) = 1/(x \ln b)$.

**43.** Clearly, $h'(t) = 0.19$ inches per month. Also, $dW/dt = (2.6)0.0024h^{1.6}0.19 = 0.72$ pounds per month (where $h = 32+0.19 \cdot 120$).

## Problem Set 3.4 - Derivatives of Trigonometric Functions

**1.** Using the sum rule, $f'(x) = \cos x - \sin x$.

**3.** Using the chain rule, $dy/dx = 2\cos 2x$.

**5.** Using the sum rule, $f'(t) = 2t - \sin t$.

**7.** Using the product rule, $dy/dx = -e^{-x}\sin x + e^{-x}\cos x = e^{-x}(\cos x - \sin x)$.

**9.** Using the chain rule, we get $f'(\theta) =$

$2\sin\theta\cos\theta = \sin 2\theta$.

**11.** Using the chain rule, we get $dy/dx = -101x^{100}\sin(x^{101})$.

**13.** Using the product rule, $p'(t) = 2t\sin t + (t^2+2)\cos t$.

**15.** Using the quotient rule, we get $q'(t) = (t\cos t - \sin t)/t^2$.

**17.** Using the quotient rule, we get $g'(x) = (1\cdot(1-\sin x) - x(-\cos x))/(1-\sin x)^2 = (1-\sin x + x\cos x)/(1-\sin x)^2$.

**19.** Using the chain rule, we obtain $y'(x) = 1/(\sin x + \cos x)\cdot(\cos x - \sin x) = (\cos x - \sin x)/(\sin x + \cos x)$.

**21.** Using the chain rule, we obtain that $f'(x) = -\cos^{-2}x(-\sin x) = \sin x/\cos^2 x = \tan x\sec x$.

**23.** Using the chain rule, we obtain $f'(x) = (-1)\tan^{-2}x(\sec^2 x) = -\csc^2 x$.

**25.** Using the quotient rule, we obtain $y'(x) = ((\sec x\tan x + \sec^2 x)(\csc x + \cot x) - (-\csc x\cot x - \csc^2 x)(\sec x + \tan x))/(\csc x + \cot x)^2$.

**27.** Using the chain rule, we get $y'(x) = (1/\sin^2 x)2\sin x\cos x = 2\cot x$.

**29. a.** The area is $g(h) = (1/2)\cos h\sin h$.

**b.** The area of the sector is $f(h) = h/2$, because $r = 1$.

**c.** The area is $k(h) = 1\cdot\tan h/2$.

**d.** When $h > 0$ small, we see that $g(h) < f(h) < k(h)$, which is the same as $\cos h\sin h < h < \tan h$. Take the reciprocal of every term, and multiply by the positive expression $\sin h$; we obtain $\cos h < \sin h/h < 1/\cos h$. The limit of the left and right sides is 1 as $h \to 0$, so by the squeeze theorem we obtain that $\lim\limits_{h\to 0}\sin h/h = 1$.

**31.** $P'(t) = -100e^{-t}\sin t + 100e^{-t}\cos t$, so $P'(2) \approx -17.94$. The population is decreasing, at a rate about 18 fish per month.

**33. a.** $P'(t) = 5000e^{\cos t + t}(-\sin t + 1)$, thus $r(t) = 1 - \sin t$.

**b.** The period is $2\pi \approx 6.28$ (hours).

**c.** $P(t)$ is never decreasing, because $r'(t) \geq 0$ and $P(t) > 0$, thus $P'(t) \geq 0$.

**35. a.** We get $T'(t) = -4\cos(\pi t/12)\pi/12$, thus $T'(12) \approx 1.047$ degrees/h. This is the rate of change of the water temperature at noon.

**b.** The temperature is increasing when $T' > 0$; this happens when $6 < t < 18$, thus between 6am and 6pm.

## Problem Set 3.5 - Linear Approximation

**1.** $y' = -\sin x$, thus the linear approximation is $y - 0 = -(x - \pi/2)$, which is $y = -x + \pi/2$; the figure shows that the linear approximation overestimates for $x < \pi/2$ and underestimates for $x > \pi/2$.

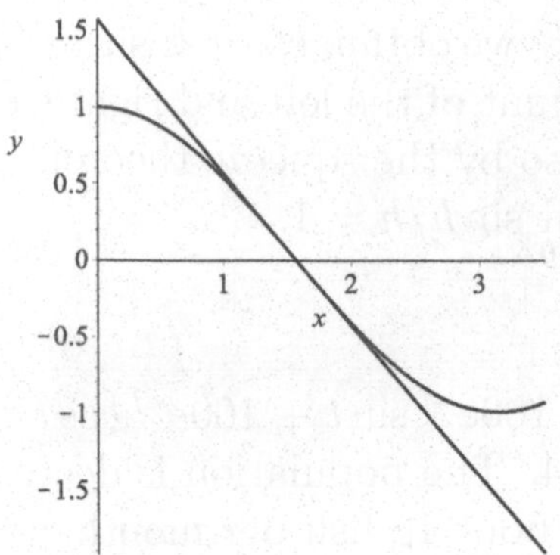

**3.** $y' = \cos x$, thus the linear approximation is $y - 1 = 0(x - \pi/2)$, which is $y = 1$; the figure shows that the linear approximation overestimates the function.

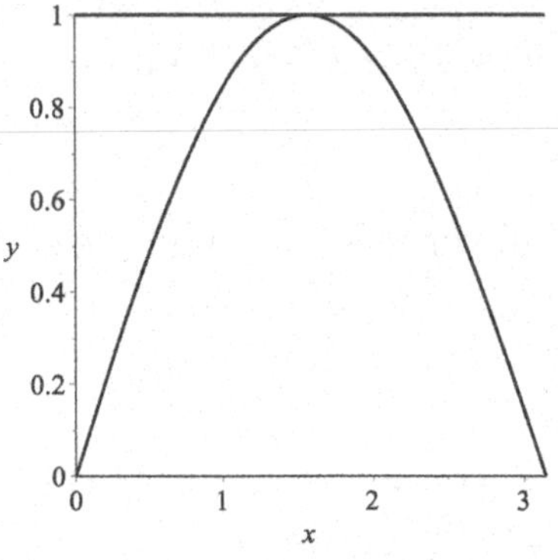

**5.** $y' = (-1)(1 + x^2)^{-2}(2x)$, thus the linear approximation is $y - 1/5 = (-4/25)(x - 2)$, which is $y = -4x/25 + 13/25$; the figure shows that the linear approximation underestimates the function.

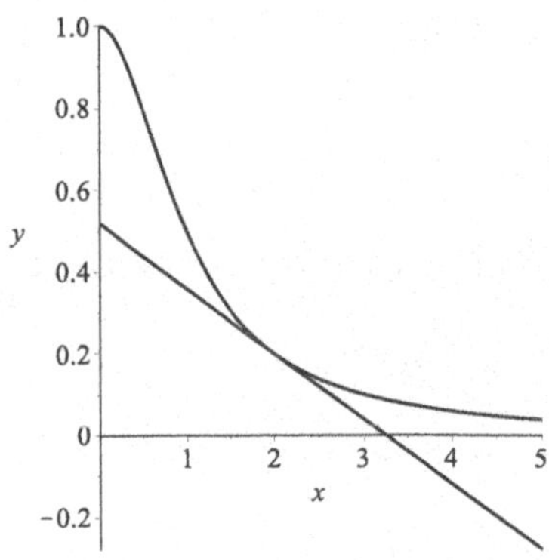

**7.** Let $f(x) = \sqrt{x}$; then $f'(x) = 1/(2\sqrt{x})$. Using linear approximation, $f(26) \approx f(25) + f'(25)(26 - 25) = 5 + 1/10 = 5.1$. The true value is $\sqrt{26} \approx 5.09902$.

**9.** Let $f(x) = \ln x$; then $f'(x) = 1/x$. Using linear approximation, $f(0.9) \approx f(1) + f'(1)(0.9 - 1) = 0 - 0.1 = -0.1$. The true value is $\ln 0.9 \approx -0.105361$.

**11.** Let $f(x) = \tan x$; then $f'(x) = \sec^2 x$. Using linear approximation, $f(0.2) \approx f(0) + f'(0)(0.2 - 0) = 0 + (1)0.2 = 0.2$. The true value is $\tan 0.2 \approx 0.20271$.

**13.** We differentiate: $y' = 1/(2\sqrt{x})$, thus the sensitivity is $1/(2\sqrt{9}) = 1/6$; $\Delta y = \Delta x/6 = 0.01/6 \approx 0.001667$.

**15.** We differentiate: $y' = 1/x$, thus the sensitivity is $1/2$; $\Delta y = \Delta x/2 = -0.1$.

**17.** We differentiate: $y' = -\sin x$, thus the sensitivity is $-\sin(\pi/2) = -1$; $\Delta y = -\Delta x = 0.01$.

**19.** The derivative is $y' = 1/(2\sqrt{x})$, thus the elasticity is $E = 9(1/(2\sqrt{9}))/\sqrt{9} = 1/2$. The error in $y$ is $1/2\%$.

**21.** The derivative is $y' = 1/x$, thus the elasticity is $E = (2/2)/\ln 2 \approx 1.44$. The error in $y$ is $1.44 \cdot 5 = 7.2\%$.

**23.** The derivative is $y' = \cos x$, so the elasticity is $E = (\pi/2)\cos(\pi/2)/\sin(\pi/2) = 0$.

**25. a.** $D'(T) = 0.021(T - 11.9)\sqrt{37 - T} + 0.021T\sqrt{37 - T} - 0.021T(T - 11.9)(1/2)(37 - T)^{-1/2}$, thus $D'(20) \approx 2.02$ and the linear approximation is $D(T) \approx D(20) + D'(20)(T - 20) = 14.03 + 2.02(T - 20) = 2.02T - 26.37$.

**b.**

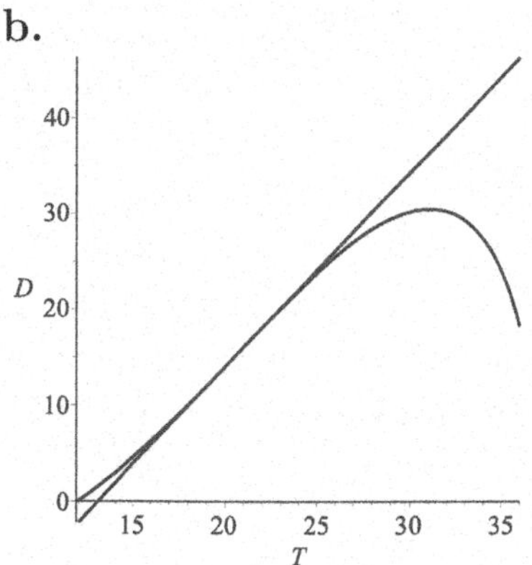


**27.** $A'(r) = 2\pi r$, thus the elasticity is $E = 2\pi r/(\pi r^2)r = 2$, and the accuracy for the calculation of $A$ is within $2 \cdot 3 = 6\%$.

**29.** $Q'(t) = 0.1t + 0.1$, and $\Delta Q \approx Q'(0)\Delta t = 0.1 \cdot (1/2) = 0.05$ ppm.

**31. a.** We differentiate to obtain $E'(T) = 4.6e^{17.3T/(T+237)}(17.3(T+237) - 17.3T)/(T+237)^2$, thus the elasticity is $TE'(T)/E(T) = 4100.1T/(T+237)^2$, which is approximately 1.725 when $T = 30$.

**b.** The percentage change is about $1.725 \cdot 5 \approx 8.63\%$.

**33. a.** We obtain that $T'(x) = -\ln 2/x^2$, thus the elasticity is $E = xT'(x)/T(x) = -x\ln 2/x^2/(\ln 2/x) = -1$.

**b.** Because $E = -1$, a change of 2% in $x$ causes a change of 2% in $T$. The answer is thus $\pm 2\%$.

**35.** $S'(R) = 2cR$, thus the elasticity is $E = R2cR/cR^2 = 2$, and then the error is $2 \cdot 1 = 2\%$.

**37. a.** $S'(t) = 0.023 - 0.008t$, thus the sensitivity in 1980 is $S'(0) = 0.023$.

**b.** The sensitivity in 2000 is $S'(20) = -0.137$.

**39. a.** $T'(y) = -(b-a)k(1+ky)^{-2}$, thus the elasticity is given by $E = yT'(y)/T(y) = y(a-b)k/((1+ky)(b+aky))$.

**b.** We obtain that the error is given by $-y/((1+y)(2+y)) \cdot 10\%$.

## Problem Set 3.6 - Higher Derivatives and Approximations

**1.** $f'(x) = e^{-x} - xe^{-x} = (1-x)e^{-x}$, thus $f''(x) = (-1)e^{-x} - (1-x)e^{-x} = (x-2)e^{-x}$.

**3.** $f'(x) = 1 + 2x + 3x^2 + 4x^3$; $f''(x) = 2 + 6x + 12x^2$; $f'''(x) = 6 + 24x$, and then $f^{(4)}(x) = 24$.

**5.** $f'(x) = 3\cos 3x$; $f''(x) = -3^2 \sin 3x$; $f'''(x) = -3^3 \cos 3x$, and $f^{(4)}(x) = 3^4 \sin 3x$. Thus every fourth derivative contains the sine function, and we multiply by 3 after every differentiation because of the chain rule. We obtain $f^{(99)}(x) = -3^{99}\cos 3x$.

**7.** $f'(t) = 2t^7 - 3t^5 - 2t$; $f''(t) = 14t^6 - 15t^4 - 2$; $f'''(t) = 84t^5 - 60t^3$, and then $f^{(4)}(t) = 420t^4 - 180t^2 = 60t^2(7t^2 - 3)$.

**9.** $f'(x) = 10(1+x)^9$; $f''(x) = 10 \cdot 9(1+x)^8$; $f'''(x) = 10 \cdot 9 \cdot 8(1+x)^7$, continuing, we obtain that $f^{(10)}(x) = 10! = 3628800$.

**11.** $f(w) = (1+w)^{-1}$, so we get $f'(w) = (-1)(1+w)^{-2}$ and $f''(w) = 2(1+w)^{-3}$.

**13.** The velocity is $v(t) = s'(t) = 2t - 3$, and the acceleration is $a(t) = v'(t) = s''(t) = 2$. The acceleration is never 0 in this case.

**15.** The velocity is given by $v(t) = s'(t) = (1/3)\sin(t/3)$, and the acceleration is $a(t) =$

$v'(t) = s''(t) = (1/9)\cos(t/3)$. The acceleration is 0 when $t = 3\pi/2$ and $t = 9\pi/2$.

**17.** The velocity is $v(t) = s'(t) = e^{-t} - te^{-t} = (1-t)e^{-t}$, and the acceleration is $a(t) = v'(t) = s''(t) = -e^{-t} - (1-t)e^{-t} = (t-2)e^{-t}$. The acceleration is 0 when $t = 2$.

**19.** $f'(x) = e^x$, thus the linear approximation is $e^0 + e^0(x-0) = 1+x$. We differentiate again to get $f''(x) = e^x$, thus $f''(0) = 1 > 0$ and the linear approximation underestimates the function.

**21.** $f'(x) = -2x$, thus the linear approximation is $(1-4)+(-4)(x-2) = 5-4x$. We differentiate again to get $f''(x) = -2$, thus $f''(0) = -2 < 0$ and the linear approximation overestimates the function.

**23.** $f'(x) = -1/(1+x)^2$, thus the linear approximation is $1/(1+2)+(-1/9)(x-2) = -x/9+5/9$. We differentiate again to get $f''(x) = 2(1+x)^{-3}$, thus $f''(2) = 2/27 > 0$ and the linear approximation underestimates the function.

**25.** $f'(x) = -1+3x^2$, $f''(x) = 6x$. This implies that $f'(x) = 0$ when $x = \pm 1/\sqrt{3}$; checking the sign of $f'$ gives that $f$ is increasing on $(-\infty, -1/\sqrt{3})$ and $(1/\sqrt{3}, \infty)$, and decreasing on $(-1/\sqrt{3}, 1/\sqrt{3})$. $f''(x) = 6x$, thus $f$ is concave down on $(-\infty, 0)$ and concave up on $(0, \infty)$. The inflection point is at $x = 0$.

**27.** $f'(x) = e^{-x} - xe^{-x} = (1-x)e^{-x}$, and $f''(x) = -e^{-x}-(1-x)e^{-x} = (x-2)e^{-x}$. Now $f'(x) = 0$ when $x = 1$; checking the sign of $f'$ gives that $f$ is increasing on $(-\infty, 1)$, and decreasing on $(1, \infty)$. $f''(x) = (x-2)e^{-x}$, thus $f$ is concave down on $(-\infty, 2)$ and concave up on $(2, \infty)$. The inflection point is at $x = 2$.

**29.** $f'(x) = (1+x-x)/(1+x)^2 = 1/(1+x)^2$, $f''(x) = -2/(1+x)^3$. The function is not defined at $x = -1$. Now $f'(x) \neq 0$; checking the sign of $f'$ gives that $f$ is increasing on $(-\infty, -1)$ and $(-1, \infty)$. Also, $f''(x) = -2/(1+x)^3$, thus $f$ is concave up on $(-\infty, -1)$ and concave down on $(-1, \infty)$. There is no inflection point.

**31.** We get $f'(x) = 12x^3 - 6x^2 - 24x + 18 = 12(x-1)^2(x+3/2)$, $f''(x) = 36x^2 - 12x - 24 = 36(x-1)(x+2/3)$. Now $f'(x) = 0$ when $x = 1$ and $x = -3/2$; checking the sign of $f'$ gives that $f$ is increasing on $(-3/2, \infty)$, and decreasing on $(-\infty, -3/2)$. $f''(x) = 36(x-1)(x+2/3)$, thus $f$ is concave down on $(-2/3, 1)$ and concave up on $(-\infty, -2/3)$ and $(1, \infty)$. The inflection points are at $x = -2/3$ and $x = 1$.

**33.** The function is not defined at $x = \pi/2 + k\pi$, $k = 0, \pm 1, \pm 2, \ldots$. $f'(x) = \sec x \tan x$, and then $f''(x) = \sec x \tan^2 x + \sec^3 x = \sec x(\tan^2 x + \sec^2 x) = \sec^3 x(\sin^2 x + 1)$. Now $f'(x) = 0$ when $x = n\pi$, where $n = 0, \pm 1, \pm 2, \ldots$. Checking the sign of $f'$, we see $f$ is increasing on the intervals $(2n\pi, 2n\pi + \pi/2)$ and $(2n\pi + \pi/2, (2n+1)\pi)$, and decreasing on the intervals $((2n+1)\pi, (2n+1)\pi + \pi/2)$ and $((2n+1)\pi + \pi/2, (2n+2)\pi)$. The second derivative shows that $y$ is concave up on the intervals $(2n\pi - \pi/2, 2n\pi + \pi/2)$ and concave down on the intervals $((2n+1)\pi - \pi/2, (2n+1)\pi + \pi/2)$. There are no inflection points.

**35.** $f'(x) = \cos x$, $f''(x) = -\sin x$; the first and second order approximations are the same: $\sin 0 + \cos 0(x-0) + \sin 0(x-0)^2/2 = x$.

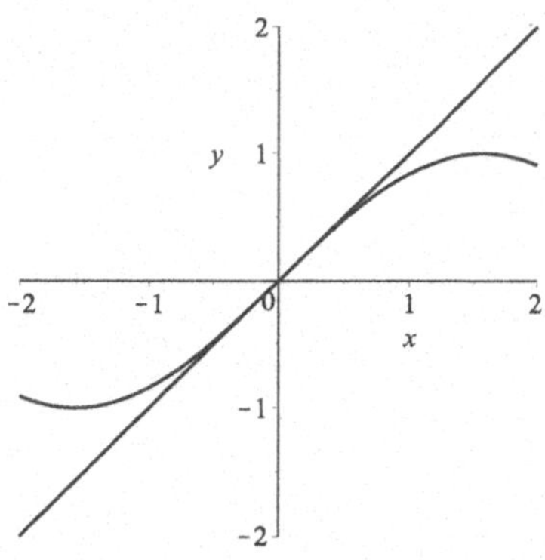


**37.** $f'(x) = e^x$, $f''(x) = e^x$; the first order approximation is $e^0 + e^0 x = 1 + x$, the second order approximation is $e^0 + e^0 x + e^0 x^2/2 = 1 + x + x^2/2$.

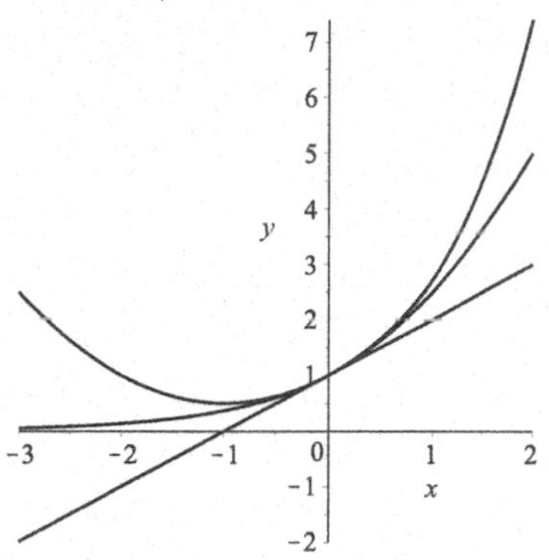


**39.** We get $f'(x) = (1/2)x^{-1/2}$, $f''(x) = (-1/4)x^{-3/2}$; the first order approximation is $4^{1/2} + (1/2)4^{-1/2}(x - 4) = x/4 + 1$, the second order approximation is $4^{1/2} + (1/2)4^{-1/2}(x-4) + (-1/4)4^{-3/2}(x-4)^2/2 = 3/4 + 3x/8 - x^2/64$.

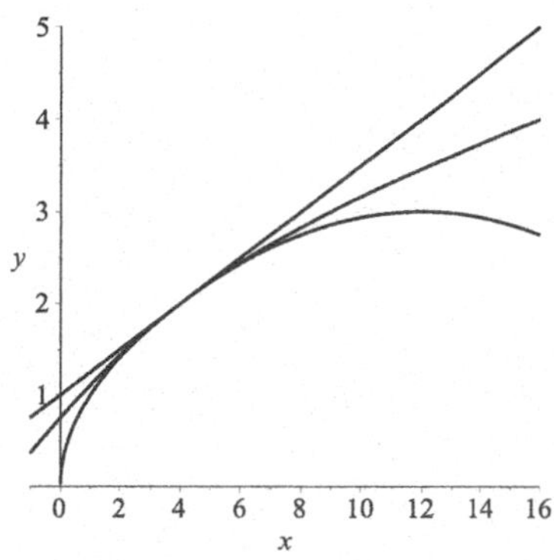


**41.** Blue: $f$, black: $f'$, red: $f''$; the derivative of a linear function is a constant, the derivative of the constant is 0.

**43.** Black: $f$, red: $f'$, blue: $f''$; the black curve is decreasing, then increasing, which is reflected in the behavior of the red curve; the red curve is decreasing-increasing-decreasing, which is reflected in the behavior of the blue curve.

**45.**

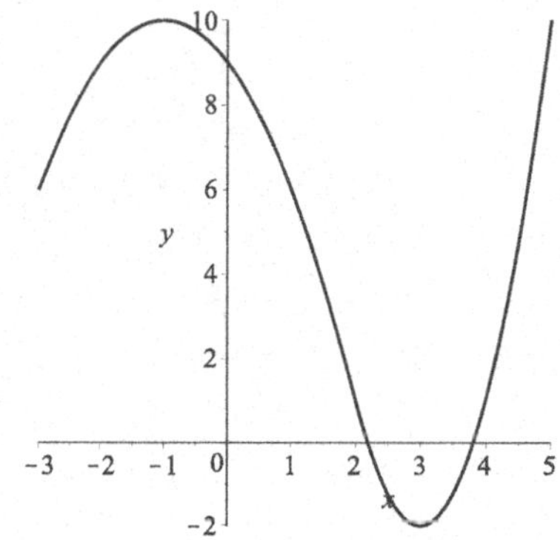


**47.** Home improvement is quality increase, so an improvement in home improvement is an increase in the quality increase, thus it is the second derivative of quality.

**49. a.** It seems the function is concave up on $(1980, 1987)$ and concave down on $(1987, 1995)$.

**b.** It means that the spread of the epidemic is slowing.

**51.** We obtain that $y' = -a^3(x^2+a^2)^{-2}2x = -2xa^3/(x^2+a^2)^2$, which means the function is increasing on $(-\infty, 0)$ and decreasing on $(0, \infty)$ (by assumption, $a > 0$). Another differentiation gives $y'' = (-2a^3(x^2 + a^2)^2 + 2xa^3 2(x^2 + a^2)2x)/(x^2 + a^2)^4 = 2a^3(3x^2 - a^2)/(x^2 + a^2)^3$. Thus the function is concave up on $(-\infty, -a/\sqrt{3})$ and $(a/\sqrt{3}, \infty)$, and concave down on $(-a/\sqrt{3}, a/\sqrt{3})$.

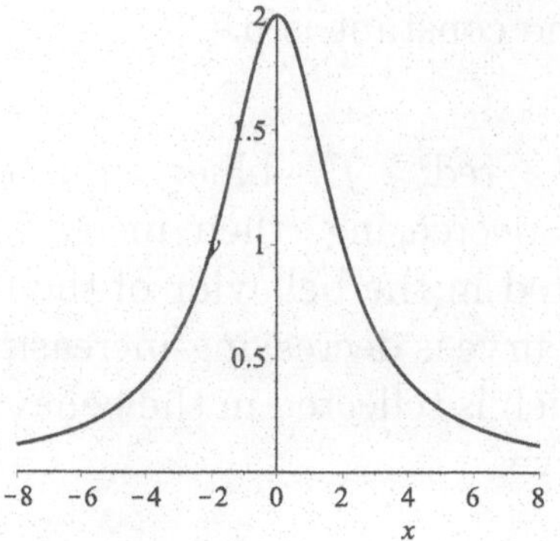


**53.** Using technology, we obtain $s(t) = -0.066t^3 + 1.245t^2 + 4.653t - 0.281$; this means $v(t) = s'(t) = -0.198t^2 + 2.49t + 4.653$ and $a(t) = v'(t) = s''(t) = -0.396t + 2.49$. The acceleration is zero at $t = 2.49/0.396 \approx 6.29$ seconds, where the maximum velocity $v(6.29) \approx 12.48$ m/s is reached. The average acceleration from this time until the end of the race is $(v(9.52) - v(6.29))/(9.52 - 6.29) = -0.64$ m/s$^2$. The figures show the distance, velocity and acceleration, respectively.

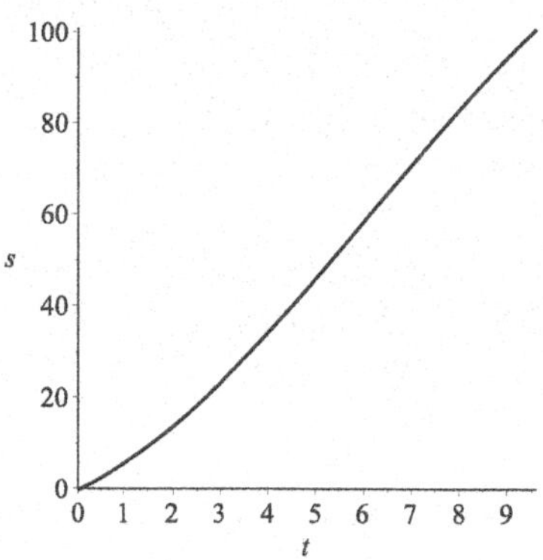


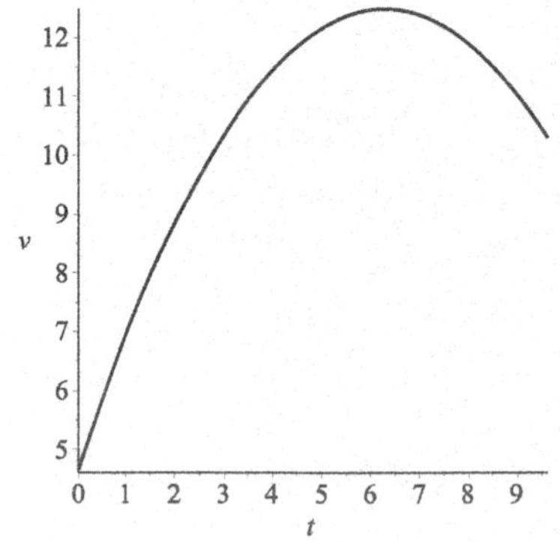


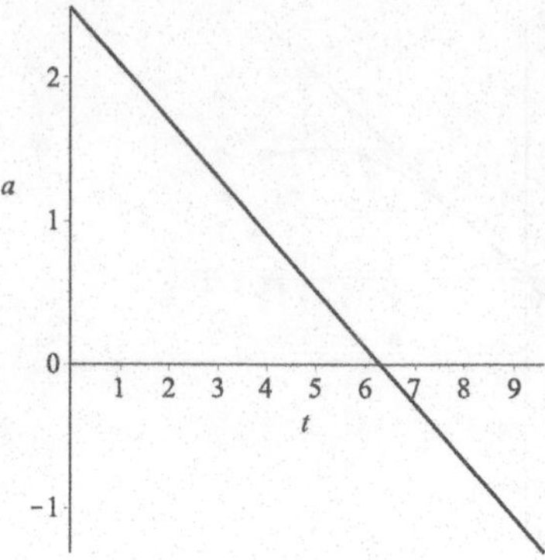


**55.** Using technology, we obtain $s(t) = -0.000571t^5 + 0.0209t^4 - 0.327738t^3 + 2.6125t^2 + 1.68505t - 0.00401$; this means $v(t) = s'(t) = -0.0028565t^4 + 0.083608t^3 - 0.983214t^2 + 5.225t + 1.68505$ and $a(t) = v'(t) = s''(t) = -0.011426t^3 + 0.25082t^2 - 1.966428t + 5.225$. The acceleration is zero at $t \approx 5.94$ seconds, where the maximum velocity $v(5.94) \approx 11.99$ m/s is reached. The average acceleration from this time until the end of the race is $(v(9.79) - v(5.94))/(9.79 - 5.94) = -0.31$ m/s$^2$. The figures show the distance, velocity and acceleration, respectively.

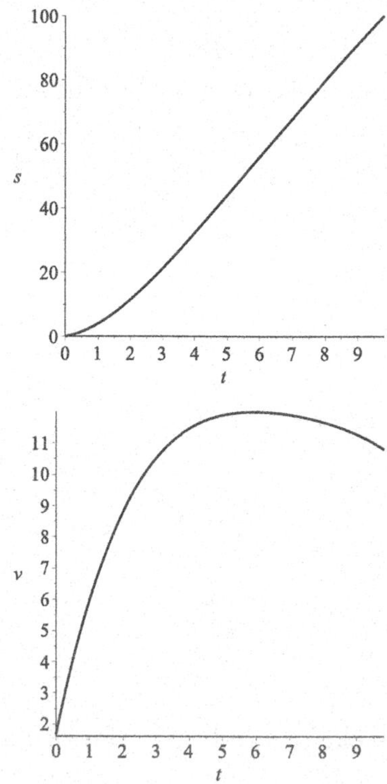

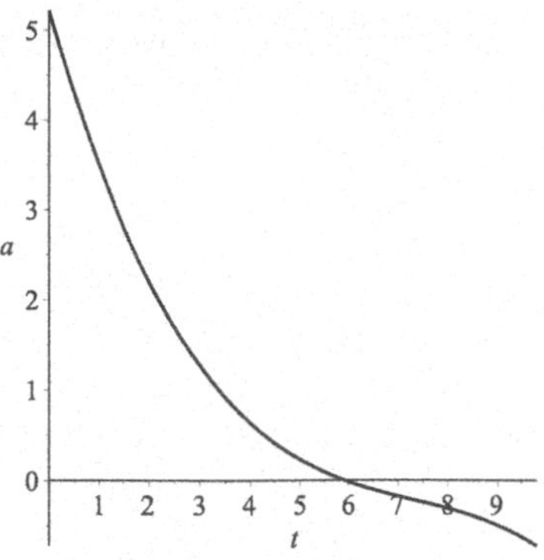


**57.** $q(a) = b$, $q'(a) = c$ and $q''(a) = 2d$ by differentiating $q(x)$ as given. Now $q(a) = f(a)$ if and only if $b = f(a)$; $q'(a) = f'(a)$ if and only if $f'(a) = c$ and $q''(a) = f''(a)$ if and only if $f''(a) = 2d$.

## Problem Set 3.7 - l'Hôpital's Rule

**1. a.** The numerator and denominator do not converge to 0; the limit (using the quotient limit law) is $\lim_{x\to\pi} \frac{1-\cos x}{x} = 2/\pi$.

**b.** The numerator and denominator do not converge to 0; the limit (using the quotient limit law) is $\lim_{x\to\pi/2} \frac{\sin x}{x} = 2/\pi$.

**3.** This is an indeterminate form; using l'Hôpital's rule, $\lim_{x\to1} \frac{x^3-1}{x^2-1} \overset{H}{=} \lim_{x\to1} \frac{3x^2}{2x} = 3/2$.

**5.** This is an indeterminate form; using l'Hôpital's rule, we get $\lim_{x\to0} \frac{1-\cos^2 x}{\sin^2 x} \overset{H}{=} \lim_{x\to0} \frac{2\cos x \sin x}{2\sin x \cos x} = 1$. Also, $1 - \cos^2 x = \sin^2 x$, thus l'Hôpital's rule is actually unnecessary here as this quotient is simply 1 for any value of $x \neq 0$.

**7.** This is an indeterminate form; using l'Hôpital's rule, $\lim_{x\to\infty} \frac{\ln x}{x^5} \overset{H}{=} \lim_{x\to\infty} \frac{1/x}{5x^4} = 0$.

**9.** This is $\lim_{x\to0^+} \frac{\sin x}{\ln x} = 0$; there is no need for l'Hôpital's rule.

**11.** Consider the natural logarithm of the expression: $2x\ln(1-3/x)$. We can use l'Hôpital's rule: $\lim_{x\to\infty} \frac{2\ln(1-3/x)}{1/x} \overset{H}{=} \lim_{x\to\infty} \frac{(3/x^2)2/(1-3/x)}{-1/x^2} = \lim_{x\to\infty} \frac{-6}{1-3/x} = -6$, thus the limit of the original expression is $e^{-6}$.

**13.** Consider the natural logarithm of the expression: $(1/x)\ln(\ln x)$. We can use l'Hôpital's rule to obtain $\lim_{x\to\infty} \frac{\ln(\ln x)}{x} \overset{H}{=} \lim_{x\to\infty} \frac{(1/\ln x)(1/x)}{1} = \lim_{x\to\infty} \frac{1}{x\ln x} = 0$, thus the limit of the original expression is $e^0 = 1$.

**15.** Consider the natural logarithm of the expression: $(1/\ln x)\ln(e^x - 1)$. We can use l'Hôpital's rule twice to get $\lim_{x\to0^+} \frac{\ln(e^x-1)}{\ln x} \overset{H}{=} \lim_{x\to0^+} \frac{(1/(e^x-1))(e^x)}{1/x} = \lim_{x\to0+} \frac{xe^x}{e^x-1} \overset{H}{=} \lim_{x\to0^+} \frac{e^x+xe^x}{e^x} = 1$, thus the limit of the original expression is $e^1 = e$.

**17.** We see easily that $\sqrt{x^2-x} - x =$
$= (\sqrt{x^2-x} - x)\frac{\sqrt{x^2-x}+x}{\sqrt{x^2-x}+x} =$
$= \frac{-x}{\sqrt{x^2-x}+x} = \frac{-1}{\sqrt{1-1/x}+1}$, thus $\lim_{x\to\infty} (\sqrt{x^2-x} - x) = -1/2$.

**19.** First, $\lim_{x\to\infty} e^{-0.01x}/x^3 = 0$. Using l'Hôpital's rule three times we obtain that $\lim_{x\to-\infty} \frac{e^{-0.01x}}{x^3} \overset{H}{=} \lim_{x\to-\infty} \frac{(-0.01)e^{-0.01x}}{3x^2} \overset{H}{=}$

$\lim_{x\to-\infty} \frac{(-0.01)^2 e^{-0.01x}}{6x} \stackrel{H}{=} \lim_{x\to-\infty} \frac{(-0.01)^3 e^{-0.01x}}{6}$ $= -\infty$. Thus the asymptote is $y = 0$ as $x \to \infty$.

**21.** The function is not defined for $x \leq 0$. Consider the natural logarithm of the expression: $(2/x)\ln(\ln\sqrt{x})$. We can use l'Hôpital's rule: $\lim_{x\to\infty} \frac{2\ln(\ln\sqrt{x})}{x} \stackrel{H}{=} \lim_{x\to\infty} \frac{2(1/\ln\sqrt{x})(1/\sqrt{x})(1/2)(1/\sqrt{x})}{1} = \lim_{x\to\infty} \frac{1}{x\ln\sqrt{x}} = 0$, thus the limit of the original expression is $e^0 = 1$. The asymptote is $y = 1$.

**23.** Let $n$ be a positive integer. If $0 < x < 1/e$, then $\ln x/x^n < -1/x^n$; the limit of the right side is $-\infty$ as $x \to 0^+$, so the limit of the original expression is also $-\infty$.

**25.** Let $n$ be a positive integer and $k > 0$. We use l'Hôpital's rule repeatedly to get $\lim_{x\to\infty} \frac{x^n}{e^{kx}} \stackrel{H}{=} \lim_{x\to\infty} \frac{nx^{n-1}}{ke^{kx}} \stackrel{H}{=} \lim_{x\to\infty} \frac{n(n-1)x^{n-2}}{k^2 e^{kx}} \stackrel{H}{=} \cdots \stackrel{H}{=} \lim_{x\to\infty} \frac{n!}{k^n e^{kx}} = 0$.

**27. a.** The growth rate is $r(x) = y'(x) = 2e^{-1/(5x)}(5x)^{-2}5 = (2/5)e^{-1/(5x)}/x^2$. $\lim_{x\to\infty} (2/5)e^{-1/(5x)}/x^2 = 0$ because the numerator converges to 1, and the denominator grows without bound.

**b.** $r'(x) = (2/5)(e^{-1/(5x)}(5x)^{-2}5x^{-2} - e^{-1/(5x)}2x^{-3}) = (2/5)e^{-1/(5x)}((1/5)x^{-4} - 2x^{-3})$, thus $r'(x) = 0$ for $x = 1/10$; $r' > 0$ for $x < 1/10$ and $r' < 0$ for $x > 1/10$. Thus the location of the maximum is at $x = 1/10$ months, about 3 days.

**c.** Clearly, $r(x) > 0$. According to part **b**, the growth accelerates on $(0, 1/10)$ and decelerates on $(1/10, \infty)$.

**d.** The figures show the biomass and the growth rate, respectively.

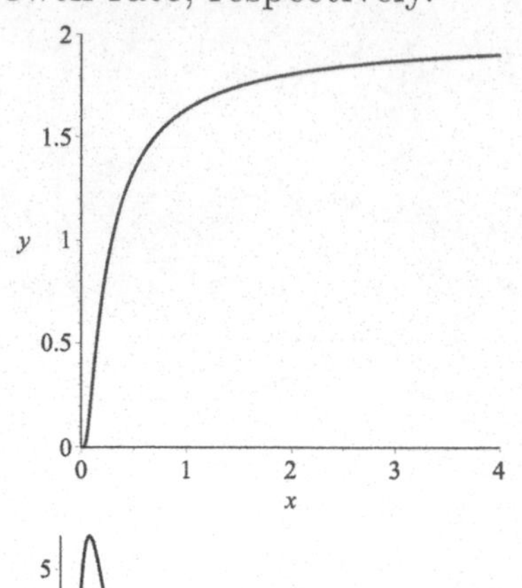


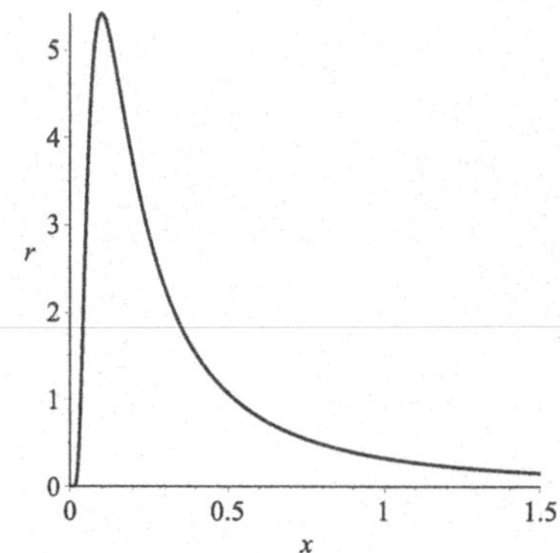


**29.** Using l'Hôpital's rule, $\lim_{x\to\infty} \frac{x^n}{\ln x} \stackrel{H}{=} \lim_{x\to\infty} \frac{nx^{n-1}}{1/x} = \lim_{x\to\infty} nx^n = \infty$, thus any power function $x^n$, $n > 0$ grows faster than the logarithmic function.

**31.** $f$ and $g$ should be differentiable on an open interval containing $a$ (where the limit is computed), $f(a) = g(a) = 0$, $g'(a) \neq 0$.

## Review Questions

**1. a.** Using the chain rule, we obtain that $dy/dx = 3x^2 + (3/2)x^{1/2} + 3\cos 3x$.

**b.** Using implicit differentiation, we get $y + xdy/dx + 3y^2dy/dx = 0$, thus $dy/dx = -y/(x + 3y^2)$.

**c.** Using the chain rule, quotient

rule and product rule we obtain that $dy/dx = (1/(x^2-1)2x(\sqrt[3]{x}(2-x)^3) - \ln(x^2-1)((1/3)x^{-2/3}(2-x)^3 - \sqrt[3]{x}3(2-x)^2))/(\sqrt[3]{x^2}(2-x)^6)$.

**d.** Using the product rule and chain rule, $dy/dx = 2xe^{-\sqrt{x}} + x^2e^{-\sqrt{x}}(-1/2)x^{-1/2}$.

**e.** Using the product rule and chain rule, $dy/dx = (3/2)x^{1/2}\cos 2x - 2x^{3/2}\sin 2x$.

**f.** Using the chain rule, we get $dy/dx = 2\sin(\pi x/4)\cos(\pi x/4)\pi/4$.

**3.** $f'(x) = 2x(2x-3)^3 + x^2 3(2x-3)^2 2 = (2x-3)^2(10x^2-6x)$, and then $f''(x) = 2(2x-3)2(10x^2-6x)+(2x-3)^2(20x-6) = 2(2x-3)(40x^2-48x+9)$.

**5.** $f(x) = e^{x^2}$; then $f'(x) = 2xe^{x^2}$ and $f''(x) = 2e^{x^2} + 4x^2e^{x^2}$. The first order approximation is $1+0\cdot x = 1$; the second order approximation is $1+0\cdot x+2\cdot x^2/2 = 1+x^2$.

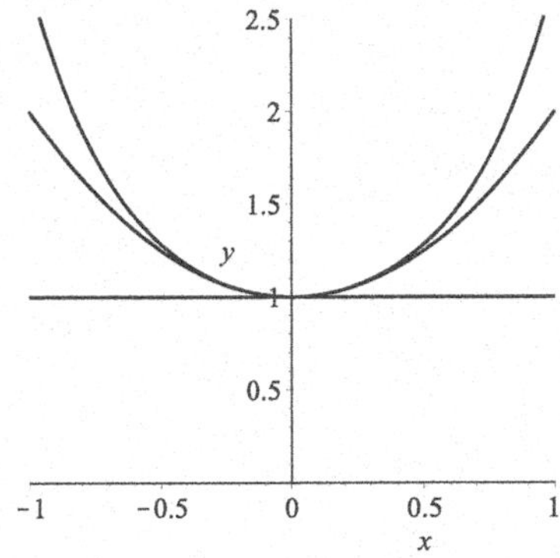


**7.** The derivative does not exist at $x = 1$.

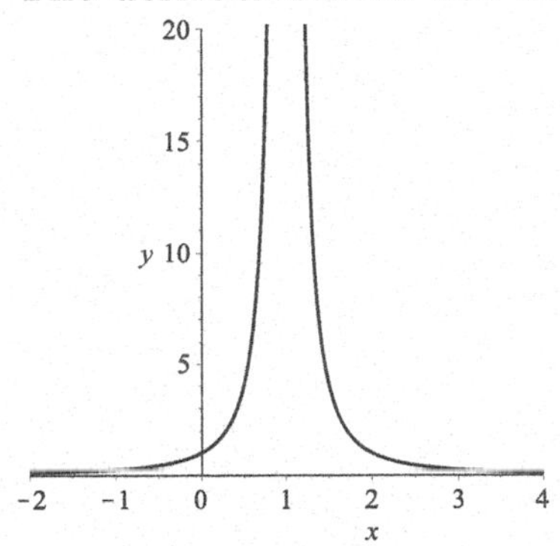


**9.** $f'(x) = 3x^2+70x-125$ and $f''(x) = 6x+70$. The function is increasing on $(-\infty,-25)$ and $(5/3,\infty)$ and decreasing on $(-25,5/3)$; it is concave down on $(-\infty,-35/3)$ and concave up on $(-35/3,\infty)$.

**11.** We get that $\lim_{t\to\pm\infty}(t^2+t+1)/(t^2+1) \overset{H}{=} \lim_{t\to\pm\infty}(2t+1)/2t \overset{H}{=} \lim_{t\to\pm\infty} 2/2 = 1$, the asymptote is $y = 1$. Also, we get $g'(t) = ((2t+1)(t^2+1)-(t^2+t+1)(2t))/(t^2+1)^2 = (1-t^2)/(t^2+1)^2$; $g''(t) = (-2t(t^2+1)^2-(1-t^2)2(t^2+1)2t)/(t^2+1)^4 = 2t(t^2-3)/(t^2+1)^3$. Thus the function is concave up on $(-\sqrt{3},0)$ and $(\sqrt{3},\infty)$ and concave down on $(-\infty,-\sqrt{3})$ and $(0,\sqrt{3})$. The inflection points are at $t = -\sqrt{3}$, $t = 0$, and $t = \sqrt{3}$.

**13. a.** $H = 100\tan\theta$.

**b.** $H(1.1) = 100\tan(1.1) \approx 196.48$ (ft).

**c.** First, $H' = 100\sec^2\theta$. Thus $E = 1.1H'(1.1)/H(1.1) \approx 2.72$. A 10% error in $\theta$ gives $2.72\cdot 10 = 27.2\%$ error in $H$.

**15.** The uptake rate is given by $f(x) = 1.2078x/(1+0.0506x) = 12$, thus $x = 19.98$. $f'(x) = 1.2078/(1+0.0506x)^2$, and then $f'(19.98) \approx 0.299$.

**17.** Differentiating with respect to $x$ both sides of the equation we get $4x^3 = 2x-2yy'$, thus $y'(x) = (x-2x^3)/y$. Now $y' = 0$ when $x = 0$ and $x = \pm 1/\sqrt{2}$. The corresponding $y$ values are $y = 0$ and $y = \pm\sqrt{x^2-x^4} = \pm 1/2$. We rule out $(0,0)$ because in $y'$ we divide by 0 there. Thus the four points we obtain are $(1/\sqrt{2},1/2)$, $(-1/\sqrt{2},1/2)$, $(1/\sqrt{2},-1/2)$, and $(-1/\sqrt{2},-1/2)$.

**19.** $y' = 100(1+\sin 2x)^{99}2\cos 2x$; thus $y'(\pi/2) = -200$. The equation of the tangent line is $y-1 = -200(x-\pi/2)$, which is $y = -200x+100\pi+1$.

## Problem Set 4.1 - Graphing Using Calculus

**1.** The function is defined and continuous everywhere, thus there are no vertical asymptotes. $\lim_{x\to\pm\infty}(x^2-x) = \lim_{x\to\pm\infty} x(x-1) = \infty$, so there are no horizontal asymptotes. $\lim_{x\to\pm\infty} y/x = \lim_{x\to\pm\infty}(x-1) = \pm\infty$, so there are no linear asymptotes. $y' = 2x-1$, $y'' = 2$. The function is decreasing when $y' < 0$, i.e. when $x < 1/2$ and increasing when $y' > 0$, i.e. when $x > 1/2$; it is also concave up for all $x$.

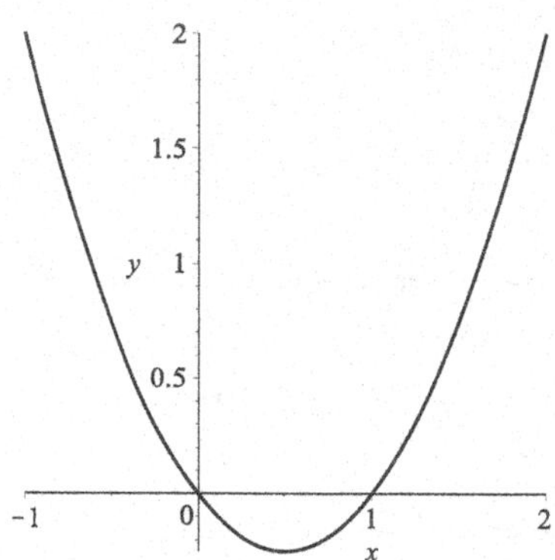

**3.** The function is defined and continuous everywhere, thus there are no vertical asymptotes. $\lim_{x\to\pm\infty} 1/(1+x^2) = 0$, so the horizontal asymptote is $y = 0$. $y' = -2x/(1+x^2)^2$, $y'' = (-2(x^2+1)^2-(-2x)2(1+x^2)2x)/(1+x^2)^4 = (6x^4+4x^2-2)/(1+x^2)^4 = (6x^2-2)/(1+x^2)^3$. The function is decreasing when $y' < 0$, i.e. when $x > 0$ and increasing when $y' > 0$, i.e. when $x < 0$; it is concave up when $y'' > 0$, i.e. when $-\infty < x < -1/\sqrt{3}$ and $1/\sqrt{3} < x < \infty$ and concave down when $y'' < 0$, i.e. when $-1/\sqrt{3} < x < 1/\sqrt{3}$. The inflection points are at $x = \pm 1/\sqrt{3}$.

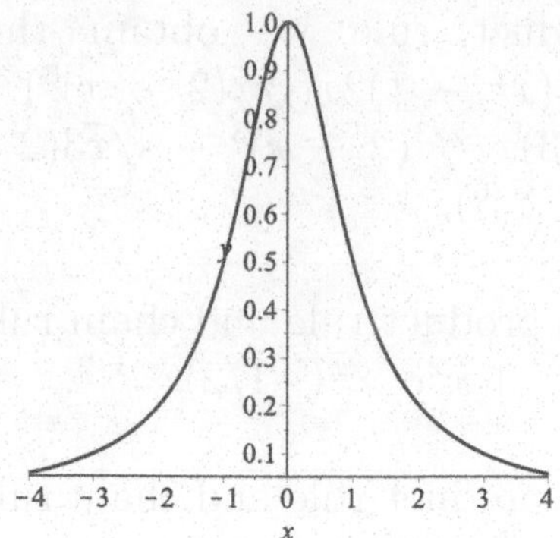

**5.** The function is not defined at $x = -2$, and there is a vertical asymptote there as $\lim_{x\to-2^-}(x+1/(2+x)) = -\infty$ and $\lim_{x\to-2^+}(x+1/(2+x)) = \infty$. $\lim_{x\to\infty}(x+1/(2+x)) = \infty$, $\lim_{x\to-\infty}(x+1/(2+x)) = -\infty$ and so there are no horizontal asymptotes. $\lim_{x\to\pm\infty} y/x = \lim_{x\to\pm\infty}(1+1/(x(2+x))) = 1$, so $y = x$ is a linear asymptote. $y' = 1-1/(2+x)^2$, $y'' = 2/(2+x)^3$. The function is decreasing when $y' < 0$, i.e. when $-3 < x < -2$ and $-2 < x < -1$, and increasing when $y' > 0$, i.e. when $-\infty < x < -3$ and $-1 < x < \infty$; it is concave up when $y'' > 0$, i.e. when $x > -2$ and concave down when $y'' < 0$, i.e. when $x < -2$.

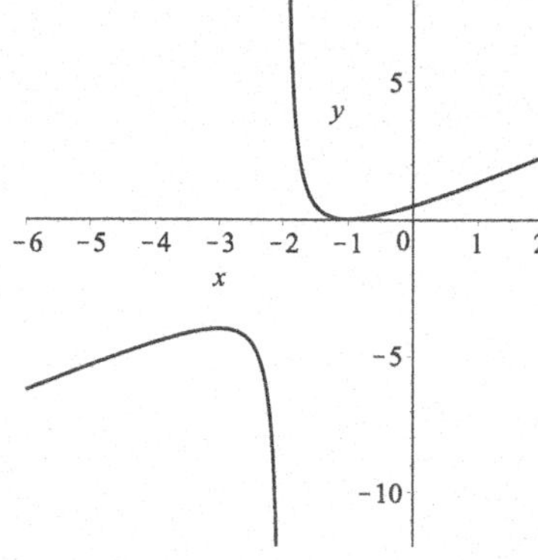

**7.** The function is defined and continuous everywhere, thus there are no vertical asymptotes. $\lim_{x\to\pm\infty}(-12x-9x^2/2+x^3) = \pm\infty$, so there are no horizontal asymptotes. $\lim_{x\to\pm\infty} y/x = \lim_{x\to\pm\infty}(x^2-9x/2-12) = \infty$, so there are no linear asymptotes. $y' = -12-$

$9x+3x^2 = 3(x-4)(x+1)$, $y'' = -9+6x$. The function is decreasing when $y' < 0$, i.e. when $-1 < x < 4$ and increasing when $y' > 0$, i.e. when $x < -1$ and $x > 4$; it is concave up when $y'' > 0$, i.e. when $x > 3/2$ and concave down when $y'' < 0$, i.e. when $x < 3/2$; there is an inflection point at $x = 3/2$.

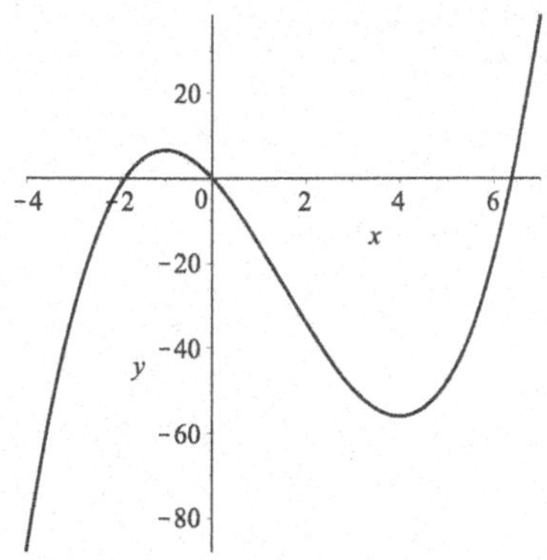


**9.** The function is defined and continuous everywhere, thus there are no vertical asymptotes. $\lim_{x\to\pm\infty}(e^x + 2e^{-x}) = \infty$, so there are no horizontal asymptotes. $\lim_{x\to\pm\infty} y/x = \lim_{x\to\pm\infty}(e^x + 2e^{-x})/x = \pm\infty$, so there are no linear asymptotes. $y' = e^x - 2e^{-x}$, $y'' = e^x + 2e^{-x}$. The function is decreasing when $y' < 0$, i.e. when $x < \ln 2/2$ and increasing when $y' > 0$, i.e. when $x > \ln 2/2$; it is concave up when $y'' > 0$, i.e. for all $x$.

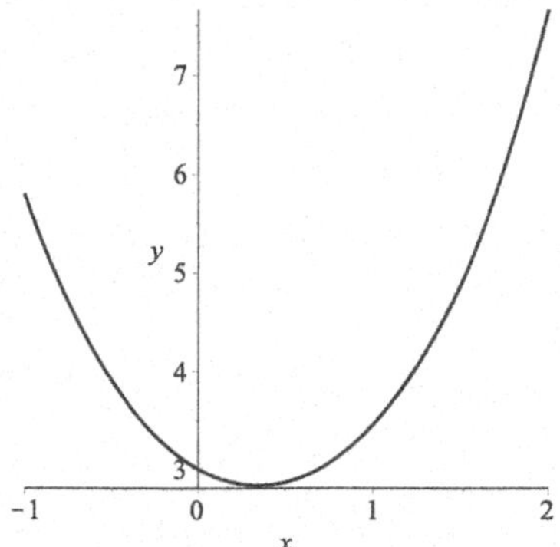


**11.** The function is defined and continuous everywhere, thus there are no vertical asymptotes. $\lim_{x\to\pm\infty}(x - x^3) = \mp\infty$, so there are no horizontal asymptotes. $\lim_{x\to\pm\infty} y/x = \lim_{x\to\pm\infty}(1 - x^2) = -\infty$, so there are no linear asymptotes. $y' = 1 - 3x^2$, $y'' = -6x$. The function is decreasing when $y' < 0$, i.e. when $x < -1/\sqrt{3}$ and $x > 1/\sqrt{3}$ and increasing when $y' > 0$, i.e. when $-1/\sqrt{3} < x < 1/\sqrt{3}$; it is concave up when $y'' > 0$, i.e. when $x < 0$ and concave down when $y'' < 0$, i.e. when $x > 0$; there is an inflection point at $x = 0$.

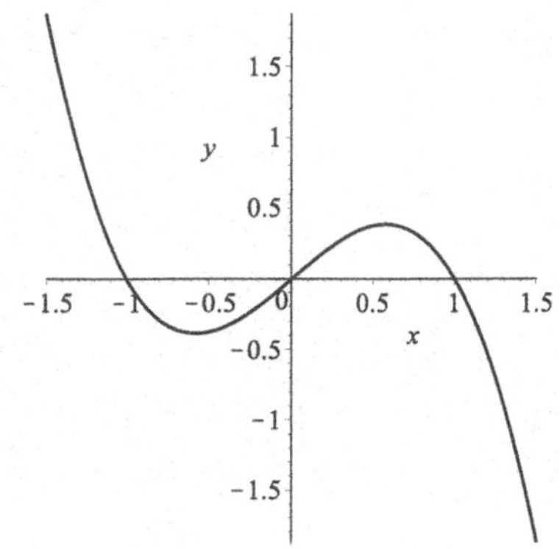


**13.** The function is not defined at $x = -1$, and there is a vertical asymptote there as $\lim_{x\to-1^-}(x-3)/(x+1) = \infty$ and $\lim_{x\to-1^+}(x-3)/(x+1) = -\infty$. $\lim_{x\to\infty}(x-3)/(x+1) = 1$, $\lim_{x\to-\infty}(x-3)/(x+1) = 1$ so the horizontal asymptote is $y = 1$. We obtain that $y' = ((x+1)-(x-3))/(x+1)^2 = 4/(x+1)^2$, $y'' = -8/(x+1)^3$. The function is increasing when $y' > 0$, i.e. when $x < -1$ and $x > -1$; it is concave up when $y'' > 0$, i.e. when $x < -1$ and concave down when $y'' < 0$, i.e. when $x > -1$.

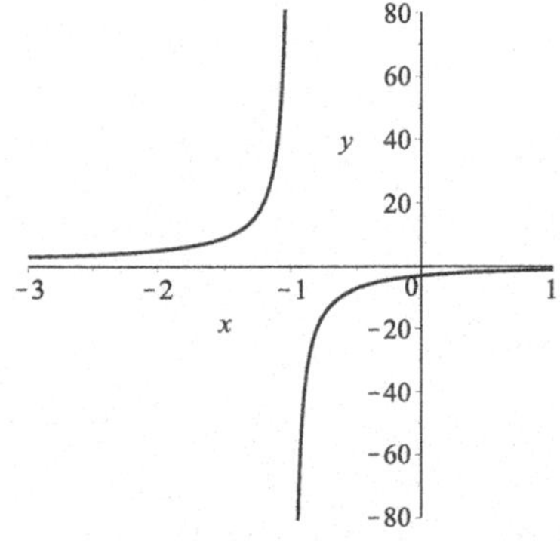


**15.** The function is defined and continuous everywhere, thus there are no vertical asymptotes. $\lim_{x\to\pm\infty}(x^4 - ax^2) = \infty$, so there are no horizontal asymptotes. $\lim_{x\to\pm\infty} y/x = \lim_{x\to\pm\infty}(x^3 - ax) = \pm\infty$, so there are no linear asymptotes. $y' = 4x^3 - 2ax = 2x(2x^2 - a)$, $y'' = 12x^2 - 2a$. If $a > 0$, then $y' < 0$ for $x < -\sqrt{a/2}$ and $0 < x < \sqrt{a/2}$ and $y' > 0$ for $-\sqrt{a/2} < x < 0$ and $x > \sqrt{a/2}$. Also, $y'' > 0$ for $x < -\sqrt{a/6}$ and $x > \sqrt{a/6}$ and $y'' < 0$ for $-\sqrt{a/6} < x < \sqrt{a/6}$. Thus the function is decreasing for $x < -\sqrt{a/2}$ and $0 < x < \sqrt{a/2}$ and increasing for $-\sqrt{a/2} < x < 0$ and $x > \sqrt{a/2}$; it is concave up for $x < -\sqrt{a/6}$ and $x > \sqrt{a/6}$ and concave down for $-\sqrt{a/6} < x < \sqrt{a/6}$. There are inflection points at $x = \pm\sqrt{a/6}$.

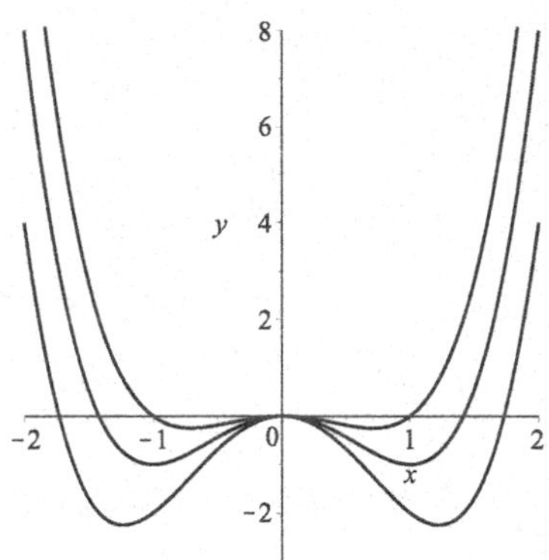


**17.** The function is defined and continuous everywhere, thus there are no vertical asymptotes. If $a > 0$, $\lim_{x\to\pm\infty}(ae^x + e^{-x}) = \infty$, so there are no horizontal asymptotes. $\lim_{x\to\pm\infty} y/x = \lim_{x\to\pm\infty}(ae^x + e^{-x})/x = \pm\infty$, so there are no linear asymptotes. We get $y' = ae^x - e^{-x}$, $y'' = ae^x + e^{-x}$. The function is decreasing when $y' < 0$, i.e. when $x < -\ln a/2$ and increasing when $y' > 0$, i.e. when $x > -\ln a/2$; it is concave up when $y'' > 0$, i.e. for all $x$.

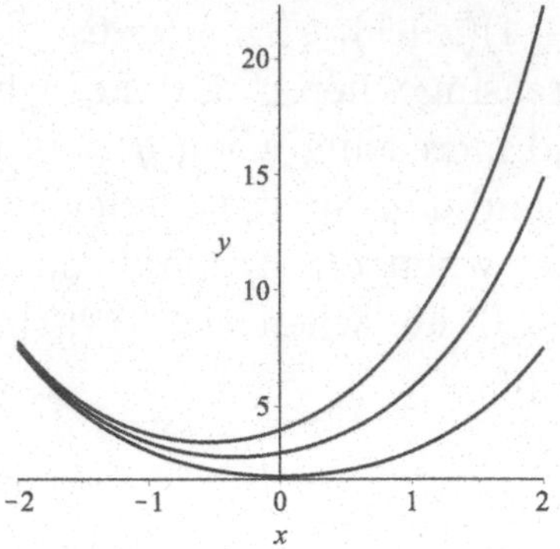


**19.** When $a = 1$, the function is just the constant function $y = 1$, not defined at $x = -1$. When $a > 1$, the function is not defined at $x = -1$, and there is a vertical asymptote there as $\lim_{x\to-1^-}(a + x)/(1 + x) = -\infty$ and $\lim_{x\to-1^+}(a + x)/(1 + x) = \infty$. $\lim_{x\to\infty}(a + x)/(1 + x) = 1$, $\lim_{x\to-\infty}(a + x)/(1 + x) = 1$ so the horizontal asymptote is $y = 1$. $y' = ((1+x)-(a+x))/(1+x)^2 = (1-a)/(1+x)^2$, $y'' = -2(1 - a)/(1 + x)^3$. The function is decreasing when $y' < 0$, i.e. when $x < -1$ and $x > -1$; it is concave up when $y'' > 0$, i.e. when $x > -1$ and concave down when $y'' < 0$, i.e. when $x < -1$. When $a < 1$, the function is not defined at $x = -1$, and there is a vertical asymptote there as $\lim_{x\to-1^-}(a + x)/(1 + x) = \infty$ and $\lim_{x\to-1^+}(a + x)/(1 + x) = -\infty$. $\lim_{x\to\infty}(a + x)/(1 + x) = 1$, $\lim_{x\to-\infty}(a + x)/(1 + x) = 1$ so the horizontal asymptote is $y = 1$. $y' = ((1+x)-(a+x))/(1+x)^2 = (1-a)/(1+x)^2$, $y'' = -2(1-a)/(1+x)^3$. The function is increasing when $y' > 0$, i.e. when $x < -1$ and $x > -1$; it is concave up when $y'' > 0$, i.e. when $x < -1$ and concave down when $y'' < 0$, i.e. when $x > -1$.

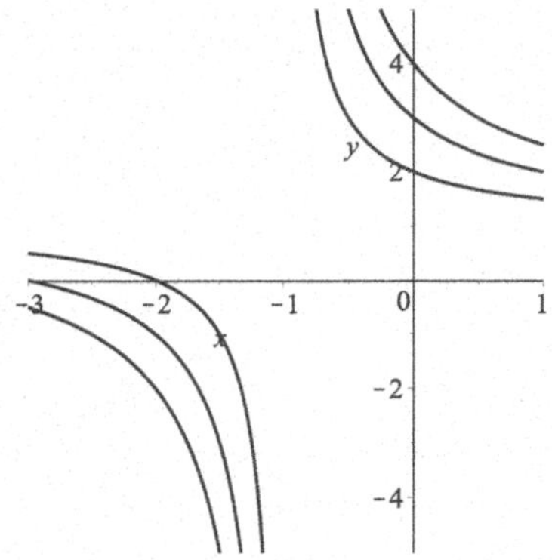


**21.**

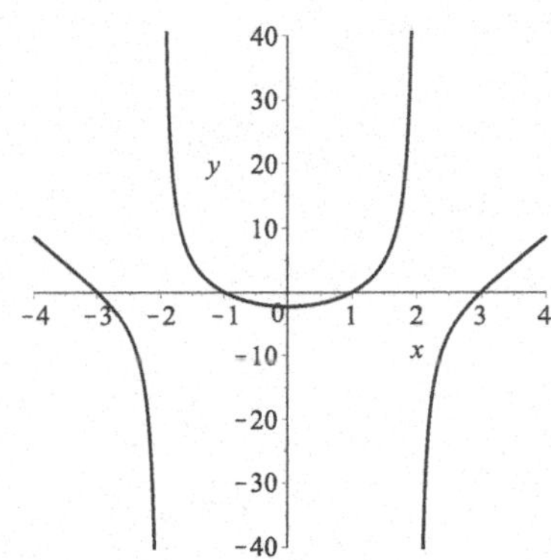


**23.** $y = ax^2 + bx + c$, thus $y' = 2ax + b$ and $y'' = 2a$. If $a \neq 0$, these are parabolas, when $a > 0$ concave up (opening upward), when $a < 0$ concave down (opening downward). If $a = 0$, we obtain lines.

**25.** We get $y' = (b-a)(-1)(1+e^x)^{-2}e^x = (a-b)e^x/(1+e^x)^2$, and $y'' = ((a-b)e^x(1+e^x)^2 - (a-b)e^x 2(1+e^x)e^x)/(1+e^x)^4 = ((a-b)e^x(1-e^x))/(1+e^x)^3$. Thus if $a \neq b$, the only point of inflection is at $x = 0$; the second derivative changes its sign there.

**27.** The domain of the function in this application is $(0, \infty)$. The horizontal asymptote is $y = 1.2078/0.0506 \approx 23.87$. $f'(x) = (1.2078(1+0.0506x) - 1.2078x(0.0506))/(1+0.0506x)^2 = 1.2078/(1+0.0506x)^2$. Thus the function is increasing for $x > 0$ (on its domain in this application).

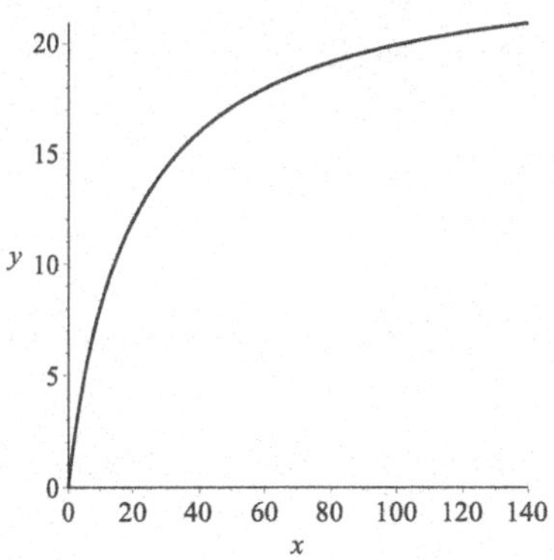


**29. a.** The domain of the function in this application is $(0, \infty)$. There is a vertical asymptote at $x = -0.76$, which is outside the domain of the function. The horizontal asymptote is $y = 58.7/1 = 58.7$.

**b.** We get $f'(x) = (58.7(0.76+x) - 58.7(x-0.03))/(0.76+x)^2 = 46.373/(0.76+x)^2$. Thus the function is increasing on its domain.

**c.** $f''(x) = -2 \cdot 46.373(0.76+x)^{-3}$. Thus the function is concave down on its domain.

**d.**

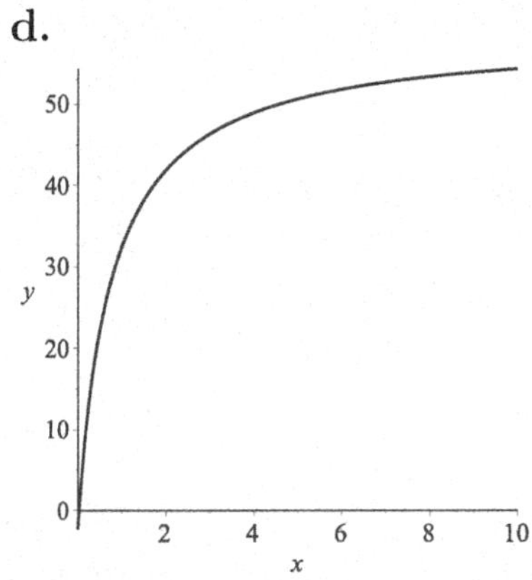


**31.** Differentiation gives that $C'(t) = 28.6(-0.3e^{-0.3t} + e^{-t}) = -8.58e^{-0.3t} + 28.6e^{-t}$, and $C''(t) = (-0.3)(-8.58)e^{-0.3t} - 28.6e^{-t}$. The inflection point is given by the equation $28.6e^{-t} = 2.574e^{-0.3t}$; thus $e^{0.7t} = 28.6/2.574$, and $t = \ln(28.6/2.574)/0.7 \approx 3.44$.

**33.** One possibility is $w(t) = 2 - e^{-t}$; we see that $w(0) = 1 > 0$, $w'(t) = e^{-t}$, which is a positive decreasing function, and $y = 2$ is the horizontal asymptote.

**35. a.** The horizontal asymptote is $y = 0$. $C'(t) = 23.725(0.7e^{-0.7t} - 0.5e^{-0.5t})$, thus the function is increasing for $0 < t < 5\ln(7/5) \approx 1.68$ and decreasing for $t > 5\ln(7/5)$.

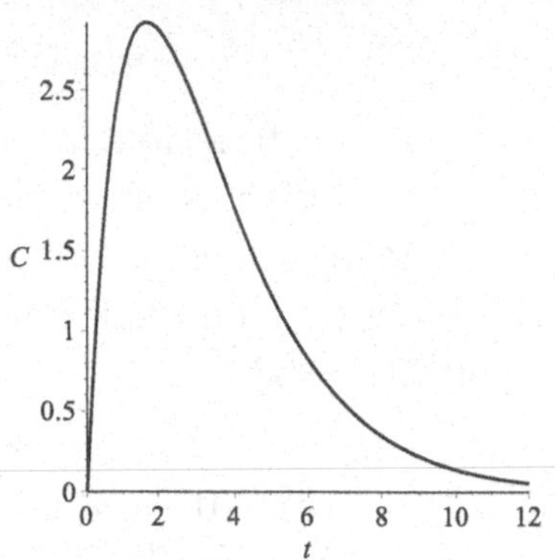


**b.** The maximum concentration is reached at $t \approx 1.68$ hours, its value is $C \approx 2.923$ micrograms/ml.

## Problem Set 4.2 - Getting Extreme

**1.** Local maxima at $x \approx -0.6$ and $x \approx 0.75$; local minima at $x = -1$, $x \approx -0.1$, and $x = 1$. Global maximum is at $x \approx 0.75$, value $y \approx 0.48$; global minimum is at $x = -1$, value $y = -0.3$.

**3.** Local maxima at $x = -2$, $x = -0.05$, $x = 0.05$, and $x = 1$; local minima at $x = -1$ and $x = 0.25$. Global maximum is at $x = -0.05$ and $x = 0.05$, value $y = 5$; global minimum is at $x = -1$, value $y = 0$.

**5.** $y' = 3 + 8x$, thus the only critical point is at $x = -3/8$; $y' < 0$ for $x < -3/8$ and $y' > 0$ for $x > -3/8$, thus $y$ has a local (and global) minimum at $x = -3/8$.

**7.** $f'(t) = 2te^{-t} - t^2e^{-t} = t(2-t)e^{-t}$. The critical points are $t = 0$ and $t = 2$. $f' < 0$ for $t < 0$ and $t > 2$ and $f' > 0$ for $0 < t < 2$. Thus $t = 0$ is a local minimum and $t = 2$ is a local maximum.

**9.** $f'(x) = ((1+x)-x)/(1+x)^2 = 1/(1+x)^2$. The only critical point is at $x = -1$ where the derivative does not exist. $f(x)$ is not defined there either.

**11.** $y' = -1 + 3x/2 + x^2 = (x+2)(x-1/2)$. The critical points are at $x = -2$ and $x = 1/2$. $y' > 0$ when $x < -2$ and $x > 1/2$, and $y' < 0$ when $-2 < x < 1/2$. Thus $x = -2$ is a local maximum and $x = 1/2$ is a local minimum.

**13.** $y' = -12 - 9x + 3x^2 = 3(x+1)(x-4)$. The critical points are at $x = -1$ and $x = 4$. $y'' = -9 + 6x$; $y''(-1) = -15$ and $y''(4) = 15$ thus $x = -1$ is a local maximum and $x = 4$ is a local minimum.

**15.** $y' = 1 - 1/(2+x)^2$; the critical points are $x = -1$, $x = -3$ and $x = -2$. At $x = -2$, the function is not defined. $y'' = 2/(2+x)^3$; so $y''(-1) = 2$ and $y''(-3) = -2$ and thus $x = -1$ is a local minimum and $x = -3$ is a local maximum.

**17.** $f'(x) = 2x - 4$; the only critical point is $x = 2$, which is inside the given interval. $f(0) = 2$, $f(3) = -1$ and $f(2) = -2$. Thus on $[0, 3]$, the global minimum is $f = -2$ at $x = 2$ and the global maximum is $f = 2$ at $x = 0$.

**19.** $f'(x) = 1 - 1/x^2$; the critical point on the given interval is $x = 1$. $f(0.1) = 10.1$, $f(10) = 10.1$ and $f(1) = 2$. Thus on $[0.1, 10]$, the global minimum is $f = 2$ at $x = 1$ and the global maximum is $f = 10.1$ at $x = 0.1$ and $x = 10$.

**21.** $f'(x) = 2x - 4$; the only critical point is $x = 2$. $\lim_{x\to\pm\infty}(x^2 - 4x + 2) = \infty$. This implies that the global minimum is $f = -2$ at $x = 2$ and there is no global maximum.

**23.** $f'(x) = 1 - 1/x^2$; critical point on $(0, \infty)$ is $x = 1$. $\lim_{x\to\infty}(x + 1/x) = \infty$, $\lim_{x\to 0^+}(x + 1/x) = \infty$. This implies that the global minimum is $f = 2$ at $x = 1$ and there is no global maximum.

**25.** Let $M = \lim_{x\to b^-} f(x)$. (We allow $M$ to be $\pm\infty$). Identify all critical points on the interval $(a, b)$. Evaluate $f$ at the critical points and at $x = a$, then use the last part of the open interval method.

**27.** $f'(x) = 2x - 4$; the only critical point is $x = 2$ (it is on $[0, \infty)$). $\lim_{x\to\infty}(x^2 - 4x + 2) = \infty$. $f(0) = 2$; these imply that the global minimum is $f = -2$ at $x = 2$ and there is no global maximum.

**29.** $f'(x) = 1 - 1/(x+2)^2$; there is no critical point on $(-1, \infty)$ (it is at $x = -1$, actually). $\lim_{x\to\infty}(x + 1/(2 + x)) = \infty$. $f(-1) = 0$; these imply that the global minimum is $f = 0$ at $x = -1$ and there is no global maximum.

**31.** If $x = c$ is a local maximum, then there is a sufficiently small $k > 0$ value such that $f(c+h) \le f(c)$ for all $h$ for which $-k < h < k$. We assumed that $f$ is differentiable at $c$, thus $\lim_{h\to 0^+}(f(c + h) - f(c))/h = f'(c) \le 0$ because the expression in the limit is always nonpositive. Also, $\lim_{h\to 0^-}(f(c+h) - f(c))/h = f'(c) \ge 0$ because the expression in the limit is always nonnegative. This implies that $f'(c) = 0$.

**33.** We have to consider the interval $[312, 324]$. The derivative is $f'(x) = 0.122463 - 3\sin(\pi x/6)\pi/6$; the critical points on the interval $[312, 324]$ are at $x \approx 312.149$ at $x \approx 317.85$. The global maximum is at $x = 324$, value is 371.93; the global minimum is at $x \approx 317.85$, value is 365.19.

**35.** First, $\operatorname{sech}^2 x = 4/(e^{2x} + 2 + e^{-2x})$. The asymptote of this function is $y = 0$, thus the asymptote of $f(t)$ is $y = 0$ as well. Now $(\operatorname{sech}^2 x)' = -4(2e^{2x} - 2e^{-2x})/(e^{2x} + 2 + e^{-2x})^2$. Thus $\operatorname{sech}^2 x$ is increasing for $x < 0$ and decreasing for $x > 0$. This means that $\operatorname{sech}^2 x$ has a global maximum at $x = 0$, where the value is 1. $f(t)$ is a shifted and stretched version of the $\operatorname{sech}^2 x$ function, thus the graph increases for $0 < t < 3.4/02 = 17$ and decreases for $t > 17$; the maximum is at $t = 17$, and the value is 890 there.

**37. a.** We get $P'(x) = Asx^{s-1}e^{-sx/r} - (s/r)Ax^s e^{-sx/r} = (1 - x/r)sAx^{s-1}e^{-sx/r}$. The critical points are $x = 0$ and $x = r$; at $x = r$ we have a maximum because the derivative changes sign there from positive to negative, and $P(0) = \lim_{x\to\infty} P(x) = 0$.

**b.** The graph shows the case $A = 2$, $s = 3$, $r = 2$. The maximum is at $x = r = 2$.

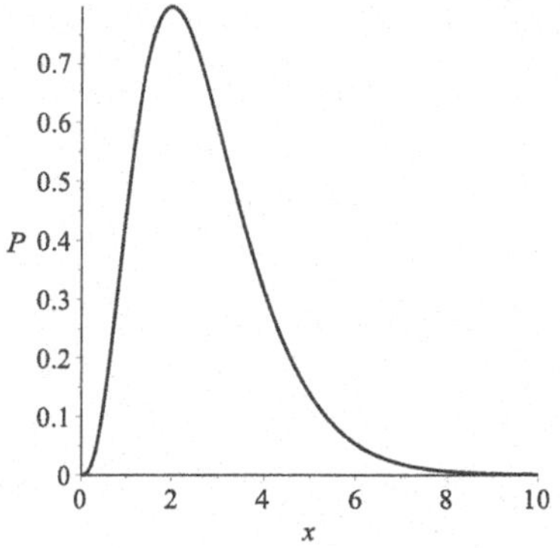


**39.** $P'(v) = -w^2/(2\rho S v^2) + (3/2)\rho A v^2$; thus $P' = 0$ when $w^2/(2\rho S v^2) = (3/2)\rho A v^2$, which is $v^4 = w^2/(3\rho^2 AS)$, and then $v = \sqrt[4]{w^2/(3\rho^2 AS)}$. This really is a minimum, because $\lim_{v\to\infty} P = \lim_{v\to 0} P = \infty$.

## Problem Set 4.3 - Optimization in Biology

**1.** The net profit is $f(x) = pY(x) - C(x) = 3(-0.1181x^2+8.525x+12.95)-6x = -0.3543x^2+19.575x+38.85$. We have to find the maximum; $f'(x) = -0.7086x + 19.575$. The critical point is at $x = 27.6249$ (thousand seeds per acre).

**3.** The net profit is $f(x) = pY(x) - C(x) = 2.2(-0.1181x^2 + 8.525x + 12.95) - 2.5x = -0.25982x^2 + 16.255x + 28.49$. We have to find the maximum; $f'(x) = -0.51964x + 16.255$. The critical point is at $x = 31.2813$ (thousand seeds per acre).

**5.** $G(x) = 0.08x(1-x/400000)$; thus $G'(x) = 0.08(1 - x/400000) - 0.08x/400000$ and the maximum is at the $x$ value where $G'(x) = 0$, i.e. $x = 200000$. The maximum harvesting rate is $H = G(200000) = 8000$ (per year).

**7.** The maximum sustainable yield is at the value $x$ where $G'(x) = 0$; $G'(x) = 2.1(1 - x/K) - 2.1x/K$, thus $x = K/2$. Then $hx = G(x)$ at this value gives $hK/2 = 2.1(K/2)(1 - 1/2)$, and thus $h = 1.05$.

**9.** Using the notation of Example 5, the function we have to minimize is $R(\theta) = (a - b\cot\theta)/r_1^4 + b\csc\theta/r_2^4$. Following the computation in Example 5, $R'(\theta) = b\csc^2\theta(1/r_2^4\cos\theta - 1/r_1^4)$. Thus the $\theta$ value we need is $\cos\theta = r_2^4/r_1^4 = (r_2/r_1)^4 = (0.04/0.06)^4 =$ and then $\theta \approx 1.3720$ radians.

**11.** Using the notation of Example 5, the function we have to minimize is $R(\theta) = (a - b\cot\theta)/r_1^4 + b\csc\theta/r_2^4$. Following the computation in Example 5, $R'(\theta) = b\csc^2\theta(1/r_2^4\cos\theta - 1/r_1^4)$. Thus the $\theta$ value we need is $\cos\theta = r_2^4/r_1^4 = (r_2/r_1)^4 = (3/4)^4 = 81/256$ and then $\theta \approx 1.2489$ radians.

**13.** Using the notation of Example 3, suppose that the distance of the ball from the shore is $d$ meters. Then the function we have to minimize is $T(x) = (15 - x)/6.4 + \sqrt{d^2 + x^2}/0.91$. Then $T'(x) = -1/6.4 + x/(0.91\sqrt{d^2 + x^2})$, and the same computation as in Example 3 gives the $x$ value $0.02(d^2 + x^2) = x^2$, thus $x^2 = 0.02d^2/0.98$, i.e. $x = d\sqrt{0.02/0.98}$. In case $d = 20$, we obtain $x \approx 2.857$, so Elvis should run $15 - 2.857 = 12.143$ meters along the shore.

**15.** Using the notation of Example 3, suppose that the distance of the ball from the shore is $d$ meters. Then the function we have to minimize is $T(x) = (15 - x)/6.4 + \sqrt{d^2 + x^2}/0.91$. Then $T'(x) = -1/6.4 + x/(0.91\sqrt{d^2 + x^2})$, and the same computation as in Example 3 gives the $x$ value $0.02(d^2 + x^2) = x^2$, thus $x^2 = 0.02d^2/0.98$, i.e. $x = d\sqrt{0.02/0.98}$. So Elvis should run $15 - 0.143d$ meters along the shore.

**17.** We get $S = \sum_{t=1}^{5}(f(t) - x_t)^2 = (3-1)^2 + (5-3)^2+(7-6)^2+(9-12)^2+(11-16)^2 = 43$.

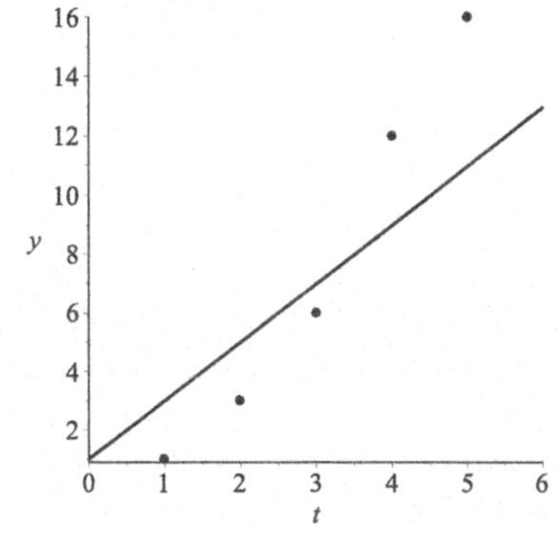


**19.** We get $S = \sum_{t=1}^{5}(f(t) - x_t)^2 \approx 42.46$.

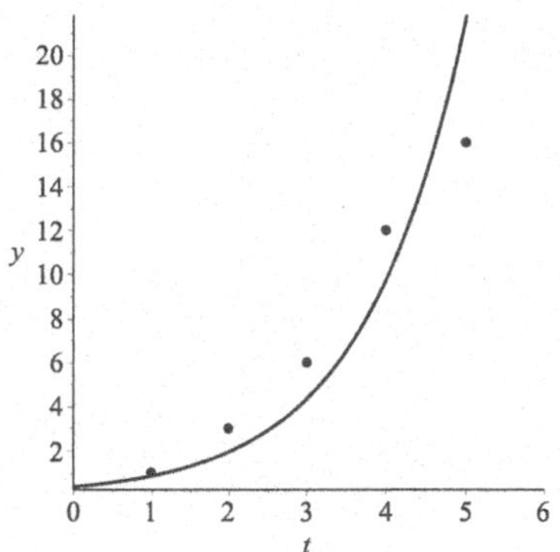


**21.** We get $S = \sum_{t=1}^{5}(f(t)-x_t)^2 = (m-1)^2 + (2m-3)^2 + (3m-6)^2 + (4m-12)^2 + (5m-16)^2 = 55m^2 - 306m + 446$.

**23.** Using the notation of Example 3, suppose that the distance of the ball from the shore is $d$ meters, and Elvis runs $k - x$ meters on the shore before jumping into the water. Then the function we have to minimize is $T(x) = (k-x)/6.4 + \sqrt{d^2+x^2}/0.91$. Then $T'(x) = -1/6.4 + x/(0.91\sqrt{d^2+x^2})$, and the same computation as in Example 3 gives the $x$ value $0.02(d^2+x^2) = x^2$, thus $x^2 = 0.02d^2/0.98$, i.e. $x = d\sqrt{0.02/0.98}$. So Elvis should run $k - 0.143d$ meters along the shore.

**25. a.** Using the notations and results of Example 4, if the doubling time is $T$, then $a = \ln 2/T$, and if the half life is $S$, then $b = \ln(1/2)/S$. Thus $V(t) = 0.44(0.9973e^{\ln(1/2)/6.24t} + 0.0027e^{\ln 2/2.9t}) = 0.4388e^{-0.111t} + 0.0012e^{0.239t}$.

**b.** $V' = -0.0487e^{-0.111t} + 0.00029e^{0.239t}$, thus $V' = 0$ at $t \approx 14.697$ days.

**c.** The model overestimates the time by about a day.

**27.** The volume is given by $v = \pi r^2 h$, thus $h = v/(\pi r^2)$. Then the total surface area is $S = 2r\pi(v/(\pi r^2)) + 2r^2\pi = 2v/r + 2r^2\pi$. Then $S'(r) = -2v/r^2 + 4r\pi$, which is zero when $4r\pi = 2v/r^2$, i.e. when $r = \sqrt[3]{v/2\pi}$.

**29.** The profit per hectare is $p(N) = pY - cN = 139.65(1.86 + 0.02741N - 0.00009N^2) - 0.71N = 259.749 + 3.11781N - 0.0125685N^2$. This is maximized when $p'(N) = 3.11781 - 0.025137N = 0$, i.e. when $N \approx 124.03$ kg per hectare.

**31.** We differentiate and obtain $G'(x) = r(1-(x/K)^\alpha) + rx(-\alpha(x/K)^{\alpha-1}/K) = r(1-(x/K)^\alpha - \alpha(x/K)^\alpha)$. Thus the maximum is at the value where $G'(x) = 0$, i.e. when $1 = (1+\alpha)(x/K)^\alpha$, which is $x = K/(1+\alpha)^{1/\alpha}$.

**33.** Assume we are standing $x$ meters away. Let the angle subtended from the ground to the top of the pedestal be $\alpha$, and the angle subtended from the ground to the top of the statue $\beta$. Then $\tan\alpha = 46/x$ and $\tan\beta = 92/x$. The angle we want to maximize is $\theta(x) = \beta(x) - \alpha(x) = \arctan(92/x) - \arctan(46/x)$. Thus $\theta'(x) = -(92/x^2)/(1+(92/x)^2) + (46/x^2)/(1+(46/x)^2)$, and this derivative is zero when $x = 46\sqrt{2} \approx 65.05$ meters.

**35.** We will assume that the number of crews is integer; i.e. we can't bring in part of a crew. Let the number of additional crews be $n$. The number of days the cleanup lasts is $200/(5(n+1)) = 40/(n+1)$. We consider two possibilities: in order to not pay a fine, we need $40/(n+1) \leq 14$, i.e. $n \geq 26/14$. For example, if we bring in 2 additional crews, the cleanup lasts 40/14 days, and we have to pay a total of $(40/14) \cdot 500 + (40/14) \cdot 2 \cdot 800 + 18{,}000 = 46{,}000$. The total cost for all $n \geq 2$ is given by $C(n) = 500(40/(n+1)) + 800n(40/(n+1)) + 18000$, thus $C'(n) = 12000/(n+1)^2 > 0$, and the function is increasing. When we assume

$n = 0$ or $n = 1$, then just the cost of the fine is higher than this amount, so the minimum is at $n = 2$.

**37.** The energy gain is $\alpha S = \alpha(2\pi r(r/2) + 2r^2\pi) = \alpha 3\pi r^2$, the loss is $\beta V = \beta r^2\pi(r/2) = \beta\pi r^3/2$. Thus the net gain is $g(r) = \alpha 3\pi r^2 - \beta\pi r^3/2$. Then $g'(r) = 6\pi\alpha r - 3\pi\beta r^2/2 = 3\pi r(2\alpha - \beta r/2)$. The net gain is maximized for $r > 0$ when $g'(r) = 0$, i.e. when $r = 4\alpha/\beta$.

**39.** We obtain that $P = 6M^{1/2}/(2M^{3/4} + 3M^{-1/4}) = 6M^{3/4}/(2M+3)$, thus $P' = (9/2M^{-1/4}(2M+3) - 12M^{3/4})/(2M+3)^2 = (-3M^{3/4} + 27/2M^{-1/4})/(2M+3)^2$. We get that $P' = 0$ when $M = 9/2$; this is a maximum because $P' > 0$ for $M < 9/2$ and $P' < 0$ for $M > 9/2$.

**41. a.** Using the result of Example 6, $a = \sum_{i=1}^{6} x_i y_i / \sum_{i=1}^{6} x_i^2 = 451679/70219 \approx 6.432$.

**b.**

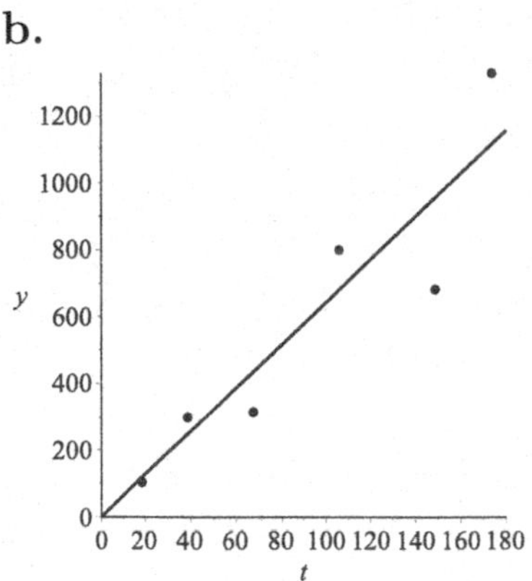


**43.** Using the assumption that $f = 0.59e^{-rt}$, we compute the quantities $\sum_{t=0}^{5} t = 15$, $\sum_{t=0}^{5} t^2 = 55$ and $\sum_{t=0}^{5} t \ln x_t = -42.842$. Thus the value of $r$ we are looking for is $r = (15\ln 0.59 + 42.842)/55 \approx 0.635$.

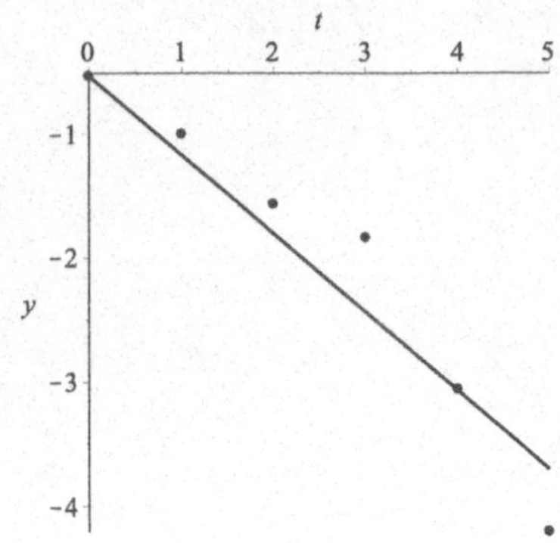


## Problem Set 4.4 - Decisions and Optimization

**1.** We want to maximize the function $R(t) = f(t)/(t+15) = 180t/((1+0.15t)(t+15)) = 180t/(0.15t^2 + 3.25t + 15)$. Then $R'(t) = (180(0.15t^2+3.25t+15) - 180t(0.3t+3.25))/(0.15t^2+3.25t+15)^2$. Thus $R'(t) = 0$ when the numerator is zero, i.e. when $-27t^2 + 2700 = 0$, which is $t = 10$ seconds (we discard the negative root).

**3.** We want to maximize the function $R(t) = f(t)/(t+10) = 360t/((1+0.5t)(t+10)) = 360t/(0.5t^2 + 6t + 10)$. Then $R'(t) = (360(0.5t^2+6t+10) - 360t(t+6))/(0.5t^2 + 6t + 10)^2$. Thus $R'(t) = 0$ when the numerator is zero, i.e. when $-180t^2 + 3600 = 0$, which is $t = 2\sqrt{5} \approx 4.5$ seconds (we discard the negative root).

**5.** We want to maximize the function $R(t) = f(t)/(t+5) = 360t/((1+0.3t)(t+5)) = 360t/(0.3t^2 + 2.5t + 5)$. Then $R'(t) = (360(0.3t^2+2.5t+5) - 360t(0.6t+2.5))/(0.3t^2+2.5t+5)^2$. Thus $R'(t) = 0$ when the numerator is zero, i.e. when $-108t^2 + 1800 = 0$, which is $t = 5\sqrt{2/3} \approx 4.1$ seconds (we discard the negative root).

**7.** Using the method of Example 5, we estimate $t \approx 14$ years.

**9.** Using the method of Example 5, we estimate $t \approx 22$ years.

**11.** For the first patch, we want to maximize $R_1(t) = 150t/((3+t)(t+2)) = 150t/(t^2 + 5t + 6)$. Then (after simplification) $R_1' = 150(6-t^2)/(t^2+5t+6)^2$, and the maximum is at $t = \sqrt{6}$, where $R_1 \approx 15.15$. For the second patch, we want to maximize $R_2(t) = 250t/((5+t)(t+3)) = 250t/(t^2+8t+15)$. Then (after simplification) $R_2' = 250(15 - t^2)/(t^2+8t+15)^2$, and the maximum is at $t = \sqrt{15}$, where $R_2 \approx 15.88$. Thus the individual should choose the second patch.

**13.** For the first patch, we want to maximize $R_1(t) = 150t/((3+t)(t+3)) = 150t/(t^2 + 6t + 9)$. Then (after simplification) $R_1' = 150(3-t)/(t+3)^3$, and the maximum is at $t = 3$, where $R_1 = 12.5$. For the second patch, we want to maximize $R_2(t) = 150t/((4+t)(t+2)) = 150t/(t^2+6t+8)$. Then (after simplification) $R_2' = 150(8 - t^2)/(t^2+6t+8)^2$, and the maximum is at $t = \sqrt{8}$, where $R_2 \approx 12.87$. Thus the individual should choose the second patch.

**15.** For the first patch, we want to maximize $R_1(t) = 250t/((5+t)(t+2)) = 250t/(t^2+7t+10)$. Then (after simplification) $R_1' = 250(10-t^2)/(t^2+7t+10)^2$, and the maximum is at $t = \sqrt{10}$, where $R_1 = 18.76$. For the second patch, we want to maximize $R_2(t) = 150t/((4+t)(t+3)) = 150t/(t^2+7t+12)$. Then (after simplification) $R_2' = 150(12-t^2)/(t^2+7t+12)^2$, and the maximum is at $t = 2\sqrt{3}$, where $R_2 \approx 10.77$. Thus the individual should choose the first patch.

**17.** We need the value $t$ where $P'(t) = P(t)/t$. We differentiate to obtain $P'(t) = (5t^{3/2}(1+t^2) - 4t^{7/2})/(1+t^2)^2 = (t^{7/2} + 5t^{3/2})/(1+t^2)^2$. Thus the equation we have to solve is $(t^{7/2}+5t^{3/2})/(1+t^2)^2 = P(t)/t = 2t^{3/2}/(1+t^2) - 1/t$. We obtain $t \approx 3.11$ (decades).

**19.** We need the value $t$ where $P'(t) = P(t)/t$. We differentiate to obtain $P'(t) = (5t^{3/2}(1+2t^2) - 8t^{7/2})/(1+2t^2)^2 = (2t^{7/2} + 5t^{3/2})/(1+2t^2)^2$. Thus the equation we have to solve is $(2t^{7/2}+5t^{3/2})/(1+2t^2)^2 = P(t)/t = 2t^{3/2}/(1+2t^2) - 1/t$. We obtain $t \approx 4.93$ (decades).

**21.** We need the value $t$ where $P'(t) = P(t)/t$. We differentiate to obtain $P'(t) = ((25/2)t^{3/2}(1+2t^2) - 20t^{7/2})/(1+2t^2)^2 = (5t^{7/2} + (25/2)t^{3/2})/(1+2t^2)^2$. Thus the equation we have to solve is $(5t^{7/2} + (25/2)t^{3/2})/(1+2t^2)^2 = P(t)/t = 5t^{3/2}/(1+2t^2) - 2/t$. We obtain $t \approx 3.72$ (decades).

**23.** From Example 6, we know that $T$ has to satisfy the equation $V'(T) = \delta e^{\delta T}V(T)/(e^{\delta T} - 1)$. Here (after simplification) $V'(T) = T^{3/2}(5+T^2)/(1+T^2)^2$ and $\delta = 0.2$, and using technology, the value of $T$ which solves the equation is $T \approx 2.43$.

**25.** From Example 6, we know that $T$ has to satisfy the equation $V'(T) = \delta e^{\delta T}V(T)/(e^{\delta T} - 1)$. Here (after simplification) $V'(T) = 7T^{3/2}(5+T^2)/6(1+T^2)^2$ and $\delta = 0.15$, and using technology, the value of $T$ which solves the equation is $T \approx 2.38$.

**27.** The function we want to minimize is $w(h) = hf(h) = 0.8h + 80h/(h-10)$. We obtain that $w'(h) = 0.8 + (80(h-10) - 80h)/(h-10)^2 = 0.8 - 800/(h-10)^2$. This means $w'(h) = 0$ when $(h-10)^2 = 1000$, i.e. when $h = 41.6$ cm (we need the positive root).

**29. a.** We have to maximize the yield, which is $Y(t) = aw(t)P(t) = 900a(5 + 400t -$

$t^2)/(900+t)$. We differentiate to obtain $Y'(t) = 900a((400-2t)(900+t)-(5+400t-t^2))/(900+t)^2$. We get that $Y'(t)=0$ when the numerator vanishes, i.e. when $-t^2-1800t+359995=0$. We find the optimum time $T=-900+\sqrt{1169995}\approx 181.66$ days.

**b.** The crop's value is given by $1 = aw(182)P(182) \approx 33006$ when harvested at 182 days; using the information that early harvest reduces the value by $10t_e\%$, we get that the expected value of the harvest is $(0+0+0+0+0+0+0+0.1aw(173)P(173)+0.2aw(174)P(174)+0.3aw(175)P(175)+0.4aw(176)P(176)+0.5aw(177)P(177)+0.6aw(178)P(178)+0.7aw(179)P(179)+0.8aw(180)P(180)+0.9aw(181)P(181)+\sum_{i=182}^{190} aw(i)P(i))/25 \approx 0.54$.

**31.** Using technology, the best fit quadratic curve is $f(x) = -92.113+490.763x-5.915x^2$. The maximum is located at the solution of $f'(x)=0$, which is $x\approx 41.485$.

**33.** Using technology, the best fit quartic curve is $f(x) = 6606.970-549.194x+50.515x^2-1.275x^3+0.010192x^4$. The maximum is located at the solution of $f'(x)=0$ on $[15,50]$; there are two of these, $x\approx 38.286$ and $x\approx 48.244$. We obtain the larger value at $x\approx 48.244$.

## Problem Set 4.5 - Linearization and Difference Equations

**1.** $x=0$ and $x=1$ are unstable; $x=1/2$ is stable.

**3.** $x=0$ and $x\approx 0.75$ are unstable equilibria.

**5.** The equilibria are given by the equation $x=2x$; thus the only equilibrium is $x=0$. We obtain that $x_1=2x_0$, then $x_2=2^2x_0$, $x_3=2^3x_0$, $\ldots x_n=2^nx_0$. If $x_0\neq 0$, $\lim_{n\to\infty} x_n=\pm\infty$, so $x^*=0$ is unstable.

**7.** The function is defined for $x\geq 0$. The equilibria are given by the equation $x=x^{1/2}$; thus the equilibria are $x=0$ and $x=1$. We obtain that $x_1=x_0^{1/2}$, then $x_2=x_0^{1/4}$, $x_3=x_0^{1/8}$, $\ldots x_n=x_0^{1/2^n}$. For $x_0>0$, $\lim_{n\to\infty} x_n=1$. Thus 1 is stable, and 0 is unstable.

**9.** The equilibria are given by the equation $x=4\sqrt{x}$; thus the equilibria are $x=0$, and $x=16$. Some algebra and mathematical induction implies that $x_n = 2^{4-(1/2^{n-2})}x_0^{1/2^n}$. Thus 0 is unstable, and 16 is stable.

**11.** The equilibria are given by the equation $x=x^2$; thus the equilibria are $x=0$, and $x=1$. In this case, $f'(x)=2x$, thus $|f'(0)|=0<1$ and $|f'(1)|=2>1$; hence 0 is stable, and 1 is unstable.

**13.** The equilibria are given by the equation $x=2x/(1+2x)$; thus the equilibria are $x=0$, and $x=1/2$. In this case, $f'(x)=2/(1+2x)^2$, thus $|f'(0)|=2>1$ and $|f'(1/2)|=1/2<1$; hence 0 is unstable, and 1/2 is stable.

**15.** The equilibria are given by the equation $x=4x(1-x)$; thus the equilibria are $x=0$, and $x=3/4$. In this case, $f'(x)=4-8x$, thus $|f'(0)|=4>1$ and $|f'(3/4)|=2>1$; hence both are unstable.

**17. a.** The equilibria are given by the equation $x=rx(1-x/100)$; thus the equilibria are $x=0$, and $x=100(1-1/r)$.

**b.** $f'(x) = r(1 - x/100) - rx/100 = r - 2rx/100$. Thus $f'(0) = r$, and 0 is stable when $(0 <)r < 1$.

**c.** The nonzero equilibrium is positive when $1 - 1/r > 0$, i.e. when $r > 1$.

**d.** From part **b**, $f'(100(1 - 1/r)) = 2 - r$, thus the nonzero equilibrium is stable if $1 < r < 3$.

**19. a.** The equilibria are given by the equation $x = rx/(1 + x)$; thus the equilibria are $x = 0$, and $x = r - 1$.

**b.** $f'(x) = (r(1 + x) - rx)/(1 + x)^2 = r/(1 + x)^2$. Thus $f'(0) = r$, and 0 is stable when $(0 <)r < 1$.

**c.** The nonzero equilibrium is positive when $r - 1 > 0$, i.e. when $r > 1$.

**d.** From part **b**, $f'(r - 1) = 1/r$, thus the nonzero equilibrium is stable if $1 < r$.

**21.** $w_1 = 1/2$, $w_2 = 1/2$, thus the function we iterate is $f(x) = (w_1x^2 + x(1 - x))/(w_1x^2 + 2x(1 - x) + w_2(1 - x)^2) = x(x - 2)/(2x^2 - 2x - 1)$. The figure shows that $x = 1/2$ is stable, the frequency approaches this value as long as $0 < x_0 < 1$. $x = 0$ and $x = 1$ are unstable equilibria.

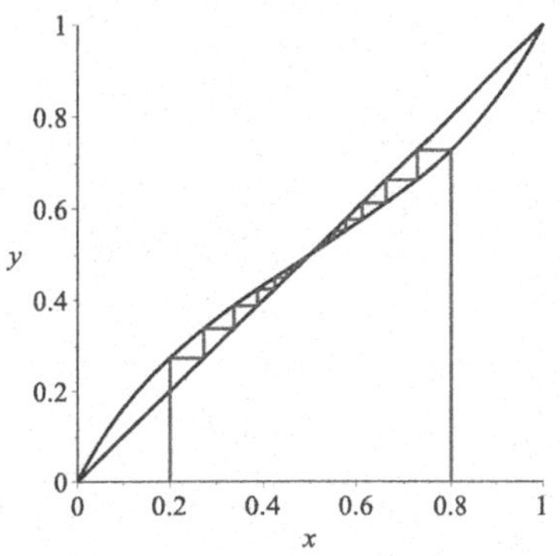


**23.** $w_1 = 1/2$, $w_2 = 2$, thus the function we iterate is $f(x) = (w_1x^2 + x(1 - x))/(w_1x^2 + 2x(1 - x) + w_2(1 - x)^2) = x/(2 - x)$. The figure shows that $x = 0$ is stable, the frequency approaches this value as long as $0 \le x_0 < 1$. $x = 1$ is an unstable equilibrium.

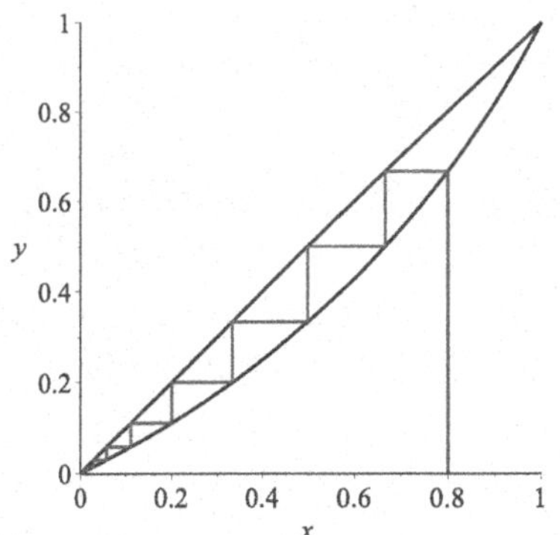


**25.** $f(x) = x^2 - 2$, $f'(x) = 2x$, thus the iteration formula is $x_{n+1} = x_n - (x_n^2 - 2)/2x_n = x_n/2 + 1/x_n$. If we start at $x_0 = 1$, we obtain that $x_{20} \approx 1.414213562$.

**27.** $f(x) = x^3 - x + 1$, $f'(x) = 3x^2 - 1$, thus the iteration formula is $x_{n+1} = x_n - (x_n^3 - x_n + 1)/(3x_n^2 - 1)$. If we start at $x_0 = -1$, we obtain that $x_{20} \approx -1.324717957$.

**29.** $f(x) = \cos x - x$, $f'(x) = -\sin x - 1$, thus the iteration formula is $x_{n+1} = x_n - (\cos x_n - x_n)/(-\sin x_n - 1)$. If we start at $x_0 = 1$, we obtain that $x_{20} \approx 0.7390851332$.

**31.** $f(x) = e^x - 5x$, $f'(x) = e^x - 5$, thus the iteration formula is $x_{n+1} = x_n - (e^{x_n} - 5x_n)/(e^{x_n} - 5)$. If we start at $x_0 = 0$, we obtain that $x_{20} \approx 0.2591711018$.

**33. a.** We can see that $\lim_{x \to \pm\infty} f(x) = -\infty$ and $f(0) = 11/8$. Thus by the intermediate value theorem, there is at least one negative and one positive root.

**b.** $f(x) = -2x^4 + 3x^2 + 11/8$, $f'(x) = -8x^3 + 6x$, thus the iteration formula is $x_{n+1} = x_n - (-2x_n^4 + 3x_n^2 + 11/8)/(-8x_n^3 + 6x_n)$. If we start at $x_0 = 2$, we obtain that

$x_{10} \approx 1.366760399$.

**c.** If $x_0 = 1/2$, then we obtain that $x_1 = -1/2$, $x_2 = 1/2$ and the iteration oscillates between these two values after that.

**35. a.** The equilibria are given by the equation $x = 2f(x) = 1.612x/(1+(0.0114x)^{7.53})$, which is satisfied by $x = 0$ and $x = (0.612)^{1/7.53}/0.0114 \approx 82.18$. We obtain that $2f'(x) = 2(0.806(1 + (0.0114x)^{7.53}) - 0.06919x(0.0114x)^{6.53})/(1 + (0.0114x)^{7.53})^2$, and then $|2f'(0)| = 1.612 > 1$ and $|2f'(82.18)| \approx 1.86 > 1$, so they are both unstable.

**b.**

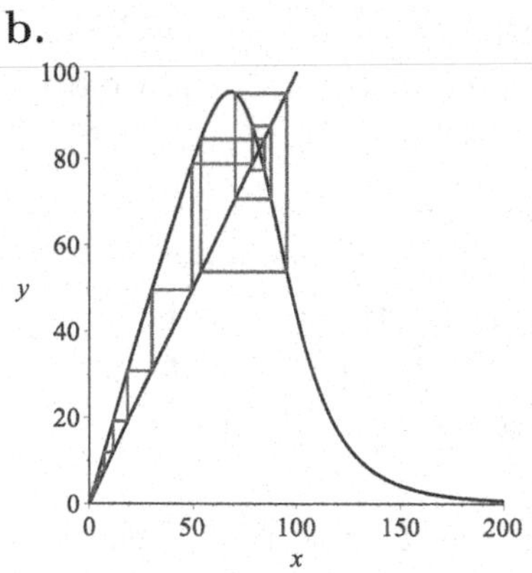


**37. a.** The equilibria are given by the equation $x = (w_1x^2 + x(1-x))/(w_1x^2 + 2x(1-x) + w_2(1-x)^2)$. Clearly, $x = 0$ is a solution. When $x \neq 0$, we obtain $1 = (w_1x + 1 - x)/(w_1x^2 + 2x(1-x) + w_2(1-x)^2)$, i.e. $w_1x+1-x = w_1x^2+2x(1-x)+w_2(1-x)^2$, which is the quadratic equation $(w_1 + w_2 - 2)x^2+(3-w_1-2w_2)x+w_2-1 = 0$. This factors into $(x-1)((w_1+w_2-2)x+1-w_2) = 0$, thus the other two roots are $x = 1$ and $x = p^* = (w_2 - 1)/(w_1 + w_2 - 2)$.

**b.** The derivative of the right side function is (after simplification) $f' = (w_1x^2 - w_2(x-1)(1+(2w_1-1)x))/(w_1x^2+2x(1-x)+w_2(1-x)^2)^2$. Thus $|f'(0)| = |w_2/w_2^2| = |1/w_2|$, and 0 is unstable when $-1 < w_2 < 1$. Also, $|f'(1)| = |w_1/w_1^2| = |1/w_1|$, and 1 is stable when $w_1 > 1$.

**c.** After a long simplification, we obtain $|f'((w_2 - 1)/(w_1 + w_2 - 2))| = |(w_1 + w_2 - 2w_1w_2)/(1 - w_1w_2)|$. When $w_1 < 1$ and $w_2 < 1$, the other two equilibria (0 and 1) are unstable by part **b**. We can check that $w_1 + w_2 - w_1w_2 - 1 = (1 - w_1)(w_2 - 1) < 0$, thus $w_1 + w_2 - 2w_1w_2 = w_1 + w_2 - w_1w_2 - 1+(1-w_1w_2) < 1-w_1w_2$, and then because $w_1w_2 < 1$, we get (after division by the right side of this inequality) that $|f'(p^*)| < 1$ and the third equilibrium is stable.

**d.** We have seen that $|f'((w_2 - 1)/(w_1 + w_2 - 2))| = |(w_1+w_2-2w_1w_2)/(1-w_1w_2)|$. When $w_1 > 1$ and $w_2 > 1$, the other two equilibria (0 and 1) are stable by part **b**. We can check that $w_1 + w_2 - w_1w_2 - 1 = (1-w_1)(w_2-1) < 0$, thus $w_1+w_2-2w_1w_2 = w_1+w_2-w_1w_2-1+(1-w_1w_2) < 1-w_1w_2$, and then because $w_1w_2 > 1$, we get (after division by the right side of this inequality) that $|f'(p^*)| > 1$ and the third equilibrium is unstable.

**39.** The volume of a spherical cap with height $H$ is given by $V = \pi H^2(3 - H)/3$ when $R = 1$, according to Problem 38. We need that this cap is one sixth of the volume of the whole sphere, thus $\pi H^2(3 - H)/3 = (4\pi/3)/6$. This gives $2 = 3H^2(3 - H)$, i.e. $3H^3-9H^2+2 = 0$. Then Newton's iteration gives $x_{n+1} = x_n - (3x_n^3 - 9x_n^2 + 2)/(9x_n^2 - 18x_n)$; let $x_0 = 1$, then $x_{20} \approx 0.518298$, and $d = 1 - H \approx 0.481702$.

**41.** The equation we want to solve is $V(x) = 2V(0) = 1$, i.e. $V(x) - 1 = 0$; $V' = 0.0012e^{0.24x} - 0.0594e^{-0.12x}$, thus Newton's iteration is given by $x_{n+1} = x_n - (V(x_n) - 1)/V'(x_n)$. If $x_0 = 25$, then $x_{20} \approx 21.925$ days.

**43.** The equation we want to solve is $V(x) = V(0) = 0.44$, i.e. $V(x) - 0.44 = 0$; $V' = 0.0010516e^{0.239x} - 0.0483516e^{-0.111x}$, thus Newton's iteration is given by $x_{n+1} = x_n - (V(x_n) - 0.44)/V'(x_n)$. If $x_0 = 20$, then $x_{20} \approx 18.715$ days.

**45.** Newton's method gives the iteration $x_{n+1} = x_n - (x_n^3 - 3x_n^2 + 2x_n + 0.4)/(3x_n^2 - 6x_n + 2)$, and $x_{n+1} = x_n - (x_n^3 - 3x_n^2 + 2x_n + 0.3)/(3x_n^2 - 6x_n + 2)$, respectively. The reason we converge to the same value in the first case is that there is only one real root there ($x \approx -0.1597$), while in the second case we have three real roots ($x \approx -0.1254$, $x \approx 1.3389$, and $x \approx 1.7865$).

**47.** The equation we want to solve is $f(x) = 400$, i.e. $f(x) - 400 = 0$. In this case, $f'(x) = 0.122463 - (\pi/2)\sin(\pi x/6)$. Thus Newton's method gives the iteration $x_{n+1} = x_n - (f(x_n) - 400)/f'(x_n)$. Choose $x_0 = 560$, then we obtain $x \approx 562.299$.

## Review Questions

**1.** $g'(x) = 3x^2 - 3$, thus $g' = 0$ at $x = \pm 1$. Now $g''(x) = 6x$, thus we have a local maximum at $x = -1$ (because $g''(-1) = -6 < 0$) and we have a local minimum at $x = 1$ (because $g''(1) = 6 > 0$). There are no global extrema.

**3.** The vertical asymptotes are at $x = 0$, $x = -1$ and $x = -2$. The horizontal asymptote is $y = \lim_{x \to \pm\infty} (x^3+3)/(x(x+1)(x+2)) = \lim_{x \to \pm\infty} (1 + 3/x^3)/((1 + 1/x)(1 + 2/x)) = 1$. Also, $f' = (3x^2x(x+1)(x+2) - (x^3+3)(3x^2 + 6x + 2))/(x(x+1)(x+2))^2 = (3x^4 + 4x^3 - 9x^2 - 18x - 6)/(x(x+1)(x+2))^2$. The zeros of the numerator are at $x \approx -0.4458$ and $x \approx 1.9773$. We can check that $f' > 0$ and $f$ is increasing on $(-\infty, -2)$, $(-2, -1)$, $(-1, -0.4458)$, and $(1.9773, \infty)$, and $f' < 0$ and $f$ is decreasing on $(-0.4458, 0)$ and $(0, 1.9973)$.

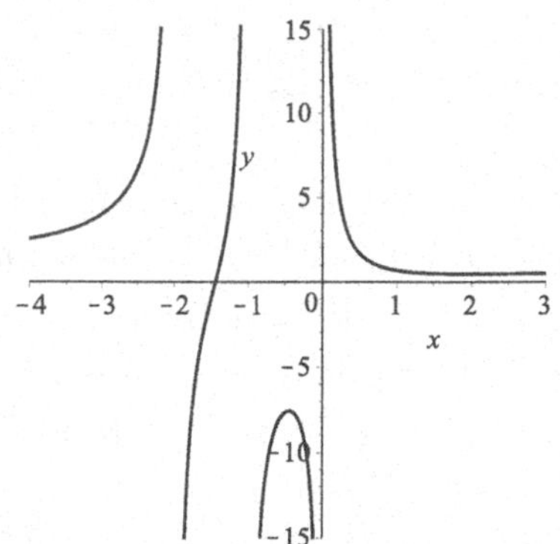


**5. a.** $f'(t) = 0.0000213t^2 - 0.0031704t + 0.1419159$, thus $f$ is increasing for all $t$ values.

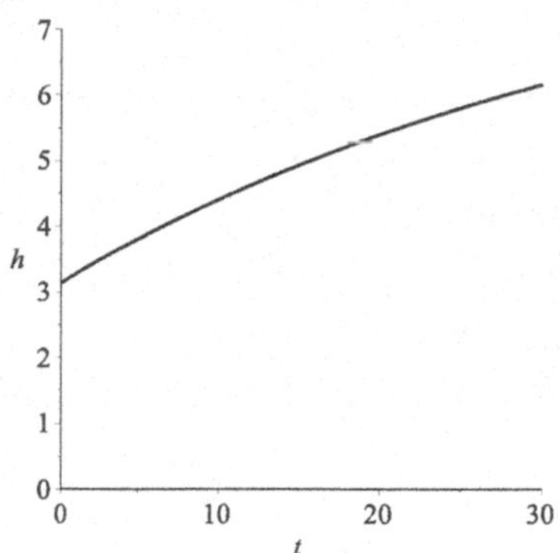


**b.** To find the time of the most rapid growth, we need the maximum of $f'(t)$, so we find $f''(t) = 0.0000426t - 0.0031704 = 0$ when $t = 74.42$; this is outside of our domain, thus the most rapid growth is at $t = 0$, and the least rapid growth is at $t = 30$.

**7.** We have to find the global maximum of $P(n)$ on $[10, 40]$. $P'(n) = -3n^2 + 55.2n + 970$; the only critical point on the interval is $n \approx 29.4$. Thus we check $n = 10$, $n = 40$, $n = 29$ and $n = 30$. We obtain the maximum at $n = 29$; the profit in this case is \$22,717.60.

**9.** The function we want to maximize is

the yield $Y(t) = P(t)S(t) = P(t)w(t)/3 = 100(1+150t-t^2/3)/(100+t)$. Then $Y'(t) = ((15000-200t/3)(100+t)-100(1+150t-t^2/3))/(100+t)^2$. To find the value of $t$ which corresponds to the maximum yield, we solve $Y'(t)=0$ and obtain $t = 134.5$ (days).

**11. a.** Following Example 7 in Section 4.5, let $x$ denote the frequency of the allele which does not cause sickle cell anemia. The assumption given in the problem means $w_1 = 0.8$ and $w_2 = 0$ in the iterating function $f(x) = (w_1x^2+x(1-x))/(w_1x^2+2x(1-x)+w_2(1-x)^2) = (0.8x^2+x(1-x))/(.8x^2+2x(1-x)+0\cdot(1-x)^2) = (x-0.2x^2)/(2x-1.2x^2) = (1-0.2x)/(2-1.2x)$ when $x \neq 0$.

**b.** The equilibria are given by the equation $x = f(x)$, i.e. $2x-1.2x^2 = 1-0.2x$, which is the same as $(1.2x-1)(x-1) = 0$, and we obtain $x=1$ and $x = 1/1.2 = 5/6$. We can check that when $x \neq 0$, $f'(x) = 0.8/(2-1.2x)^2$, so $f'>0$ and $f$ is increasing on this interval. Also, $f'(1) = 1/0.8 > 1$, and $f'(5/6) = 0.8 < 1$ and 1 is unstable and 5/6 is stable.

**c.** In the long run, the proportion of the non-sickle cell allele converges to 5/6.

**13.** Assume the bird spends $x$ minutes on the first tree, and $5-x$ on the second tree. The total energy intake is given by $E(x) = 200(1-e^{-x})+100(1-e^{-(5-x)})$. Then $E'(x) = 200e^{-x}-100e^{x-5}$, thus $E'=0$ when $2 = e^{2x-5}$, and then $x = (5+\ln 2)/2$.

**15.** We differentiate: $G'(N) = 2.61(1-(N/148)^\theta)+2.61N(-\theta(N/148)^{\theta-1})(1/148) = 2.61-2.61(1+\theta)(N/148)^\theta$. The maximum is reached when $G'(N)=0$, i.e. when $1 = (1+\theta)(N/148)^\theta$, which is $N = 148/(1+\theta)^{1/\theta}$.

**17. a.** The equilibria are given by the equation $x = 3x^2/(1+x^2)$, i.e. $x(1+x^2) = 3x^2$. One solution is $x=0$; when $x \neq 0$, the other solutions are given by $1+x^2 = 3x$, thus $x = (3-\sqrt{5})/2$ and $x = (3+\sqrt{5})/2$.

**b.** $f'(x) = (6x(1+x^2)-3x^2 2x)/(1+x^2)^2 = 6x/(1+x^2)^2$, thus $f'>0$ and $f$ is increasing for all $x>0$.

**c.** $f'(0) = 0$, $f'((3-\sqrt{5})/2) = 1+\sqrt{5}/3$ and $f'((3+\sqrt{5})/2) = 1-\sqrt{5}/3$. Thus 0 and $(3+\sqrt{5})/2$ are stable, and $(3-\sqrt{5})/2$ is unstable.

**19. a.**

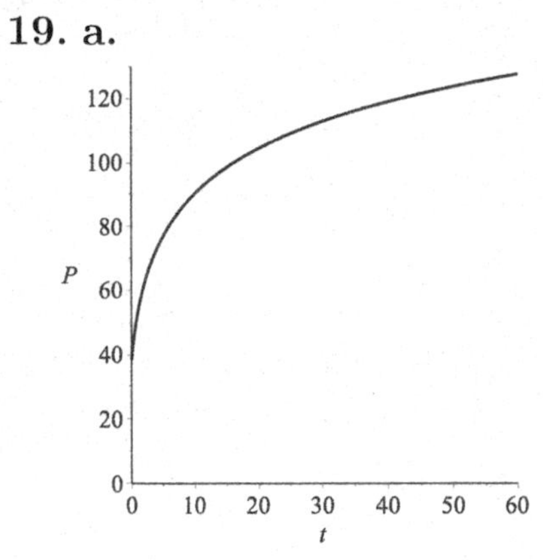


**b.** $P'(t) = 0.98 \cdot 21.8/(0.98t+1) = 21.364/(0.98t+1)$.

**Problem Set 5.1 - Antiderivatives**

**1.** We get $F(x) = 2x+C$, because $F'(x) = 2$ by the power rule.

**3.** We obtain $F(x) = x^2+3x+C$, because $F'(x) = 2x+3$ by the power rule.

**5.** We obtain $F(x) = 6x^5/5+C$, because $F'(x) = 6x^4$ by the power rule.

**7.** We get $F(x) = 2x^3/3 - 5x + C$, because $F'(x) = 2x - 5$ by the power rule.

**9.** We get $F(t) = 2t^4 + 15t^2/2 + C$, because $F'(t) = 8t^3 + 15t$ by the power rule.

**11.** We obtain $F(x) = -5x^{-1} + C$, because $F'(x) = 5/x^2$ by the power rule.

**13.** We obtain $F(x) = \sin x + C$, because $F'(x) = \cos x$.

**15.** We get $F(x) = -(3/2\pi)\cos(2\pi x) + C$, because $F'(x) = 3\sin(2\pi x)$ by the chain rule.

**17.** We get $F(x) - 3e^x + C$, because $F'(x) = 3e^x$.

**19.** We get $F(x) = (2/5)x^{5/2} + (2/3)x^{3/2} + \ln x + C$, because $F'(x) = x^{3/2} + x^{12} + x^{-1}$ by the power rule.

**21.** We get $F(u) = 3u^2 + 3\sin u + C$, because $F'(u) = 6u + 3\cos u$.

**23.** We get $F(x) = 2x + C$, because $F'(x) = 2$. Also, $1 = F(0) = C$, thus $F(x) = 2x + 1$.

**25.** We get $F(x) = x^2 + 3x + C$, because $F'(x) = 2x + 3$. Also, $0 = F(-3) = 9 - 9 + C$, thus $F(x) = x^2 + 3x$.

**27.** We get $F(x) = (6/5)x^5 + C$, because $F'(x) = 6x^4$. Also, $-2 = F(1) = (6/5) + C$, thus $F(x) = (6/5)x^5 - 16/5$.

**29. a.** $F(x) = x - 2x^2 + C$, because $F'(x) = 1 - 4x$. Also, $0 = F(1) = 1 - 2 + C$, thus $F(x) = x - 2x^2 + 1$.

**b.**

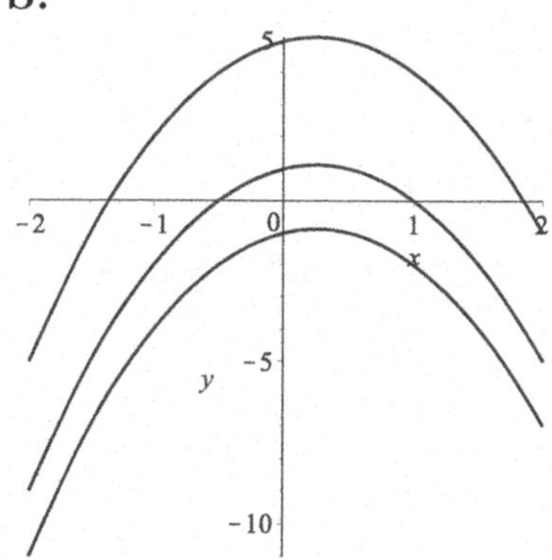


**c.** The maximum of $F(x) + C$ is where $F'(x) = 1 - 4x = 0$, i.e. at $x = 1/4$ ($F''(x) = -4 < 0$, so it is a maximum). The value of the maximum is $(1/4) - 2(1/4)^2 + 1 + C$, thus we need that $C = 2(1/4)^2 - 1/4 - 1 = -9/8$.

**31.** The function is $F(x) = x^3/3 + (3/2)x^2$, because $F'(x) = x^2 + 3x$ and $F(0) = 0$.

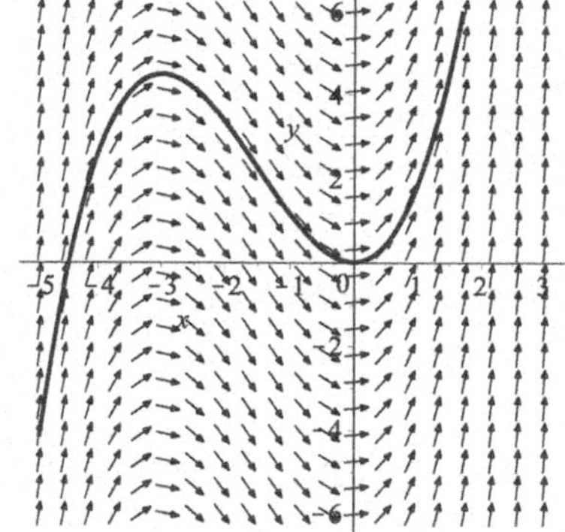

**33.** The function is $F(x) = x^2/2 + e^x + 1$, because $F'(x) = x + e^x$ and $F(0) = 2$.

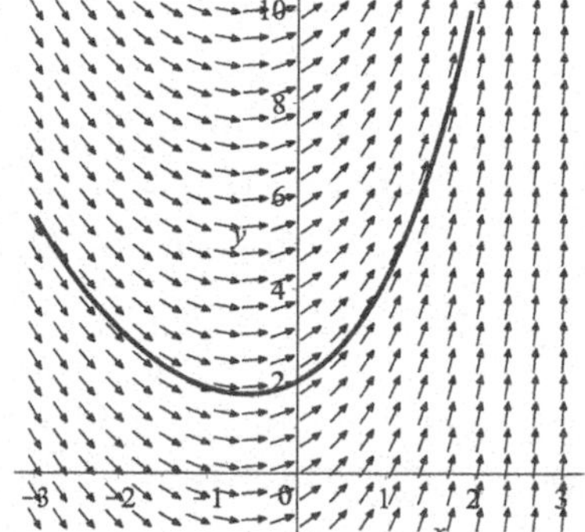

**35.** The function is $F(x) = x^2/2 + 1$, because

$F'(x) = x$ and $F(0) = 1$.

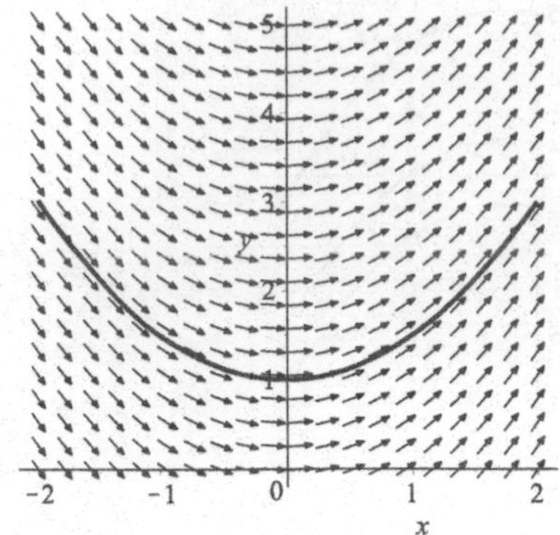

**37.** The function is $F(x) = \sin x + 1$, because $F'(x) = \cos x$ and $F(0) = 1$.

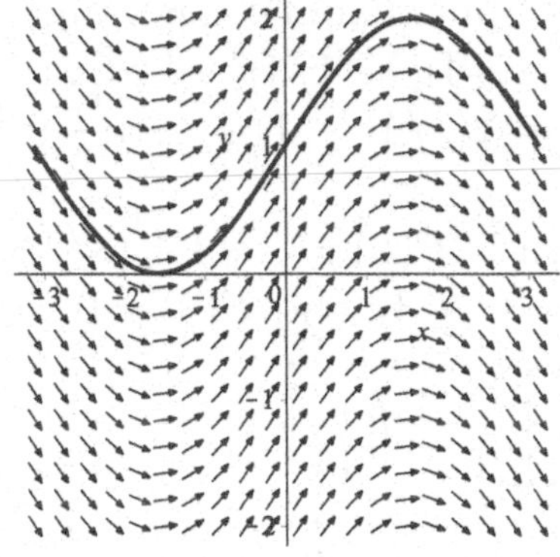

**39.** We get $F(x) = -(1/a)\cos(ax) + C$, because $F'(x) = \sin ax$ by the chain rule.

**41. a.** Using the given $T(x)$, we get $F'(x) = -0.06075 + 0.00112(80 + 10\cos(2\pi x)) = 0.02885 + 0.0112\cos(2\pi x)$. Thus $F(x) = 0.028855x + 0.0112\sin(2\pi x)/2\pi + C$, and also $0 = F(0) = 0 + 0 + C$; so we obtain $F(x) = 0.02885x + 0.0112\sin(2\pi x)/2\pi$.

**b.** We have to find $x$ such that $F(x) = 1$, i.e. $0.02885x + 0.0112\sin(2\pi x)/2\pi = 1$. Using technology, we obtain that $x \approx 34.72$ (days).

**43. a.** Using the given $T(t)$, we get $F'(t) = -0.0582211 + 0.00417376(30 + 10\sin(2\pi t)) = 0.0669917 + 0.0417376\sin(2\pi t)$. Thus $F(t) = 0.0669917t - 0.0417376\cos(2\pi t)/2\pi + C$, and also $0 = F(0) = 0 - 0.0417376/2\pi + C$; so we obtain $F(t) = 0.0669917t - 0.0417376\cos(2\pi t)/2\pi + 0.0417376/2\pi$.

**b.** We have to find $t$ such that $F(t) = 1$, i.e. $0.0669917t - 0.0417376\cos(2\pi t)/2\pi + 0.0417376/2\pi = 1$. Using technology, we obtain that $t \approx 14.91$ (days).

**45.** Using the notations and ideas of Example 8, the distance fallen is given by $s(t) = 4.9t^2 = 305$, thus $t = \sqrt{305/4.9} \approx 7.89$ (seconds). The velocity reached is $v(t) = 9.8t = 9.8 \cdot 7.89 = 77.3$ (m/sec).

**47.** Let $s(t)$ be the distance fallen, $v(t) = s'(t)$ be the velocity, and $a(t) = v'(t) = 5.2$ ft/s$^2$ is the acceleration. Then $v(t) = 5.2t + C$, $0 = v(0) = 0 + C$; so $v(t) = 5.2t$. Finding the antiderivative of this, we obtain that $s(t) = 2.6t^2 + C$; now $0 = s(0) = 0 + C$. Thus $s(t) = 2.6t^2$ and we have to find the time $t$ such that $2.6t^2 = 4$; we obtain $t = \sqrt{4/2.6} \approx 1.24$ seconds. When $a = 32$ ft/s$^2$, the same computation yields $t = \sqrt{4/16} = 0.5$ seconds.

**49.** The population $P(t)$ is the antiderivative of $4 + 5t^{2/3}$; thus $P(t) = 4t + 3t^{5/3} + C$. We know that $2000 = P(0) = C$, thus $P(8) = 2000 + 4 \cdot 8 + 3 \cdot 8^{5/3} = 2128$ people.

**51.** The antiderivative is $v(r) = ar^2/2 + C$, and we know that $0 = v(R) = aR^2/2 + C$. Thus $v(r) = ar^2/2 - aR^2/2 = (a/2)(r^2 - R^2)$.

**53.** First, she travels $0.7 \cdot 88 = 61.6$ feet before hitting the breaks. At this time (denote it by $t = 0$), the deceleration is $a = -28$ ft/s$^2$. Thus the antiderivative is $v(t) = -28t + C$, where $88 = v(0) = 0 + C$. We obtained that $v(t) = -28t + 88$. This gives that the car stops when $v(t) = 0$, i.e. at $t = 88/28$ seconds. The distance traveled is the antiderivative of $v(t)$: $s(t) = -14t^2 + 88t + C$, where $0 = s(0) = 0 + 0 + C$. Thus she travels an additional $s(88/28) \approx 138.29$ feet. The total distance before stopping is $138.29 + 61.6 = 199.89$ feet. The cow lives.

## Problem Set 5.2 - Accumulated Change and Area under a Curve

**1.** $\Delta x = (1 - 0)/4 = 1/4$; this gives $R_4 = \sum_{k=1}^{4} f(k/4)(1/4) = \sum_{k=1}^{4}(2(k/4) + 1)(1/4) = 9/4 = 2.25$.

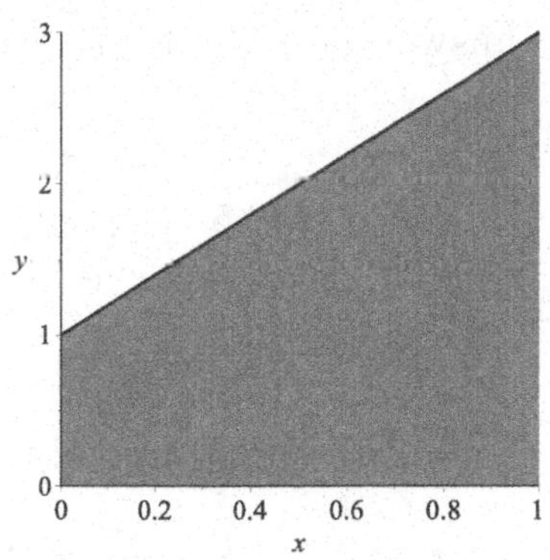

**3.** $\Delta x = (2 - 0)/4 = 1/2$; this gives $R_4 = \sum_{k=1}^{4} f(k/2)(1/2) = \sum_{k=1}^{4}(k/2)^2(1/2) = 15/4 = 3.75$.

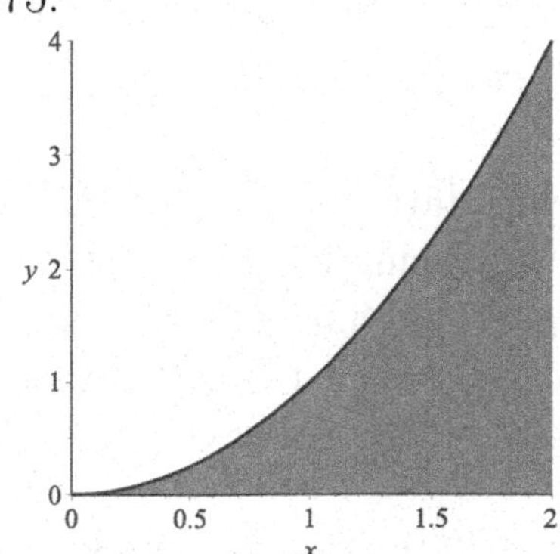

**5.** $\Delta x = (3 - 1)/4 = 1/2$; this gives $R_4 = \sum_{k=1}^{4} f(1 + k/2)(1/2) = \sum_{k=1}^{4}(1 + k/2)^3(1/2) =$ 27.

**7.** $\Delta x = (1 - 0)/4 = 1/4$; this gives $R_4 = \sum_{k=1}^{4} f(k/4)/4 = \sum_{k=1}^{4}((k/4)^2 + (k/4)^3)/4 = 55/64 \approx 0.859$.

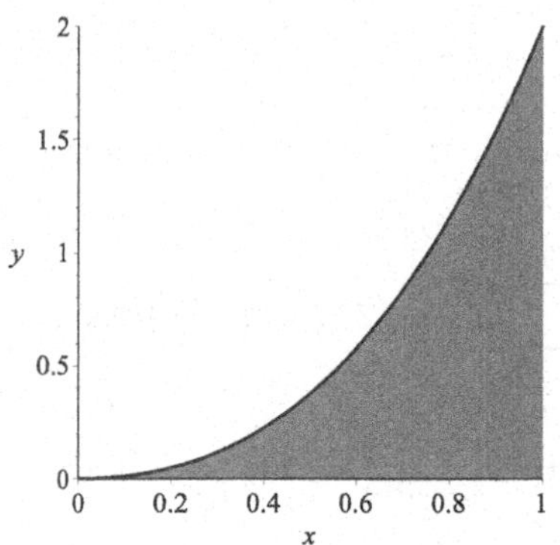

**9.** $\Delta x = (2 - 1)/4 = 1/4$; this gives $R_4 = \sum_{k=1}^{4} f(1 + k/4)/4 = \sum_{k=1}^{4}(1/(1 + k/4))/4 = 533/840 \approx 0.635$.

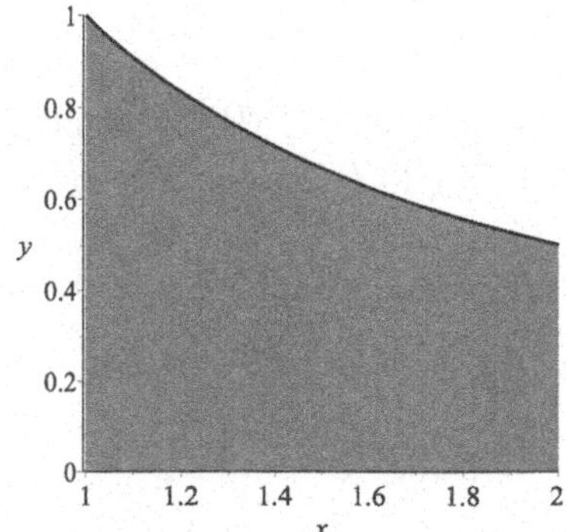

**11.** $\Delta x = (0 - (-\pi/2))/4 = \pi/8$; thus $R_4 =$

$\sum_{k=1}^{4} f(-\pi/2+k\pi/8)(\pi/8) = \sum_{k=1}^{4} \cos(-\pi/2+k\pi/8)(\pi/8) \approx 1.183$.

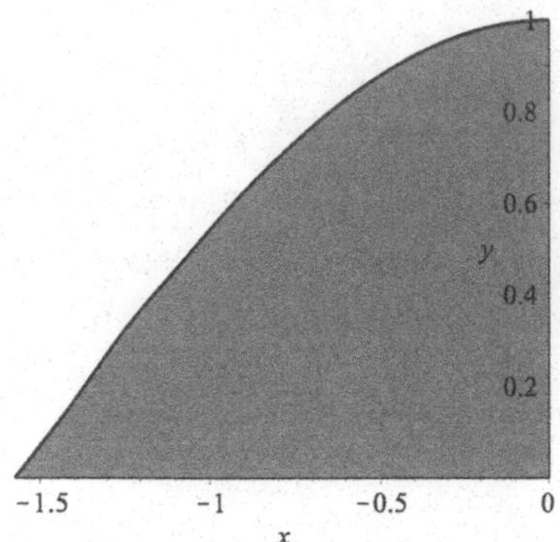


**13.** $\Delta x = (1-0)/10 = 1/10$; thus $R_{10} = \sum_{k=1}^{10} f(k/10)(1/10) = \sum_{k=1}^{10}(4k/10)(1/10) = 22/10 = 2.2$.

**15.** $\Delta x = (0-(-\pi/2))/10 = \pi/20$; this gives that $R_{10} = \sum_{k=1}^{10} f(-\pi/2+k\pi/20)(\pi/20) = \sum_{k=1}^{10} \cos(-\pi/2+k\pi/20)(\pi/20) \approx 1.076$.

**17.** $\Delta x = (3-0)/10 = 3/10$; then we obtain that $R_{10} = \sum_{k=1}^{10} f(3k/10)(3/10) = (3/10)\sum_{k=1}^{10} \ln((3k/10)^2+1) \approx 3.7557$.

**19.** In this case, $L_n = (0/n)^2(1/n) + (1/n)^2(1/n) + (2/n)^2(1/n) + \ldots + ((n-1)/n)^2(1/n) = (1/n^3)(0^2+1^2+2^2+\ldots+(n-1)^2) = (1/n^3)((n-1)(n-1+1)(2(n-1)+1)/6) = (1-1/n)(2-1/n)/6$. Thus $\lim_{n\to\infty} L_n = 1/3$.

**21.** $\Delta x = (4-0)/n = 4/n$; the right endpoints are thus $4k/n$, $k = 1, 2, \ldots, n$. Using these endpoints in the Riemann sum, $R_n = \sum_{k=1}^{n}(4k/n)(4/n) = (16/n^2)(1+2+\ldots+n) = (16/n^2)(n(n+1)/2) = 8(1+1/n)$. Thus $\lim_{n\to\infty} R_n = 8$.

**23.** $\Delta x = (2-0)/n = 2/n$; the left endpoints are thus $2k/n$, $k = 0, 1, 2, \ldots, n-1$. Using these endpoints in the Riemann sum, $L_n = \sum_{k=0}^{n-1}(2k/n)^3(2/n) = (16/n^4)(0^3+1^3+2^3+\ldots+(n-1)^3) = (16/n^4)((n-1)^2n^2/4) = 4(1-1/n)^2$. Thus $\lim_{n\to\infty} L_n = 4$.

**25.** We have to solve $(75-50)x = 25x = 1587$; thus the answer is $x = 1587/25 = 63.48$ (days).

**27.** We have to solve $(68.5-41)x = 27.5x = 1365.5$; thus the answer is $x = 1365.5/27.5 \approx 49.65$ (days).

**29.** Using the right endpoints, we obtain $\sum_{k=1}^{10} 890\ \mathrm{sech}^2(0.2\cdot 3k - 3.4)\cdot 3 \approx 8858$. We expect this answer to be more accurate, because we use more intervals.

**31.** Using the left endpoints, we estimate $0\cdot 2 + 100\cdot 2 + 400\cdot 2 + 900\cdot 2 + 1900\cdot 2 + 2500\cdot 2 + 1500\cdot 2 = 14,600$ cases.

**33.** For the given data, $m_i \le k < M_i$ for $i = 1, 2, \ldots 14$, thus using the given formula, we obtain $1.45+1.79+3.56+5.96+6.72+0.86+2.57+3.76+5.65+3.13+2.03+2.40+2.00+3.50 = 45.38$ degree-days.

**35.** The estimate is $(79-44)/4+(44-43)/2+(79-43)^2/(4(79-40)) \approx 17.558$ degree-days.

**37.** The estimate is $\sum_{k=1}^{5}(T(2k)-40)\cdot 2 \approx$

298.7 degree-days.

**39.** The right endpoint estimate for the distance traveled is $5(0.5) + 9.5(0.5) + 15.1(0.5) + 21(0.5) + 25(0.5) = 37.8$ meters.

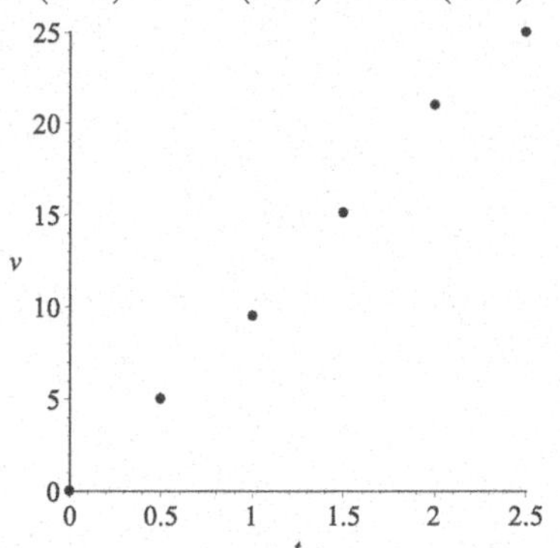


**41.** Using right endpoints, an estimate is $25{\cdot}10+20{\cdot}10+20{\cdot}10+15{\cdot}10+15{\cdot}10+0{\cdot}10 = 950$ square feet.

## Problem Set 5.3 - The Definite Integral

**1.** $\lim_{n\to\infty} \sum_{i=1}^{n} \frac{i}{n^2} = \lim_{n\to\infty} \sum_{i=1}^{n} \frac{i}{n} \cdot \frac{1}{n} = \int_0^1 x\,dx.$

**3.** We compute: $\lim_{n\to\infty} \sum_{i=1}^{n} \left(-2 + \frac{3i}{n}\right) \cdot \frac{3}{n} = \lim_{n\to\infty} \sum_{i=1}^{n} \left(-6 + 9\frac{i}{n}\right) \cdot \frac{1}{n} = \int_0^1 (-6+9x)\,dx.$

**5.** $\lim_{n\to\infty} \sum_{i=1}^{n} \left(1 - \frac{i^2}{n^2}\right) \cdot \frac{1}{n} = \int_0^1 (1-x^2)\,dx.$

**7.** Dividing the interval $[1,2]$ into $n$ equal parts, the right endpoints of the intervals are $1+i/n$, $i = 1, 2, \ldots, n$, and the length of the subintervals is $1/n$. Thus $\int_1^2 x^4\,dx = \lim_{n\to\infty} \sum_{i=1}^{n} \left(1+\frac{i}{n}\right)^4 \cdot \frac{1}{n}.$

**9.** Dividing the interval $[0,1]$ into $n$ equal parts, the right endpoints of the intervals are $i/n$, $i = 1, 2, \ldots, n$, and the length of the subintervals is $1/n$. Thus $\int_0^1 e^x\,dx = \lim_{n\to\infty} \sum_{i=1}^{n} e^{i/n} \cdot \frac{1}{n}.$

**11.** Dividing the interval $[-1,1]$ into $n$ equal parts, the right endpoints of the intervals are $-1+2i/n$, $i = 1, 2, \ldots, n$, and the length of the subintervals is $2/n$. Thus $\int_{-1}^1 |x|\,dx = \lim_{n\to\infty} \sum_{i=1}^{n} \left|-1+\frac{2i}{n}\right| \cdot \frac{2}{n}.$

**13.** The area is given by $(4+1/2)(1+8)/2 - (3-1/2)(6-1)/2 = 14.$

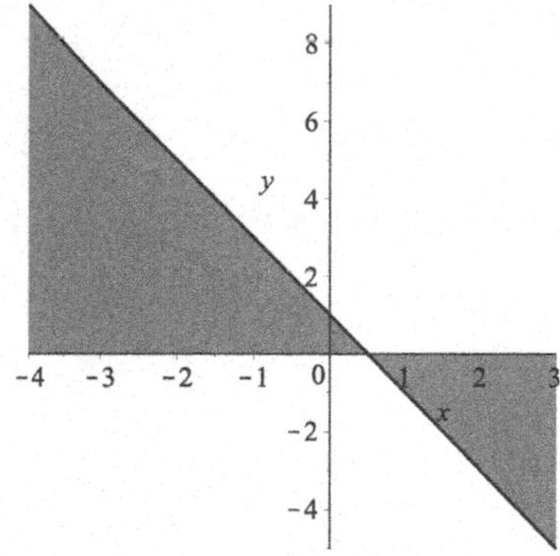


**15.** The area is $16\pi/4 = 4\pi$, a quarter of the circle with radius 4.

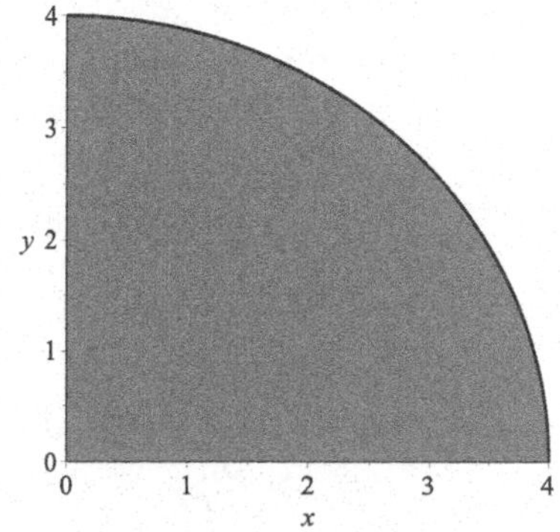


**17.** By the opposite rule, $\int_0^{-1} f(x)\,dx =$

$- \int_{-1}^{0} f(x)\,dx = -1/3.$

**19.** By the sum and scalar rule, $\int_{-1}^{2} [2f(x) - 3g(x)]\,dx = 2\int_{-1}^{2} f(x)\,dx - 3\int_{-1}^{2} g(x)\,dx = 6 - 9/2 = 3/2.$

**21.** By the splitting property, $\int_{-1}^{2} g(x)\,dx = \int_{-1}^{0} g(x)\,dx + \int_{0}^{2} g(x)\,dx$, thus $\int_{-1}^{0} g(x)\,dx = \int_{-1}^{2} g(x)\,dx - \int_{0}^{2} g(x)\,dx = 3/2 - 2 = -1/2.$

**23.** We know that $\sin x \leq 1$ for all $x$ values, thus by the bounding property $\int_{0}^{\pi} \sin x\,dx \leq 1 \cdot (\pi - 0) = \pi.$

**25.** For $-1 \leq x \leq 1$, $1 \leq \sqrt{1+x^2} \leq \sqrt{2}$. Then by the bounding property $(1)(1-(-1)) \leq \int_{-1}^{1} \sqrt{1+x^2}\,dx \leq \sqrt{2}(1-(-1)).$

**27.** The function is symmetric on the $y$-axis, thus $\int_{-1}^{1} x^2\,dx = 2(1/3) = 2/3.$

**29.** Let $\int_{-2}^{4} f(x)\,dx = A$ and $\int_{-2}^{4} g(x)\,dx = B$. Then by the sum and scalar rules the assumptions are $5A+2B = 7$ and $3A+B = 10$. These imply that $-A = 5A + 2B - (6A + 2B) = 7 - 20 = -13$, thus $A = 13$; and then $B = 10 - 3A = -29$.

**31.** By the splitting property, $\int_{1}^{3} f(x)\,dx = \int_{1}^{2} f(x)\,dx + \int_{2}^{3} f(x)\,dx$, thus $\int_{1}^{2} f(x)\,dx = 5-(-2) = 7$. Then we obtain $\int_{-1}^{2} f(x)\,dx = \int_{-1}^{1} f(x)\,dx + \int_{1}^{2} f(x)\,dx = 3 + 7 = 10.$

**33.** The length of the subintervals is given by $(0-(-2))/5 = 2/5$; the right endpoints are given by $-2 + 2k/5$, $k = 1, 2, \ldots 5$. The approximation is $R_5 = \sum_{k=1}^{5}(-2+2k/5)^2(2/5) = 48/25 = 1.92$; this is less than the integral because the rectangles are under the graph of the function.

**35.** The length of the subintervals is given by $(4-1)/5 = 3/5$; the right endpoints are $1 + 3k/5$, $k = 1, 2, \ldots 5$. We obtain $R_5 = \sum_{k=1}^{5}(1/(1+3k/5))(3/5) \approx 1.188$; this is less than the integral because the rectangles are under the graph of the function.

**37.** $\Delta x = (3-0)/n = 3/n$; the right endpoints are thus $3k/n$, $k = 1, 2, \ldots, n$. Using these endpoints in the Riemann sum, $R_n = \sum_{k=1}^{n}[(3k/n)^3 - 3](3/n) = (81/n^4)(1^3 + 2^3 + \ldots + n^3) - 3(3/n)(1 + 1 + \ldots + 1) = (81/n^4)(n^2(n+1)^2/4) - 9 = (81/4)(1 + 1/n)^2 - 9$. Thus $\lim_{n\to\infty} R_n = 45/4 = 11.25$.

**39.** True; the area of the rectangle is given by $C(b-a)$.

**41.** False; consider for example the interval $[0,1]$ and the function $f(x) = x$. Then $[f(x)]^2 = x^2 \leq x$, and by the dominance property the area is not greater.

**43.** The formula is $\int_{0}^{15}(60 + 10\sin 2\pi x)\,dx - \int_{0}^{15} 50\,dx = \int_{0}^{15} 10\,dx + 10\int_{0}^{15} \sin 2\pi x\,dx = 150$ degree-days. This is $(150/3000) \cdot 100 = 5\%$ of its ripening.

**45.** The formula is $\int_0^{25} 4x\,dx + \int_{25}^{50} (200 - 4x)\,dx$. The area under the function is the sum of the areas of two triangles: $(1/2)25(100) + (1/2)25(100) = 2500$.

**47.** Using the splitting property, $\int_{-a}^{a} f(x)\,dx = \int_{-a}^{0} f(x)\,dx + \int_0^a f(x)\,dx = 0$, because the graph of the function is symmetric with respect to the origin, thus the first integral is $-1$ times the second integral in the above expression.

**49.** We use the splitting property twice, first for $a \leq c \leq b$, then for $c \leq d \leq b$: $\int_a^b f(x)\,dx = \int_a^c f(x)\,dx + \int_c^b f(x)\,dx = \int_a^c f(x)\,dx + \int_c^d f(x)\,dx + \int_d^b f(x)\,dx$.

**51.** We obtain $\Delta x = (\pi - 0)/n = \pi/n$; the right endpoints are thus $k\pi/n$, $k = 1, 2, \ldots, n$. Using these endpoints in the Riemann sum, $R_n = \sum_{k=1}^{n} \sin(k\pi/n)(\pi/n) = (\pi/n)\sum_{k=1}^{n} \sin(k\pi/n)$. Using technology, we get that $\sum_{k=1}^{n} \sin(k\pi/n) = \cot(\pi/2n)$. We also know that $\lim_{x\to 0} \sin x/x = 1$, thus $\lim_{n\to\infty} R_n = \lim_{n\to\infty} 2(\pi/2n)\cos(\pi/2n)/\sin(\pi/2n) = 2$.

## Problem Set 5.4 - The Fundamental Theorem of Calculus

**1. a.** Using the evaluation theorem, we get $\int_{-10}^{10} 6\,dx = 6x\big|_{-10}^{10} = 60 - (-60) = 120$.

**b.** Using the evaluation theorem, we get $\int_3^5 2x + a\,dx = x^2 + ax\big|_{-3}^{5} = (25 + 5a) - (9 - 3a) = 16 + 8a$.

**3. a.** Using the evaluation theorem, we get $\int_0^4 x^2 - 1\,dx = x^3/3 - x\big|_0^4 = 4^3/3 - 4 = 52/3$.

**b.** Using the evaluation theorem, we get $\int_0^\pi \sin x + x\,dx = -\cos x + x^2/2\big|_0^\pi = (-(-1) + \pi^2/2) - (-1) = \pi^2/2 + 2$.

**5. a.** Using the evaluation theorem, we get $\int_0^9 \sqrt{x}\,dx = (2/3)x^{3/2}\big|_0^9 = (2/3)9^{3/2} = 18$.

**b.** Using the evaluation theorem, we get $\int_0^1 5u^7 + \pi^2\,du = (5/8)u^8 + \pi^2 u\big|_0^1 = \pi^2 + 5/8$.

**7. a.** Using the evaluation theorem, we get $\int_1^2 (2x)^\pi\,dx = 2^\pi(1/(\pi+1))x^{\pi+1}\big|_1^2 = 2^{2\pi+1}/(\pi+1) - 2^\pi/(\pi+1)$.

**b.** Using the evaluation theorem, we get $\int_{-1}^{1} e^{x+1}\,dx = e^{x+1}\big|_{-1}^{1} = e^2 - e^0 = e^2 - 1$.

**9. a.** Using the power rule for differentiation, $\int 4t^3 + 3t^2\,dt = t^4 + t^3 + C$.

**b.** Using the power rule for differentiation, $\int -8t^3 + 15t^5\,dt = -2t^4 + (5/2)t^6 + C$.

**11. a.** $\int -3\cos u\,du = -3\sin u + C$.

**b.** Using the power rule for differentiation, $\int 5t^3 - \sqrt{t}\,dt = (5/4)t^4 - (2/3)t^{3/2} + C$.

**13. a.** We compute: $\int \sqrt{x}(x+1)\,dx = \int x^{3/2}+x^{1/2}\,dx = (2/5)x^{5/2}+(2/3)x^{3/2}+C.$

**b.** We compute: $\int \sqrt{t}(t-\sqrt{t})\,dt = \int t^{3/2}-t\,dt = (2/5)t^{5/2}-t^2/2+C.$

**15. a.** We compute: $\int (x^2-4)/(x-2)\,dx = \int x+2\,dx = x^2/2+2x+C.$

**b.** We compute: $\int (x^2-1)/(x+1)\,dx = \int x-1\,dx = x^2/2-x+C.$

**17.** Using the fundamental theorem, $\frac{d}{dx}\int_{-1}^{x}\sqrt{1+u^2}\,du = \sqrt{1+x^2}.$

**19.** Using the fundamental theorem, $\frac{d}{dx}\int_{4}^{x}\frac{dt}{2+\sin t^2} = \frac{1}{2+\sin x^2}.$

**21.** Using the opposite rule and the fundamental theorem, $\frac{d}{dx}\int_{x}^{2}\frac{e^u}{u}\,du = -\frac{d}{dx}\int_{2}^{x}\frac{e^u}{u}\,du = -\frac{e^x}{x}.$

**23.** By plugging in $x=0$, we get $0=1+a$, thus $a=-1$. Differentiating both sides with respect to $x$, the fundamental theorem gives $f(x)=-2\sin 2x$.

**25. a.** First, $F(x) = \int x^{-1/2}-4\,dx = 2x^{1/2}-4x+C$; we also know that $0 = F(1) = 2-4+C$, thus $C=2$. So $F(x) = 2\sqrt{x}-4x+2$.

**b.**

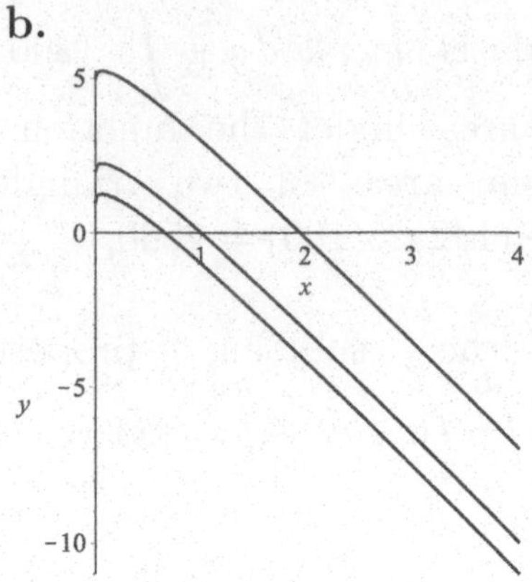


**27. a.** $G'(x)=g(x)$; thus we have local maxima and minima of $G(x)$ when $g(x)$ changes sign. At $x=1/2$ we have a local maximum ($g$ is changing from positive to negative), at $x=3/2$ we have a local minimum ($g$ is changing from negative to positive), at $x=5/2$ we have a local maximum ($g$ is changing from positive to negative), at $x=7/2$ we have a local minimum ($g$ is changing from negative to positive). At the endpoint $x=0$ we have a local minimum, because $G(0)=0$ and then $G$ starts to increase (because $g>0$), and at the endpoint $x=9/2$ we have a local maximum, because $G$ is increasing for $7/2<x<9/2$ (because $g>0$ there).

**b.** $G''(x)=g'(x)$, so $G$ is concave up on $(0,1/4)$, $(1,2)$ and $(3,4)$ (because $g$ is increasing there, so $g'>0$), and $G$ is concave down on $(1/4,1)$, $(2,3)$ and $(4,9/2)$ (because $g$ is decreasing there, so $g'<0$).

**c.** The global maximum is at $x=9/2$, because the integral adds more area for the second and third positive bumps for the function than the preceding negative bumps subtract; the global minimum is at $x=7/2$, because the first and second negative bumps subtract more area than the preceding positive bumps add.

**d.**

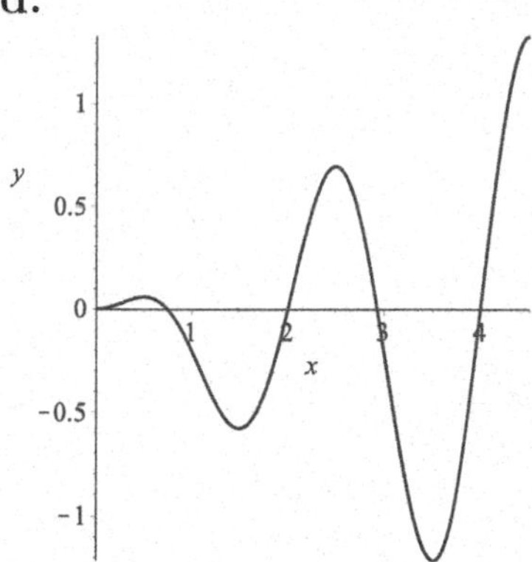


**29.** The increase is $\int_5^9 0.1762x^2 - 3.986x + 22.68\,dx = 0.1762x^3/3 - 3.986x^2/2 + 22.68x\big|_5^9 \approx 14.59$ cm.

**31. a.** The expression for the accumulated degree-days is $\int_0^x (68 + 17\sin(2\pi s))\,ds - \int_0^x 50\,ds = 1597$.

**b.** Using technology, we obtain $x \approx 88.4$ days.

**33.** Let $\int_a^x f(u)\,du = F(x)$. We see that $F(a) = 0$. By the fundamental theorem, $F(x)$ is an antiderivative of $f(x)$, i.e. $F'(x) = f(x)$. Let $\tilde{F}(x)$ be any antiderivative of $f(x)$; then $(\tilde{F}(x) - F(x))' = \tilde{F}'(x) - F'(x) = f(x) - f(x) = 0$. Thus $\tilde{F}(x) - F(x) = C$, and then $\tilde{F}(b) - \tilde{F}(a) = F(b) + C - (F(a) + C) = F(b) - F(a) = F(b) = \int_a^b f(u)\,du$ and we are done.

**Problem Set 5.5 - Substitution**

**1. a.** Using the evaluation theorem, we get $\int_0^4 2t + 4\,dt = t^2 + 4t\big|_0^4 = 16 + 16 = 32$.

**b.** Using the substitution $u = 2t + 4$, $du = 2dt$ and we obtain $\int_0^4 (2t+4)^{-1/2}\,dt = \int_4^{12} (1/2)u^{-1/2}\,du = u^{1/2}\big|_4^{12} = \sqrt{12} - 2$.

**3. a.** Using the evaluation theorem, we get $\int_0^{\pi/2} \cos t\,dt = \sin t\big|_0^{\pi/2} = 1 - 0 = 1$.

**b.** Using the substitution $u = t^2$, we get $du = 2tdt$ and we obtain $\int_0^{\pi} t\cos(t^2)\,dt = \int_0^{\pi^2} \cos u\,/2du = \sin u\,/2\big|_0^{\pi^2} = \sin(\pi^2)/2$.

**5. a.** Using the evaluation theorem, we get $\int_0^{16} \sqrt[4]{x}\,dx = (4/5)x^{5/4}\big|_0^{16} = (4/5)16^{5/4} - 0 = 128/5$.

**b.** Using the substitution $u = x + 2$, we get $du = dx$ and we obtain $\int_0^1 \sqrt[4]{x+2}\,dx = \int_2^3 \sqrt[4]{u}\,du = (4/5)u^{5/4}\big|_2^3 = (4/5)3^{5/4} - (4/5)2^{5/4}$.

**7. a.** We obtain that $\int x^2\sqrt{2x^3}\,dx = \int \sqrt{2}x^{7/2}\,dx = \sqrt{2}(2/9)x^{9/2} + C$.

**b.** We substitute $u = 2x^3 - 5$, $du = 6x^2dx$ and we obtain $\int 6x^2\sqrt{2x^3 - 5}\,dx = \int u^{1/2}\,du = 2u^{3/2}/3 + C = 2(2x^3 - 5)^{3/2}/3 + C$.

**9.** Let $u = 2x + 3$, then $du = 2dx$ and $\int (2x+3)^4\,dx = \int (1/2)u^4\,du = (1/10)u^5 + C = (1/10)(2x+3)^5 + C$.

**11.** Let $u = x^2 + 4$, then $du = 2xdx$ and $\int x\sqrt{x^2+4}\,dx = \int (1/2)\sqrt{u}\,du = (1/3)u^{3/2} + C = (1/3)(x^2+4)^{3/2} + C$.

**13.** Let $u = \sin x$, then $du = \cos x dx$ and $\int \cot x\,dx = \int (\cos x/\sin x)\,dx = \int (1/u)\,du = \ln u + C = \ln(\sin x) + C$.

**15.** Let $u = \ln x$, then $du = (1/x)dx$ and $\int (\ln x/x)\,dx = \int u\,du = u^2/2 + C = (1/2)\ln^2 x + C$.

**17.** Let $u = 5x^2 - x$, then $du = (10x-1)dx$ and $\int_{-1}^{2} (5x^2-x)^2(10x-1)\,dx = \int_6^{18} u^2\,du = (1/3)u^3\big|_6^{18} = (18^3 - 6^3)/3 = 1872$.

**19.** Let $u = 1/x$, then $du = -(1/x^2)dx$ and $\int_1^2 (e^{1/x}/x^2)\,dx = -\int_1^{1/2} e^u\,du = e^u\big|_{1/2}^1 = e - e^{1/2}$. (We used the opposite rule.)

**21.** Let $u = 1 + e^{0.2x}$; $du = 0.2e^{0.2x}dx$ and we obtain $\int_0^1 (0.58e^{0.2x}/(1+e^{0.2x}))\,dx = \int_2^{1+e^{0.2}} (0.58/0.2)(1/u)\,du = 2.9\ln u\big|_2^{1+e^{0.2}} = 2.9(\ln(1+e^{0.2}) - \ln 2) \approx 0.3045$.

**23.** Let $u = x - 1$, then $du = dx$ and we obtain that $\int_1^2 x\sqrt{x-1}\,dx = \int_0^1 (u+1)\sqrt{u}\,du = \int_0^1 u^{3/2} + u^{1/2}\,du = (2/5)u^{5/2} + (2/3)u^{3/2}\big|_0^1 = 2/5 + 2/3 = 16/15$.

**25.** Let $u = \sin x$; then $du = \cos x\,dx$ and $\int e^{\sin x}\cos x\,dx = \int e^u\,du = e^u + C = e^{\sin x} + C$.

**27.** Let $u = x/2$; then $du = (1/2)dx$ and $\int_2^{12} f(x)\,dx = \int_1^6 2f(2u)\,du = 2(-3) = -6$.

**29.** We get that $F(t) = \int 10e^{0.3t}\,dt = (10/0.3)e^{0.3t} + C$; also, $10 = F(0) = 100/3 + C$, thus $C = 10 - 100/3$. This means that one day from now we will have $F(24) = (100/3)e^{0.3\cdot 24} + 10 - 100/3 \approx 44624$ dust mites.

**31.** On the left side, let $u = \ln(N/b)$, so $du = (1/(N/b))(1/b)\,dN = (1/N)\,dN$ and then $\int dN/(N\ln(N/b)) = \int 1/u\,du = \ln u = \ln(\ln(N/b))$. On the right side, $-\int at\,dt = -at^2/2 + C$. This gives $N = be^{ce^{-at^2/2}}$; using the constants $a = 1$, $b = 10$ and the fact that $N(0) = 5$ we obtain $N = 10e^{-(\ln 2)e^{-t^2/2}}$.

**33.** The number of moose killed is given by $\int_0^3 3.36\cdot 0.1e^{0.2t}/(0.42 + 0.1e^{0.2t})\,dt$. Let $u = 0.42 + 0.1e^{0.2t}$; $du = 0.02e^{0.2t}dt$, and we get $\int_0^3 3.36\cdot 0.1e^{0.2t}/(0.42+0.1e^{0.2t})\,dt = (0.336/0.02)\int_{0.52}^{0.42+0.1e^{0.6}} 1/u\,du = 16.8\ln u\big|_{0.52}^{0.42+0.1e^{0.6}} \approx 2.4659$.

**Problem Set 5.6 - Integration by Parts and Partial Fractions**

**1.** Let $u = x$, $dv = e^{-x}dx$. Then $du = dx$, $v = -e^{-x}$ and we obtain that $\int xe^{-x}\,dx = -xe^{-x} + \int e^{-x}\,dx = -xe^{-x} - e^{-x} + C$.

**3.** Let $u = \ln x$, $dv = xdx$. Then $du = (1/x)dx$, $v = x^2/2$ and we obtain that $\int x\ln x\,dx = (x^2/2)\ln x - \int (x^2/2)(1/x)\,dx = (x^2/2)\ln x - x^2/4 + C$.

**5.** Using the substitution $t = \sqrt{x}$, $dt = ((1/2)/\sqrt{x})dx$, thus $\int (\ln\sqrt{x})/\sqrt{x}\,dx = \int 2\ln t\,dt$. Let $u = 2\ln t$, $dv = dt$. Then $du = (2/t)dt$ and $v = t$ and we obtain that $\int 2\ln t\,dt = 2t\ln t - \int (2/t)t\,dt = 2t\ln t - 2t + C$. So we get that $\int (\ln\sqrt{x})/\sqrt{x}\,dx = 2\sqrt{x}\ln\sqrt{x} - 2\sqrt{x} + C$.

**7.** Let $I = \int e^{2x}\sin 3x\,dx$. Choose $u = e^{2x}$, $dv = \sin 3x\,dx$; then $du = 2e^{2x}dx$, and $v = -(1/3)\cos 3x$. We obtain that $I = \int e^{2x}\sin 3x\,dx = -(1/3)e^{2x}\cos 3x + (2/3)\int e^{2x}\cos 3x\,dx$. For the second integration by parts, let $u = e^{2x}$, and $dv = \cos 3x\,dx$; then $du = 2e^{2x}\,dx$, and $v = (1/3)\sin 3x$. Continuing, we get that $I = \int e^{2x}\sin 3x\,dx = -(1/3)e^{2x}\cos 3x + (2/3)\int e^{2x}\cos 3x\,dx = -(1/3)e^{2x}\cos 3x + (2/3)((1/3)e^{2x}\sin 3x - (2/3)\int e^{2x}\sin 3x\,dx) = e^{2x}((2/9)\sin 3x - (1/3)\cos 3x) - 4I/9$. Now rearranging this equation we obtain that $I = e^{2x}((2/9)\sin 3x - (1/3)\cos 3x)/(13/9) + C = e^{2x}((2/13)\sin 3x - (3/13)\cos 3x) + C$.

**9.** We compute: $\sin x\cos x = (1/2)\sin 2x$, thus $\int x\sin x\cos x\,dx = (1/2)\int x\sin 2x\,dx$; as in problem **4**, let $u = x$, $dv = \sin 2x\,dx$. Then $du = dx$, $v = -(1/2)\cos 2x$ and we obtain that $(1/2)\int x\sin 2x\,dx = -(x/4)\cos 2x + \int (1/4)\cos 2x\,dx = -(x/4)\cos 2x + (1/8)\sin 2x + C$.

**11.** Let $u = x$, $dv = e^{-x}\,dx$. Then $du = dx$, $v = -e^{-x}$ and we obtain $\int_0^4 xe^{-x}\,dx = -xe^{-x}\big|_0^4 + \int_0^4 e^{-x}\,dx = -4e^{-4} + (-e^{-x})\big|_0^4 = -5e^{-4} + 1 \approx 0.9084$.

**13.** Let $u = 3[\ln(3x)]^2$, $dv = dx$. Then $du = 3[6\ln(3x)]/(3x)dx$, $v = x$ and we obtain $\int_{1/3}^e 3[\ln(3x)]^2\,dx = 3x[\ln(3x)]^2\big|_{1/3}^e - \int_{1/3}^e 6\ln(3x)\,dx = 3e[\ln(3e)]^2 - \int_{1/3}^e 6\ln(3x)\,dx$. For this second integral, let $u = 6\ln(3x)$, $dv = dx$. Then $du = 3(6/3x)dx$, $v = x$ and we get $\int_{1/3}^{e} 6\ln(3x)\,dx = 6x\ln(3x)\Big|_{1/3}^{e} - \int_{1/3}^{e} 3(6/3x)x\,dx = 6e\ln(3e) - 6x\Big|_{1/3}^{e} = 6e\ln(3e) - (6e-2) = 2 + 6e\ln 3$; thus the original integral is $3e[\ln(3e)]^2 - 6e\ln 3 - 2 \approx 15.9973$.

**15.** Let $u = x$, $dv = (\sin x + \cos x)\,dx$. Then $du = dx$, $v = -\cos x + \sin x$ and we obtain $\int_0^{\pi} x(\sin x + \cos x)\,dx = x(-\cos x + \sin x)\Big|_0^{\pi} - \int_0^{\pi}(-\cos x + \sin x)\,dx = \pi - (-\sin x - \cos x)\Big|_0^{\pi} = \pi - 2 \approx 1.1416$.

**17.** The partial fraction decomposition is $A/N + B/(1000 - N) = 1/(N(1000 - N))$; this gives $A(1000 - N) + BN = 1$. Substituting $N = 0$ we get $A = 1/1000$, $N = 1000$ we get $B = 1/1000$. Thus $\int \frac{dN}{N(1000 - N)} = \frac{1}{1000}\int \frac{1}{N} + \frac{1}{1000 - N}\,dN = \frac{1}{1000}(\ln N - \ln(1000 - N)) + C = (1/1000)\ln(N/(1000 - N)) + C$.

**19.** We simplify by $x$ to get $\int \frac{x}{x(x-1000)}\,dx = \int \frac{1}{x - 1000}\,dx = \ln(x - 1000) + C$.

**21.** The partial fraction decomposition is $A/x + B/(x+1) + C/(x-2) = 1/(x(x+1)(x-2))$; this gives $A(x+1)(x-2) + Bx(x-2) + Cx(x+1) = 1$. Substituting $x = 0$ we get $A = -1/2$, $x = -1$ we get $B = 1/3$, and $x = 2$ we get $C = 1/6$. Thus $\int \frac{dx}{x(x+1)(x-2)} = \int \frac{-1/2}{x} + \frac{1/3}{x+1} + \frac{1/6}{x-2}\,dx = -(1/2)\ln x + (1/3)\ln(x+1) + (1/6)\ln(x-2) + C$.

**23.** Substitute $t = \ln x$; then $x = e^t$ and $dx = e^t\,dt$, so $\int \cos(\ln x)\,dx = \int e^t \cos t\,dt$. Now let $I = \int e^t \cos t\,dt$. Choose $u = e^t$, $dv = \cos t\,dt$; then $du = e^t dt$, and $v = \sin t$. We obtain that $I = \int e^t \cos t\,dt = e^t \sin t - \int e^t \sin t\,dt$. For the second integration by parts, let $u = e^t$, and $dv = \sin t\,dt$; then $du = e^t\,dt$, and $v = -\cos t$. Continuing, we get that $I = \int e^t \cos t\,dt = e^t \sin t - \int e^t \sin t\,dt = e^t \sin t + e^t \cos t - \int e^t \cos t\,dt = e^t(\sin t + \cos t) - I$. Now rearranging this equation we obtain that $I = e^t(\sin t + \cos t)/2 + C$. Thus the original integral is $\int \cos(\ln x)\,dx = e^{\ln x}(\sin(\ln x) + \cos(\ln x))/2 + C = x(\sin(\ln x) + \cos(\ln x))/2 + C$.

**25.** Substitute $t = \ln x$; then $dt = (1/x)dx$ and $\int \frac{\ln x \sin(\ln x)}{x}\,dx = \int t \sin t\,dt$. Let $u = t$, $dv = \sin t\,dt$. Then $du = dt$, $v = -\cos t$ and we obtain that $\int t\sin t\,dt = -t\cos t + \int \cos t\,dt = -t\cos t + \sin t + C$. Thus the original integral is $\int \frac{\ln x \sin(\ln x)}{x}\,dx = -\ln x \cos(\ln x) + \sin(\ln x) + C$.

**27.** Substitute $t = e^x$; then $dt = e^x\,dx$ and $\int \frac{e^{2x}dx}{e^{2x}+3e^x+2} = \int \frac{t}{t^2+3t+2}\,dt = \int \frac{t}{(t+1)(t+2)}\,dt$. The partial fraction decomposition is $A/(t+1) + B/(t+2) = t/((t+1)(t+2))$; this gives $A(t+2)+B(t+1) = t$. Substituting $t = -1$ we get $A = -1$, $t = -2$ we get $B = 2$. Thus $\int \frac{t}{(t+1)(t+2)}\,dt = \int \frac{-1}{t+1} + \frac{2}{t+2}\,dt = -\ln(t+1) + 2\ln(t+2) + C$. Thus the original integral is $\int \frac{e^{2x}dx}{e^{2x}+3e^x+2} = -\ln(e^x+1) + 2\ln(e^x+2) + C$.

**29. a.** Substitute $t = x^2 - 1$; then $dt = 2x\,dx$ and $\int \frac{x^3}{x^2-1}\,dx = (1/2)\int \frac{t+1}{t}\,dt = (1/2)\int 1 + (1/t)\,dt = t/2 + (1/2)\ln t + C = x^2/2 + (1/2)\ln(x^2-1) + C$.

**b.** For partial fractions, we have to simplify the expression first: $\int \frac{x^3}{x^2-1}\,dx = \int \frac{x^3-x+x}{x^2-1}\,dx = \int x + \frac{x}{(x-1)(x+1)}\,dx$. The partial fraction decomposition is $A/(x-1) + B/(x+1) = x/((x-1)(x+1))$; this gives $A(x+1) + B(x-1) = x$. Substituting $x = 1$ we get $A = 1/2$, $x = -1$ we get $B = 1/2$. Thus $\int x + \frac{x}{(x-1)(x+1)}\,dx = \int x + \frac{1/2}{x-1} + \frac{1/2}{x+1}\,dx = x^2/2 + (1/2)\ln(x-1) + (1/2)\ln(x+1) + C$.

**31.** A possible table is

| $D$ | $I$ |
|---|---|
| $\ln x$ | $1$ |
| $1/x$ | $x$ |

. Then $\int \ln x\,dx = x\ln x - \int (1/x)x\,dx$. We get that $\int \ln x\,dx = x\ln x - x + C$.

**33.** A possible table is

| $D$ | $I$ |
|---|---|
| $e^x$ | $\cos x$ |
| $e^x$ | $\sin x$ |
| $e^x$ | $-\cos x$ |

. Then $\int e^x\cos x\,dx = e^x\sin x - (-e^x\cos x) + \int e^x(-\cos x)\,dx$. Rearranging this we get that $\int e^x\cos x\,dx = e^x(\sin x + \cos x)/2 + C$.

**35.** Generally, $p(x)/((x-r_1)(x-r_2)\dots(x-r_n)) = A_1/(x-r_1) + A_2/(x-r_2) + \dots + A_n/(x-r_n)$. If we multiply both sides by the denominator of the left side, we get $p(x) = A_1(x-r_2)\dots(x-r_n) + A_2(x-r_1)\dots(x-r_n)+\dots+A_n(x-r_1)\dots(x-r_{n-1})$, i.e. on the right side the $A_i$'s are multiplied by all the $x-r_j$ except the one with the same index as $A_i$, so $x-r_i$. Now plugging in this $r_i$ value to both sides, all terms on the right side disappear except the one containing $A_i$; so we get that $A_i = p(r_i)/((r_i-r_1)\dots(r_i-r_n))$ - but this is exactly Heaviside's method.

**37.** The total contribution during the first five weeks is $\int_0^5 2000te^{-0.2t}\,dt$ dollars. Let $u = 2000t$, $dv = e^{-0.2t}$; then $du = 2000dt$, $v = -5e^{-0.2t}$ and we get that $\int_0^5 2000te^{-0.2t}\,dt = -10000te^{-0.2t}\big|_0^5 +$

$10000\int_0^5 e^{-0.2t}\,dt = -10000te^{-0.2t} - 50000e^{-0.2t}\big|_0^5 \approx 13,212.10$.

**39.** The population growth during the 8th year is given by $\int_7^8 5(t+1)\ln\sqrt{t+1}\,dt = \int_7^8 (5/2)(t+1)\ln(t+1)\,dt$. Let $u = \ln(t+1)$ and $dv = (t+1)\,dt$; then $du = (1/(t+1))\,dt$ and $v = (t+1)^2/2$ and we get $\int_7^8 (5/2)(t+1)\ln(t+1)\,dt = (5/2)[((t+1)^2/2)\ln(t+1)\big|_7^8 - \int_7^8 (t+1)/2\,dt] = (5/2)[((t+1)^2/2)\ln(t+1) - (t+1)^2/4\big|_7^8] \approx 45.49$ thousand individuals.

**41.** The number of individuals captured and released is $\int_0^7 10 + 5\sin(2\pi t)e^{-0.01t}\,dt$. Using integration by parts twice, rearranging the equation obtained and solving for the integral we obtain that the antiderivative is $10t - 500e^{-t/100}(200\pi\cos(2\pi t) + \sin(2\pi t))/(1+40000\pi^2)$. The integral thus gives 70.05 individuals.

## Problem Set 5.7 - Numerical Integration

**1. a.** $\Delta x = (2-1)/4 = 1/4$; thus $L_4 = \sum_{k=0}^{3} f(1+k/4)(1/4) = \sum_{k=0}^{3}(1+k/4)^2(1/4) = 63/32 \approx 1.9688$.

**b.** $\Delta x = (2-1)/4 = 1/4$; thus $R_4 = \sum_{k=1}^{4} f(1+k/4)(1/4) = \sum_{k=1}^{4}(1+k/4)^2(1/4) = 87/32 \approx 2.7188$.

**c.** $\Delta x = (2-1)/4 = 1/4$; thus $M_4 = \sum_{k=1}^{4} f(7/8+k/4)/4 = \sum_{k=1}^{4}(7/8+k/4)^2/4 = 149/64 \approx 2.3281$.

**d.** $\Delta x = (2-1)/4 = 1/4$; thus $S_4 = (1/12)(f(1)+4f(5/4)+2f(3/2)+4f(7/4)+f(2)) = 7/3 \approx 2.3333$.

**3. a.** $\Delta x = (1-0)/4 = 1/4$; thus we obtain $L_4 = \sum_{k=0}^{3} f(k/4)(1/4) = \sum_{k=0}^{3}\cos(2k/4)/4 \approx 0.6222$.

**b.** $\Delta x = (1-0)/4 = 1/4$; thus we obtain $R_4 = \sum_{k=1}^{4} f(k/4)(1/4) = \sum_{k=1}^{4}\cos(2k/4)/4 \approx 0.2681$.

**c.** $\Delta x = (1-0)/4 = 1/4$; thus we obtain that $M_4 = \sum_{k=1}^{4} f(-1/8+k/4)(1/4) = \sum_{k=1}^{4}\cos(2(-1/8+k/4))(1/4) \approx 0.4594$.

**d.** $\Delta x = (1-0)/4 = 1/4$; thus $S_4 = (1/12)(f(0)+4f(1/4)+2f(1/2)+4f(3/4)+f(1)) \approx 0.4548$.

**5. a.** $\Delta x = (1-0)/4 = 1/4$; thus $L_4 = \sum_{k=0}^{3} f(k/4)(1/4) = \sum_{k=0}^{3} 1/(1+(k/4)^2)(1/4) \approx 0.8453$.

**b.** $\Delta x = (1-0)/4 = 1/4$; thus $R_4 = \sum_{k=1}^{4} f(k/4)(1/4) = \sum_{k=1}^{4} 1/(1+(k/4)^2)(1/4) \approx 0.7203$.

**c.** $\Delta x = (1-0)/4 = 1/4$; thus we obtain that $M_4 = \sum_{k=1}^{4} f(-1/8 + k/4)(1/4) = \sum_{k=1}^{4} 1/(1 + (-1/8 + k/4)^2)(1/4) \approx 0.7867.$

**d.** $\Delta x = (1-0)/4 = 1/4$; thus $S_4 = (1/12)(f(0)+4f(1/4)+2f(1/2)+4f(3/4)+f(1)) \approx 0.7854.$

**7. a.** $\Delta x = (2-0)/6 = 1/3$; thus we obtain that $L_6 = \sum_{k=0}^{5} f(k/3)(1/3) = \sum_{k=0}^{5}(k/3)\cos(k/3)\,(1/3) \approx 0.5111.$

**b.** $\Delta x = (2-0)/6 = 1/3$; thus we obtain that $R_6 = \sum_{k=1}^{6} f(k/3)(1/3) = \sum_{k=1}^{6}(k/3)\cos(k/3)\,(1/3) \approx 0.2337.$

**c.** $\Delta x = (2-0)/6 = 1/3$; thus we obtain that $M_6 = \sum_{k=1}^{6} f(-1/6 + k/3)(1/3) = \sum_{k=1}^{6}(-1/6 + k/3)\cos(-1/6 + k/3)\,(1/3) \approx 0.4175.$

**d.** $\Delta x = (2-0)/6 = 1/3$; thus $S_6 = (1/9)(f(0) + 4f(1/3) + 2f(2/3) + 4f(1) + 2f(4/3) + 4f(5/3) + f(2)) \approx 0.4029.$

**9. a.** $\Delta x = (1-0)/4 = 1/4$; thus $L_4 = \sum_{k=0}^{3} f(k/4)(1/4) = \sum_{k=0}^{3} 1/(1+(k/4)^3)(1/4) \approx 0.8942.$

**b.** $\Delta x = (1-0)/4 = 1/4$; thus $R_4 = \sum_{k=1}^{4} f(k/4)(1/4) = \sum_{k=1}^{4} 1/(1+(k/4)^3)(1/4) \approx 0.7692.$

**c.** $\Delta x = (1-0)/4 = 1/4$; thus we obtain that $M_4 = \sum_{k=1}^{4} f(-1/8 + k/4)(1/4) = \sum_{k=1}^{4} 1/(1 + (-1/8 + k/4)^3)(1/4) \approx 0.8376.$

**d.** $\Delta x = (1-0)/4 = 1/4$; thus $S_4 = (1/12)(f(0)+4f(1/4)+2f(1/2)+4f(3/4)+f(1)) \approx 0.8358.$

**11. a.** $\Delta x = (2-(-2))/6 = 2/3$; thus we obtain that $L_6 = \sum_{k=0}^{5} f(-2 + 2k/3)(2/3) = \sum_{k=0}^{5} \cos(-2 + 2k/3)^2\,(2/3) \approx 1.1607.$

**b.** $\Delta x = (2-(-2))/6 = 2/3$; thus we obtain that $R_6 = \sum_{k=1}^{6} f(-2 + 2k/3)(2/3) = \sum_{k=1}^{6} \cos(-2 + 2k/3)^2\,(2/3) \approx 1.1607.$

**c.** $\Delta x = (2-(-2))/6 = 2/3$; thus we obtain that $M_6 = \sum_{k=1}^{6} f(-7/3 + 2k/3)(2/3) = \sum_{k=1}^{6} \cos(-7/3 + 2k/3)^2\,(2/3) \approx 0.7995.$

**d.** $\Delta x = (2-(-2))/6 = 2/3$; thus $S_6 = (2/9)(f(-2)+4f(-4/3)+2f(-2/3)+4f(0)+2f(2/3)+4f(4/3)+f(2)) \approx 1.0356.$

**13.** We know that $E_L \le K_1(b-a)^2/(2n) = K_1(2-0)^2/(2n) = 2K_1/n$. Now $f'(x) = \sqrt{4-x} - (1/2)x/\sqrt{4-x}$; the maximum of this function on the interval $[0,2]$ is at $x = 0$, where $f'(0) = 2$. Thus we need $4/n < 0.01$, i.e. $n > 400$. We obtain $L_{401} \approx 3.24655$.

**15.** We get $E_S \leq K_4(b-a)^5/(180n^4) = K_4(0-(-2))^5/(180n^4) = (32/180)K_4/n^4$. Now $f^{(4)}(x) = e^x$; the maximum of the absolute value of this function on the interval $[-2, 0]$ is 1. Thus we need $(32/180)/n^4 < 0.01$, i.e. $n > 2.05$. We obtain $S_4 \approx 0.8649$.

**17.** We get $E_M \leq K_2(b-a)^3/(24n^2) = K_2(\pi/2-0)^3/(24n^2) = (\pi^3/192)K_2/n^2$. Now $f''(x) = -2\cos 2\theta$; the maximum of the absolute value of this function on the interval $[0, \pi/2]$ is 2. Thus we need $2(\pi^3/192)/n^2 < 0.01$, i.e. $n > 5.68$. We obtain $M_6 \approx 0.7854$.

**19.** We know that $E_R \leq K_1(b-a)^2/(2n) = K_1(1-0)^2/(2n) = K_1/(2n)$. Now $f'(x) = \cos x e^{\sin x}$; the maximum of the absolute value of this function on the interval $[0, 1]$ is less than $3/2$. Thus we need $(3/2)/(2n) < 0.001$, i.e. $n > 750$. We obtain $R_{751} \approx 1.63275$.

**21.** We get $E_S \leq K_4(b-a)^5/(180n^4) = K_4(1-0)^5/(180n^4) = (1/180)K_4/n^4$. It can be shown that the maximum of the absolute value of the fourth derivative of this function on the interval $[0, 1]$ is less than 260. Thus we need $(260/180)/n^4 < 0.02$, i.e. $n > 2.9$. We obtain $S_4 \approx 0.2326$.

**23. a.** We get $E_M \leq K_2(b-a)^3/(24n^2) = K_2(4-1)^3/(24n^2) = (27/24)K_2/n^2$. Now $f''(x) = 2/x^3$; the maximum of the absolute value of this function on the interval $[1, 4]$ is 2. Thus we need $2(27/24)/n^2 < 5 \cdot 10^{-5}$, i.e. $n \geq 213$.

**b.** We get $E_S \leq K_4(b-a)^5/(180n^4) = K_4(4-1)^5/(180n^4) = (243/180)K_4/n^4$. Now $f^{(4)}(x) = 24/x^5$ and the maximum of the absolute value of this function on the interval $[1, 4]$ is 24. Thus we need $24(243/180)/n^4 < 5 \cdot 10^{-5}$, i.e. $n \geq 30$ ($n$ has to be even).

**25. a.** We get $E_M \leq K_2(b-a)^3/(24n^2) = K_2(2-0)^3/(24n^2) = (8/24)K_2/n^2$. Now $f''(x) = -\cos x$; the maximum of the absolute value of this function on the interval $[0, 2]$ is 1. Thus we need $(1/3)/n^2 < 5 \cdot 10^{-5}$, i.e. $n \geq 82$.

**b.** We get $E_S \leq K_4(b-a)^5/(180n^4) = K_4(2-0)^5/(180n^4) = (32/180)K_4/n^4$. Now $f^{(4)}(x) = \cos x$ and the maximum of the absolute value of this function on the interval $[0, 2]$ is 1. Thus we need $(32/180)/n^4 < 5 \cdot 10^{-5}$, i.e. $n \geq 8$ ($n$ has to be even).

**27. a.** We get $E_M \leq K_2(b-a)^3/(24n^2) = K_2(4-(-1))^3/(24n^2) = (125/24)K_2/n^2$. Now $f''(x) = 6x+4$; the maximum of the absolute value of this function on the interval $[-1, 4]$ is 28. Thus we need $28(125/24)/n^2 < 5 \cdot 10^{-5}$, i.e. $n \geq 1708$.

**b.** We get $E_S \leq K_4(b-a)^5/(180n^4) = K_4(4-(-1))^5/(180n^4) = (3125/180)K_4/n^4$. Now $f^{(4)}(x) = 0$ and the maximum of the absolute value of this function on the interval $[-1, 4]$ is 0. Thus $n = 2$ gives the exact result here.

**29.** $L_6 \approx (1/2)(3/4 + 1/4 + 0 + 1/4 + 3/4 + 3/2) = 7/4$; $R_6 \approx (1/2)(1/4+0+1/4+3/4+3/2+7/4) = 9/4$; $S_6 \approx (1/6)(3/4+4(1/4)+2(0) + 4(1/4) + 2(3/4) + 4(3/2) + 7/4) = 2$.

**31.** The estimate is $(4/3)(3762 + 4 \cdot 5604 + 2 \cdot 8828 + 4 \cdot 10901 + 2 \cdot 10210 + 4 \cdot 13511 + 2 \cdot 10670 + 4 \cdot 8521 + 2 \cdot 4606 + 4 \cdot 2610 + 2 \cdot 4529 + 4 \cdot 6986 + 2 \cdot 10133 + 4 \cdot 13279 + 2 \cdot 13872 + 4 \cdot 18694 + 2 \cdot 15875 + 4 \cdot 11498 + 2 \cdot 9644 + 4 \cdot 6899 + 2 \cdot 5044 + 4 \cdot 2151 + 668) = 791{,}824$.

**33. a.** The integral representing the number

of deaths is $\int_0^{30} 890\,\text{sech}^2(0.2t-3.4)\,dt$.

**b.** With $n=10$, $\Delta x=3$ and $S_{10}\approx 1(3.96+4(13.07)+2(42.65)+4(133.97)+2(373.78)+4(761.52)+2(855.33)+4(497.56)+2(192.38)+4(62.88)+19.42)=8827.66$.

**35. a.** We obtain $R_{24}=(1/24)(16.7+15.3+16.0+19.6+20.2+19.1+18.7+18.7+18.5+18.7+19.3+20.2+22.1+25.4+27.9+29.7+30.8+31.0+30.6+29.2+26.5+25.6+23.9+23.0)=22.7792$ degree-days.

**b.** It would take $1587/22.7792\approx 69.67$ days.

**37.** We obtain $S_8=(2/3)(0+4(100)+2(400)+4(900)+2(1900)+4(2500)+2(1500)+4(400)+0)\approx 15467$ cases.

**39.** The answer should be close to 314 $\text{cm}^2$.

## Problem Set 5.8 - Applications of Integration

**1.** According to the survival-renewal equation, the number of patients at the clinic after 15 months is given by $e^{-15/20}300+\int_0^{15} e^{-(15-t)/20}20\,dt = 300e^{-15/20}+400e^{(t-15)/20}\Big|_0^{15}\approx 352.76$. The number is larger than in Example 1.

**3.** According to the survival-renewal equation, the number of patients at the clinic after 15 months is given by $e^{-15/10}300+\int_0^{15} e^{-(15-t)/10}20\,dt = 300e^{-15/10}+200e^{(t-15)/10}\Big|_0^{15}\approx 222.31$.

**5.** According to the survival-renewal equation, the number of patients at the clinic after 15 months is given by $e^{-15/20}300+\int_0^{15} e^{-(15-t)/20}(10+t)\,dt = 300e^{-15/20}+(20t-200)e^{(t-15)/20}\Big|_0^{15}\approx 336.18$.

**7.** According to the survival-renewal equation (and using technology), we obtain $\int_0^5 e^{-0.625(5-t)}100(1250+250\sin(2\pi t))\,dt\approx 187{,}446$.

**9.** According to the survival-renewal equation (and using technology), we get $\int_0^5 \frac{100(1250+250\sin(2\pi t))}{0.25+(5-t)^2}\,dt\approx 349{,}823$.

**11.** According to the survival-renewal equation (and using technology), we obtain $\int_0^5 e^{-1.25(5-t)}100(1250+250\sin(2\pi t))e^{-0.2t}\,dt\approx 42{,}149$.

**13.** The amount she will have is $\int_0^{40} e^{0.1(40-t)}4000\,dt\approx 2{,}143{,}926$.

**15.** The amount she will have is $\int_0^{50} e^{0.1(50-t)}1000\,dt\approx 1{,}474{,}132$.

**17.** The work done is $W=Fd=90\cdot 3=270$ ft-lb.

**19.** The work done is $W=Fd=850\cdot 15=12{,}750$ ft-lb.

**21.** According to the survival-renewal equation (and using technology), we obtain $5e^{-2\cdot 24}+\int_0^{24} e^{-2(24-t)}10\,dt=5$ mg.

**23.** **a.** The amount we will have is $\int_0^{43} e^{0.09(43-t)}1000\,dt \approx \$521,582$. The total amount we paid is $43 \cdot 1000 = \$43,000$.

**b.** The amount we will have is $\int_0^{33} e^{0.09(33-t)}2000\,dt \approx \$410,932$. The total amount we paid is $33 \cdot 2000 = \$66,000$.

**25.** Using Simpson's rule, $\int_0^{22} c(t)\,dt = (1/3)(0.00+4\cdot0.20+2\cdot0.7+4\cdot1.6+2\cdot2.5+4\cdot3.5+2\cdot4.8+4\cdot5.5+2\cdot6.0+4\cdot6.3+2\cdot6.3+4\cdot5.5+2\cdot4.5+4\cdot3.5+2\cdot2.5+4\cdot1.8+2\cdot1.10+4\cdot0.60+2\cdot0.50+4\cdot0.2+2\cdot0.2+4\cdot0.1+0.0) = 57.8$, so $F = 5/57.8 \approx 0.0865$ L/s, or 5.19 L/min.

**27.** **a.** The amount accumulated in the brain is approximated by $F(2.5/3)(4(0.031-0.012)+2(0.039-0.027)+4(0.041-0.034)+0.044-0.042) \approx 0.1083F$.

**b.** Using the previous part, we get $F \approx 58.8/0.1083 \approx 542.936$ cc/min.

**29.** The work done is $W = \int_0^{20} 0.4x\,dx = 80$ ft-lb.

**31.** 100 men in seven days eat $7 \cdot 100 \cdot 2$ lbs of pasta, which is $1400 \cdot 16 = 22400$ ounces. This contains 2240000 calories; with 10% efficiency this is 224000 calories.

**33.** The volume of a slice with thickness $\Delta x$ at depth $x$ in the conical hole can be found by $(3 - 3x/5)^2\pi\Delta x$, so the work required to lift this slice is $\rho g(3 - 3x/5)^2\pi x\Delta x$, where $\rho$ is density of the dirt and $g$ is the standard gravity. Thus the work needed to dig the hole is $W = \int_0^5 \rho g(3 - 3x/5)^2\pi x\,dx = 577857$ J. With 5% efficiency, we need $(577857/4184)/(200 \cdot 0.05) \approx 13.8$ servings of pasta.

## Review Questions

**1.** $F(x) = 2x^{1/2} + C$, because $F'(x) = x^{-1/2}$.

**3.** Let $u = t^2$, $dv = \sin 2t\,dt$. Then $du = 2t\,dt$, $v = -(1/2)\cos 2t$ and we obtain $\int_0^{\pi/2} t^2 \sin 2t\,dt = -(1/2)t^2 \cos 2t\big|_0^{\pi/2} + \int_0^{\pi/2} t\cos 2t\,dt = \pi^2/8 + \int_0^{\pi/2} t\cos 2t\,dt$. Now let $u = t$, $dv = \cos 2t\,dt$; then $du = dt$ and $v = (1/2)\sin 2t$ and $\int_0^{\pi/2} t\cos 2t\,dt = (1/2)t\sin 2t\big|_0^{\pi/2} - \int_0^{\pi/2} (1/2)\sin 2t\,dt = 0 - \int_0^{\pi/2} (1/2)\sin 2t\,dt = 1/4\cos 2t\big|_0^{\pi/2} = -1/4 - (1/4) = -1/2$. Thus the original integral is $\pi^2/8 - 1/2$.

**5.** Using partial fractions, we get $A/(x+3) + B/(x+2) = (x+1)/((x+3)(x+2))$; this implies $A(x+2) + B(x+3) = x+1$. Substituting $x = -3$, $A = 2$; $x = -2$, we get $B = -1$. Thus $\int_{-1}^{1} (x+1)/((x+3)(x+2))\,dx = \int_{-1}^{1} 2/(x+3) - 1/(x+2)\,dx = 2\ln(x+3) - \ln(x+2)\big|_{-1}^{1} = 2\ln 4 - \ln 3 - (2\ln 2 - \ln 1) = \ln(4/3)$.

**7.** We obtain $F(x) = \int (x+1)/x^2\,dx = \int 1/x + 1/x^2\,dx = \ln x - 1/x + C$. Now $-2 = F(1) = \ln 1 - 1 + C$, thus $C = -1$ and $F(x) = \ln x - 1/x - 1$.

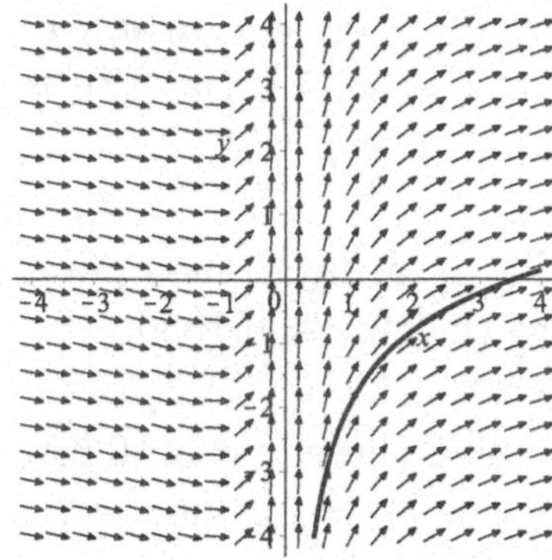

**9.** Using the fundamental theorem and the chain rule, $dy/dx = 2\sin(4x^2)$.

**11.** $T = 12$, $N_0 = 450$, $r(t) = 150$ (per month). The survival-renewal equation gives that the number of individuals is $s(T)N_0 + \int_0^T s(T-t)r(t)\,dt$, so the left endpoint Riemann sum is $0.1 \cdot 450 + (0.2 \cdot 150) \cdot 3 + (0.3 \cdot 150) \cdot 3 + (0.5 \cdot 150) \cdot 3 + (1.0 \cdot 150) \cdot 3 = 945$ patients.

**13. a.**

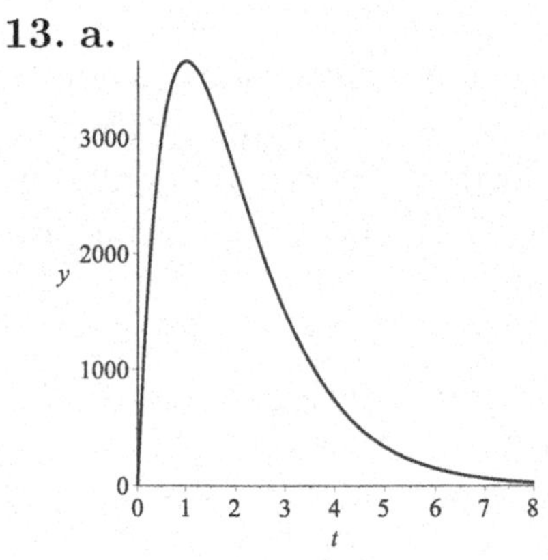


**b.** The number of people infected by the disease by time $T$ is $\int_0^T 10000te^{-t}\,dt$; using integration by parts with $u = t$, $dv = e^{-t}\,dt$ we have $du = dt$, $v = -e^{-t}$ and $10000\int_0^T te^{-t}\,dt = 10000(-te^{-t}\big|_0^T + \int_0^T e^{-t}\,dt) = 10000(-Te^{-T} + (-e^{-t}\big|_0^T)) = 10000(1-(1+T)e^{-T})$.

**c.** We have to solve the equation $1/2 = 1-(1+T)e^{-T}$; technology gives $T \approx 1.68$ (months).

**15.** $\Delta x = (7-3)/n = 4/n$, thus $R_n = \sum_{k=1}^{n} \tan(3+4k/n)(4/n)$, and the integral is $\lim_{n\to\infty} R_n$.

**17.** Let $t = x+1$, then $dt = dx$ and $\int_1^2 x/\sqrt{x+1}\,dx = \int_2^3 (t-1)/\sqrt{t}\,dt = \int_2^3 \sqrt{t} - 1/\sqrt{t}\,dt = (2/3)t^{3/2} - 2t^{1/2}\big|_2^3 = (2/3)3^{3/2} - 2(3)^{1/2} - ((2/3)2^{3/2} - 2(2)^{1/2}) = (2/3)\sqrt{2}$.

**19.** Using the graph, $\int_{-1}^{2} 1-|x|\,dx = 1/2 + 1/2 - 1/2 = 1/2$.

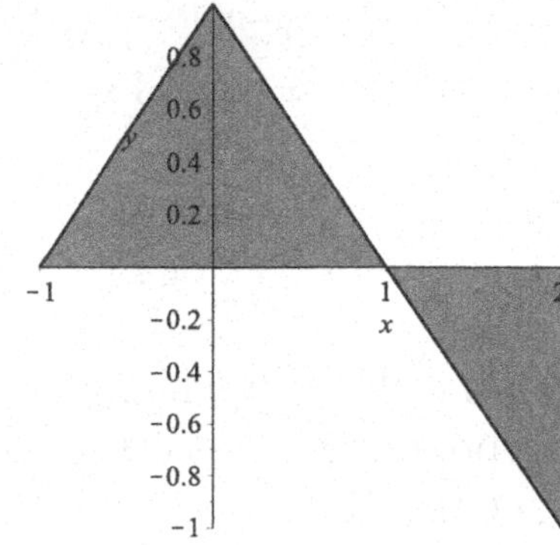

**Problem Set 6.1 - A Modeling Introduction to Differential Equations**

**1.** $N'(t) = kN(t)$, $k > 0$. $N(t)$ is the number of bacteria, $k$ is the proportionality constant.

**3.** $T'(t) = -k(T(t) - A)$, $k > 0$. $T(t)$ is the body's temperature, $A$ is the ambient temperature, $k$ is the proportionality constant.

**5.** $N'(t) = ae^{-bt}N(t)$, $b > 0$.

**7.** $y'(t) = k(P - y(t))y(t)$.

**9.** We have to solve $2 = e^{0.64t}$; we obtain $t = \ln 2/0.64 \approx 1.083$ years.

**11.** We have to solve $1/2 = e^{-0.41t}$; we obtain $t = \ln(1/2)/(-0.41) \approx 1.691$ months.

**13.** We have to solve $1/2 = e^{-0.03t}$; we obtain $t = \ln(1/2)/(-0.03) \approx 23.1049$ centuries, or 2310.49 years.

**15.** We have to solve $4 = e^{0.2t}$; we obtain $t = \ln 4/0.2 \approx 6.931$ years.

**17.** We have to solve $0.1 = e^{-0.0025t}$; we obtain $t = \ln(0.1)/(-0.0025) \approx 921.03$ years.

**19.** The equilibria are given by $dP/dt = P(100-P) = 0$, i.e. when $P = 0$ or $P = 100$.

**21.** $dP/dt = P(100 - P) < 0$ when $P < 0$ or $P > 100$; the population density is not negative, so we get $P > 100$.

**23.** The equilibria are given by $dP/dt = P(P-1)(100-P) = 0$, i.e. when $P = 0$, $P = 1$ or $P = 100$.

**25.** $dP/dt = P(P-1)(100-P) < 0$ when $0 < P < 1$ or $P > 100$.

**27.** The equilibria are given by $dp/dt = -p^3 + 3p^2 - 2p = -p(p-1)(p-2) = 0$, i.e. when $p = 0$, $p = 1$ or $p = 2$.

**29.** If the half-life is 5 days, then $N_0/2 = N_0e^{-\lambda \cdot 5}$, thus $\lambda = (\ln 2)/5 \approx 0.1386$.

**31.** We have to solve the equation $0.6N_0 = N_0e^{-\lambda t}$; we obtain $t = \ln 0.6/(-\lambda) = \ln 0.6/(-0.00012449) \approx 4103$ years.

**33.** The rate of decay is $6.08/6.68 \approx 0.91$ times as much as living wood, thus the age is $t = \ln 0.91/(-\lambda) = \ln 0.91/(-0.00012449) \approx 756$ years.

**35. a.** The equilibrium solutions are given by the equation $10N(1 - N/10000) = 25000$, i.e. $N = 5000$. If the initial population size is less than 5000, then the population dies out; if the initial population size is larger than 5000, then it decreases toward 5000.

**b.** The equilibrium solutions are given by the equation $10N(1 - N/10000) = 5N$, i.e. $N = 0$ or $N = 5000$. If the initial population size is larger than 0, then the population size approaches 5000.

**c.** The equilibrium solutions are given by the equation $10N(1 - N/10000) = 12N$, i.e. $N = 0$ or $N = -2000$. The population goes extinct for all initial population sizes.

**37. a.** The solution corresponding to the model is $y(t) = y(0)e^{rt}$; we obtain $1.1 = y(2) = 0.3e^{2r}$, thus $r = \ln(1.1/0.3)/2 \approx 0.6496$.

**b.** The value of $K$ is about 72.

**39. a.** The solution corresponding to the model is $y(t) = y(0)e^{rt}$; we obtain $8 = y(1) = 2e^{r}$, thus $r = \ln 4 \approx 1.386$.

**b.** The value of $K$ is about 750.

## Problem Set 6.2 - Separable Equations

**1.** Differentiating both sides of $t^2 + y^2 = 7$ with respect to $t$ we obtain $2t + 2y dy/dt = 0$, thus $dy/dt = -t/y$.

**3.** Differentiating both sides of $y = C/t$ with respect to $t$ we obtain $dy/dt = -C/t^2 = -(C/t)/t = -y/t$.

**5.** Differentiation gives $dy/dt = e^{\sin t} \cos t = y \cos t$.

**7.** Differentiation gives $dy/dt = 2e^{-t} = 100 - y$.

**9.** $y' = (1/2)(\cos t + \sin t)$ by differentiation; $\sin t - y = \sin t - (1/2)(\sin t - \cos t) = (1/2)(\cos t + \sin t)$, so the given $y$ is a solution.

**11.** $y' = \cos t + \sin t$ by differentiation; $\sin t - y = \sin t - (\sin t - \cos t) = \cos t$, so the given $y$ is not a solution.

**13.** First, differentiation gives $y' = (e^t(1 - e^t) - (-e^t)(1 + e^t))/(1 - e^t)^2 = 2e^t/(1 - e^t)^2$; also, $(y^2 - 1)/2 = ((1 + e^t)^2/(1 - e^t)^2 - 1)/2 = (1/2)((1 + 2e^t + e^{2t}) - (1 - 2e^t + e^{2t}))/(1 - e^t)^2 = 2e^t/(1 - e^t)^2$, so the given $y$ is a solution.

**15.** First, differentiation gives $y' = -e^t$; also, $(y^2 - 1)/2 = (1/2)((2 - e^t)^2 - 1) = (1/2)(3 - 4e^t + e^{2t})$, so the given $y$ is not a solution.

**17.** Clearly, $y = 0$ is a solution. Separating the variables, $y^{-3}dy = dt$, thus integration gives $-(1/2)y^{-2} = t + \tilde{C}$ and then $y = \pm\sqrt{1/(C - 2t)}$.

**19.** Integrating both sides, we obtain $y = \sin t + C$.

**21.** Separating the variables, $e^y dy = dt$, thus integration gives $e^y = t + C$ and then $y = \ln(t + C)$.

**23.** Clearly, $y = 0$ is a solution. Separating the variables, $(1/y)dy = 3x dx$, thus integration gives $\ln y = 3x^2/2 + \tilde{C}$ and then $y = Ce^{3x^2/2}$.

**25.** Clearly, $y = 0$ is a solution. Separating the variables, $(1/y)dy = (2x/\sqrt{1 + x^2})dx$, thus integration gives $\ln y = 2\sqrt{1 + x^2} + \tilde{C}$ and then $y = Ce^{2\sqrt{1+x^2}}$.

**27.** Clearly, $y = 0$ is a solution. Separating the variables, $(1/\sqrt{y})dy = \sqrt{x}\, dx$, thus integration gives $2\sqrt{y} = (2/3)x^{3/2} + \tilde{C}$ and then $y = (x^{3/2}/3 + C)^2$.

**29.** Separating the variables, $(1 + y)^{-2}dy = dt$, thus integration gives $-1/(1 + y) = t + C$. The initial condition shows that $-1/(1 + 2) = 0 + C$, thus $C = -1/3$ and then $y = 1/(1/3 - t) - 1$.

**31.** Separating the variables, $y dy = te^{-t}dt$, thus integration gives $y^2/2 = -(1 + t)e^{-t} + C$. The initial condition shows that $9/2 = -1 + C$, thus $C = 11/2$ and then we get $y = \sqrt{-2(1 + t)e^{-t} + 11}$.

**33.** Separating the variables, $(y + e^y)dy = (t + 1)dt$, thus integration gives $y^2/2 + e^y = t^2/2 + t + C$. The initial condition shows that $8 + e^4 = 9/2 + 3 + C$, thus $C = e^4 + 1/2$ and then $y^2/2 + e^y = t^2/2 + t + e^4 + 1/2$.

**35.** Separating the variables, we obtain $1/(y(y - 1))dy = (1/(y - 1) - 1/y)dy = dt$, thus integration gives $\ln|y - 1| - \ln y = t + C$. The initial condition shows that $0 = 0 + C$, thus $C = 0$ and then $\ln(|y-1|/y) = t$, which gives $y = 1/(1+e^t)$.

**37.** If $e^t$ is a solution, then $e^t = 5-t+g(e^t)$, thus $g(e^t) = e^t - 5 + t$, and then $g(y) = y - 5 + \ln y$.

**39.** Separating the variables, we obtain $P^{-2}dP = 0.4873dt$, thus $-P^{-1} = 0.4873t + C$. Now $P(0) = 0.2$, thus $-5 = C$ and then $P(t) = 1/(5-0.4873t)$. Doomsday is at time $t = 5/0.4873 \approx 10.26$, thus around 2026 AD.

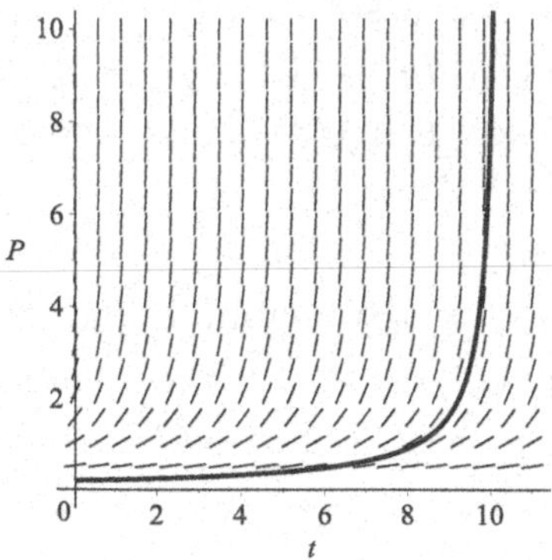

**41. a.** Separating the variables, $dy/(y\ln(y/K)) = -adt$, thus by substituting $u = \ln(y/K)$, $du = (1/K)/(y/K)dy = (1/y)dy$, thus $dy/(y\ln(y/K)) = (1/u)du$, and we obtain $\ln u = \ln(\ln(y/K)) = -at + C$, thus $y = Ke^{ce^{-at}}$. If $K = 100$, $a = 1/2$ and $y(0) = 1$, we get that $1 = 100e^c$, thus $c = \ln(1/100) = -\ln 100$. The solution is $y = 100e^{(-\ln 100)e^{-t/2}} = 100^{1-e^{-t/2}}$.

**b.**

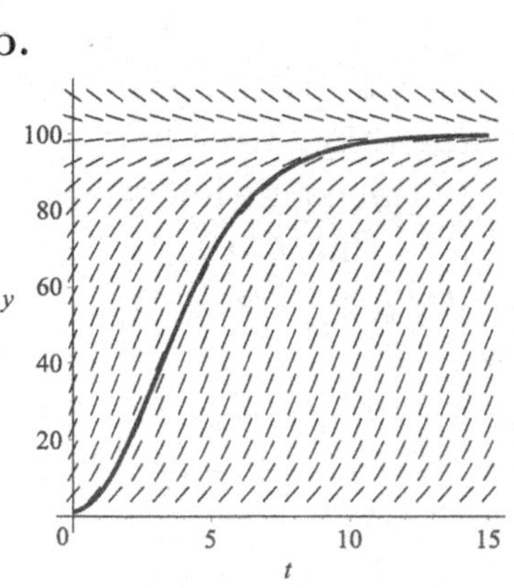

**43.** If $a = b$, then the equation is $dy/dt = k(a-y)^2$. Separating the variables, we get $(a-y)^{-2}dy = kdt$, thus $(a-y)^{-1} = kt + C$. Also, $y(0) = 0$, thus $a^{-1} = C$. Then $a - y = 1/(kt + (1/a))$ and $y = a - a/(akt+1) = a^2kt/(akt+1)$. The graph shows the case $k = 1$, $a = 1$.

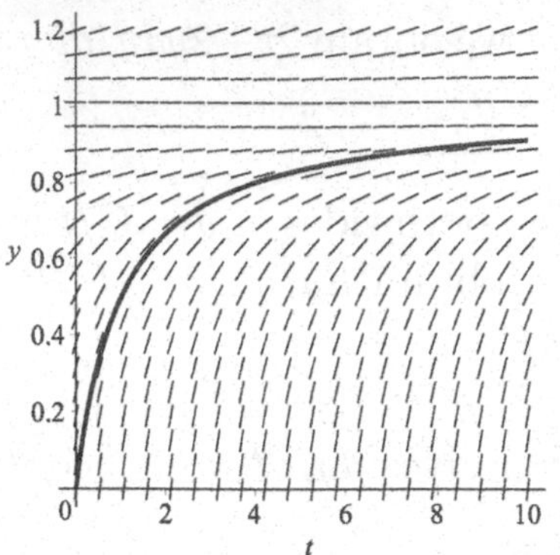

## Problem Set 6.3 - Linear Models in Biology

**1.** Using the result of Example 1, $1/2 = e^{-b\cdot(2.4/24)}$, thus $b = 10\ln 2 \approx 6.9315$. Also, $0 = a - bV$, thus $a = bV \approx 6.9315\cdot 1.89\cdot 10^5 \approx 1,310,000$.

**3.** Using the result of Example 1, $1/2 = e^{-b\cdot(4/24)}$, thus $b = 6\ln 2 \approx 4.1589$. Also, $0 = a - bV$, thus $a = bV \approx 4.1589\cdot 2.25\cdot 10^5 \approx 935,700$.

**5.** Using the result of Example 1, $1/2 = e^{-b\cdot(6/24)}$, thus $b = 4\ln 2 \approx 2.7726$. Also, $0 = a - bV$, thus $a = bV \approx 2.7726\cdot 3.15\cdot 10^5 \approx 873,400$.

**7.** From Example 2, $y = 84(1 - e^{-0.17t})$. We have to solve $y = 5 \cdot 5.6$, and we obtain $1 - e^{-0.17t} = 1/3$, thus $t = -\ln(2/3)/0.17 \approx 2.385$ hours.

**9.** From Example 2, $y = 84(1 - e^{-0.17t})$. We have to solve $y = 12 \cdot 5.6$, and we obtain

$1 - e^{-0.17t} = 4/5$, thus $t = -\ln(1/5)/0.17 \approx$ 9.467 hours.

**11.** From Example 2, $y = 84(1-e^{-0.17t})$. We have to solve $y = 14.5 \cdot 5.6$, and we obtain $1-e^{-0.17t} = 29/30$, thus $t = -\ln(1/30)/0.17 \approx$ 20.007 hours.

**13.** Using the notation and results of Example 3, $y' = 10 - by$, and the solution of this differential equation when $y(0) = 0$ is given by $y(t) = 10(1 - e^{-bt})/b$. To find $b$, we have to numerically solve $1.6 \cdot 5 = y(1) = 10(1 - e^{-b})/b$; we obtain $b \approx 0.4642$.

**15.** Using the notation and results of Example 3, $y' = 12 - by$, and the solution of this differential equation when $y(0) = 0$ is given by $y(t) = 12(1 - e^{-bt})/b$. To find $b$, we have to numerically solve $2 \cdot 5 = y(1) = 12(1 - e^{-b})/b$; we obtain $b \approx 0.3764$.

**17.** Using the notation and results of Example 3, $y' = 20 - by$, and the solution of this differential equation when $y(0) = 0$ is given by $y(t) = 20(1 - e^{-bt})/b$. To find $b$, we have to numerically solve $2 \cdot 5 = y(1) = 20(1 - e^{-b})/b$; we obtain $b \approx 1.5936$.

**19.** Using the notation and results of Example 4, $y' = 2 - (y/50) \cdot 23 = 2 - 23y/50$. The solution of this equation when $y(0) = 0$ is given by $y = (100/23)(1 - e^{-23t/50})$, and the limiting amount is $100/23$ km$^3$. The 2% level is 1 km$^3$, and it is reached when $1 = (100/23)(1 - e^{-23t/50})$, i.e. when $t \approx 0.5682$. Let the new inflow rate be $p$; then $y' = p - 23y/50$, thus the limiting value will be 2% when $p = 23/50$ km$^3$/year.

**21.** Using the notation and results of Example 4, $y' = 2 - (y/100) \cdot 23 = 2 - 23y/100$. The solution of this equation when $y(0) = 0$ is given by $y = (200/23)(1 - e^{-23t/100})$, and the limiting amount is $200/23$ km$^3$. The 2% level is 2 km$^3$, and it is reached when $2 = (200/23)(1 - e^{-23t/100})$, i.e. when $t \approx 1.1364$. Let the new inflow rate be $p$; then $y' = p - 23y/100$, thus the limiting value will be 2% when $p = 23 \cdot 2/100 = 0.46$ km$^3$/year.

**23.** Using the notation and results of Example 4, $y' = 2 - (y/50) \cdot 18 = 2 - 18y/50$. The solution of this equation when $y(0) = 0$ is given by $y = (50/9)(1 - e^{-9t/25})$, and the limiting amount is $50/9$ km$^3$. The 2% level is 1 km$^3$, and it is reached when $1 = (50/9)(1-e^{-9t/25})$, i.e. when $t \approx 0.5513$. Let the new inflow rate be $p$; then $y' = p - 18y/50$, thus the limiting value will be 2% when $p = 18/50$ km$^3$/year.

**25.** Assume that the GDP is $y(t)$, $t = 0$ corresponds to the year 1990. The model is $y' = 0.0508y$, thus we obtain $y(t) = y_0 e^{0.0508t} = 5464e^{0.0508t}$. The predicted GDP in 2003 is $y(13) = 5464e^{0.0508 \cdot 13} \approx 10575.9$ billion.

**27.** Assume that the number of divorces is $y(t)$, $t = 0$ corresponds to the year 1990. The model is $y' = 0.047y$, thus we obtain $y(t) = y_0 e^{0.047t} = 1175e^{0.047t}$ (in thousands). The predicted number of divorces in 2004 is $y(14) = 1175e^{0.047 \cdot 14} \approx 2268.8$ thousand.

**29.** Separating the variables, $dx/(a - bx) = dt$, thus $-(1/b)\ln(a - bx) = t + C$; also, $x(0) = 0$, thus $-(1/b)\ln a = C$. Then we get that $\ln(a - bx) = -bt + \ln a$, thus $a - bx = e^{-bt+\ln a} = ae^{-bt}$, and then $x = (a/b)(1-e^{-bt})$. The limiting value is thus $a/b$ as $t \to \infty$, assuming naturally that $b > 0$. The amount is equal to half its limiting value when $a/2b = (a/b)(1 - e^{-bt})$, i.e. when $e^{-bt} = 1/2$, which gives $t = (\ln 2)/b$.

**31.** Let $y$ denote the amount of drug in the

patient's body. The model is $y' = a - by$, and according to Problem 29, the limiting value is $a/b$. We need that this limiting value is $5.6 \cdot 2 = 11.2$ mg. Like in Example 2, $b = (\ln 2)/2.7$, thus we need that the infusion rate $a = 11.2(\ln 2)/2.7 \approx 2.88$ mg/h.

**33.** Let $y$ denote the amount of drug in the patient's body. The model is $y' = a - by$, and according to Problem 29, the solution of this equation is $y(t) = (a/b)(1 - e^{-bt})$. The amount to be reached is $5 \cdot 12 = 60$ mg, and $b = 2$. Thus we need that $60 = y(0.3) = (a/2)(1 - e^{-2 \cdot 0.3})$. We obtain $a \approx 265.964$ mg/h.

**35.** Let $y$ denote the amount of drug in the patient's body. The model is $y' = a - by$, and according to Problem 29, the solution of this equation is $y(t) = (a/b)(1 - e^{-bt})$, where $a = 250$. The amount at $t = 4$ is $50 \cdot 5.5 = 275$ mg. Thus we have $275 = y(4) = (250/b)(1 - e^{-4b})$. A numerical solver gives $b \approx 0.882444$ per hour.

**37.** Let $y(t)$ denote the amount of dye in the tank, measured in grams. Time $t$ is measured in minutes. The model is $y' = -2(y/300)$, and $y(0) = 2 \cdot 300 = 600$ g. The solution can be obtained by separation of variables; we get $y = 600e^{-t/150}$ g.

**39.** Let the temperature of the coffee be $T(t)$. Newton's law of cooling gives the model $T' = -k(T - 20)$. Separation of variables gives $dT/(T - 20) = -kdt$, thus $\ln(T - 20) = -kt + C$, and then $T = 20 + Ce^{-kt}$. We know that $T(0) = 95$, thus $C = 75$ and $T = 20 + 75e^{-kt}$. Also, $45 = T(5) = 20 + 75e^{-5k}$, which means $e^{-5k} = 1/3$. We have to solve the equation $22 = 20 + 75(1/3)^{t/5}$, which gives $t \approx 16.4951$ minutes.

**41.** From Example 6, the von Bertalanffy curve is $L(t) = L_\infty(1 - e^{-k(t-t_0)})$. Half the limiting size is reached when $L_\infty/2 = L(t) = L_\infty(1 - e^{-k(t-t_0)})$, i.e. when $e^{-k(t-t_0)} = 1/2$ or $t = t_0 + (\ln 2)/k$. For females, this gives $t = -0.27 + (\ln 2)/(0.75) \approx 0.6542$; for males, $t = -0.395 + (\ln 2)/0.537 \approx 0.8958$ years.

## Problem Set 6.4 - Slope Fields and Euler's Method

**1.**

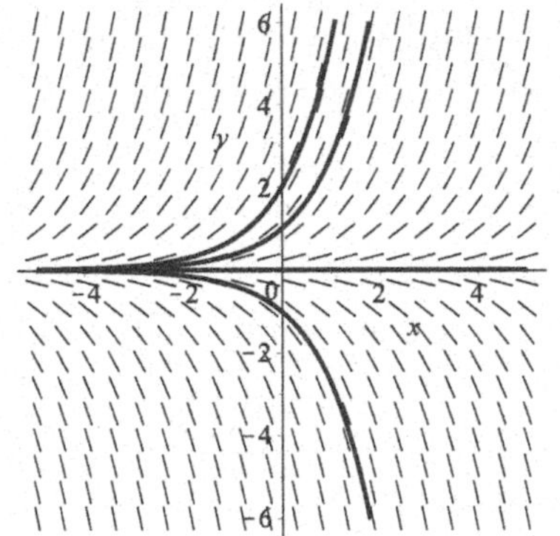

**3.**

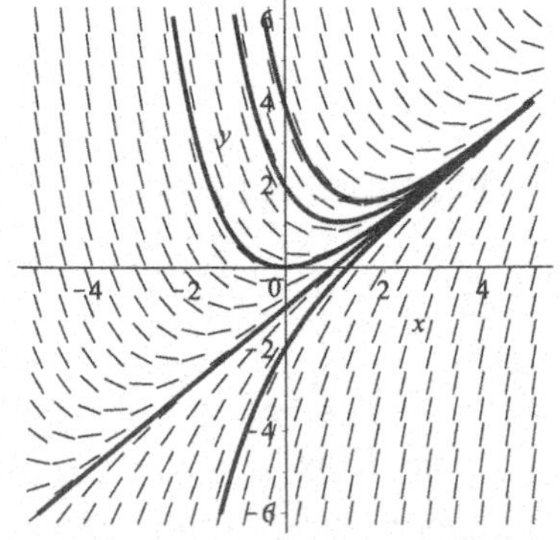

**5.**

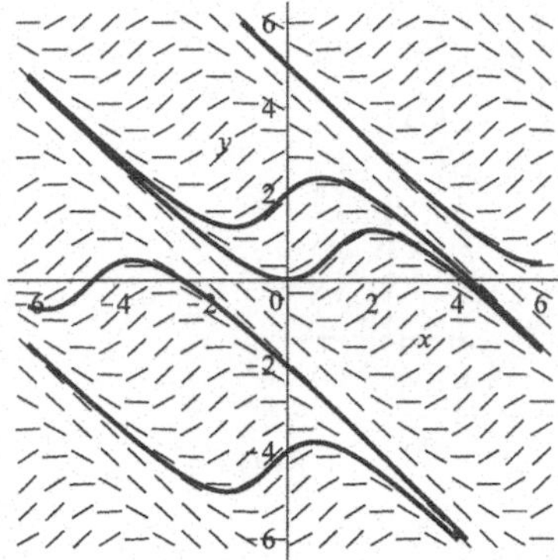

**7.**

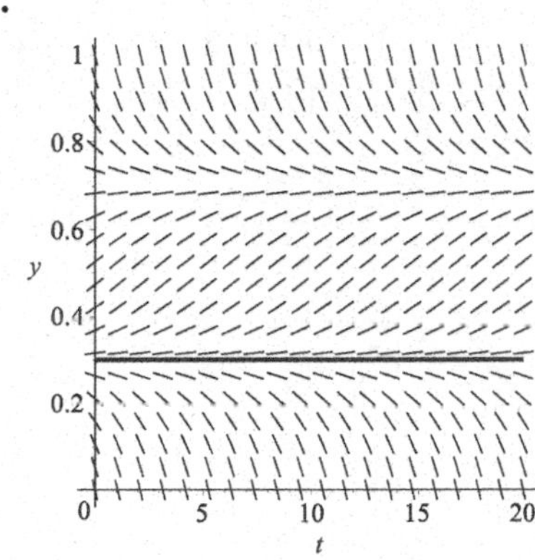


**9.**

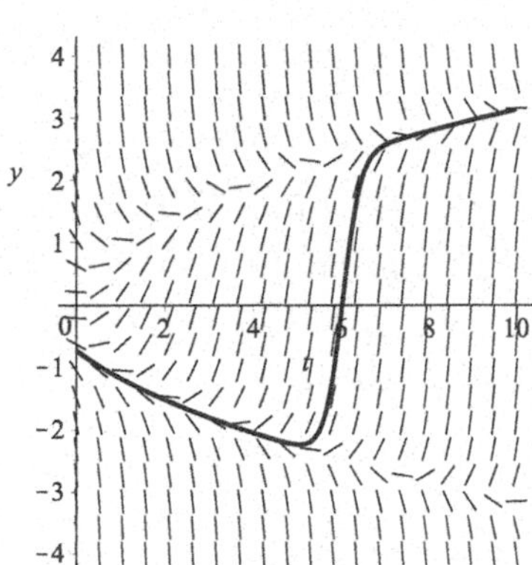


**11.a.** D - the slope depends only on $t$ for this graph.

**b.** B - the slope depends on $t$ and $y$ and is zero for $y = \pi$.

**c.** C - the slope depends only on $y$ for this graph.

**d.** A - the slope depends on $t$ and $y$ and is zero for $t = \pi$.

**13.**

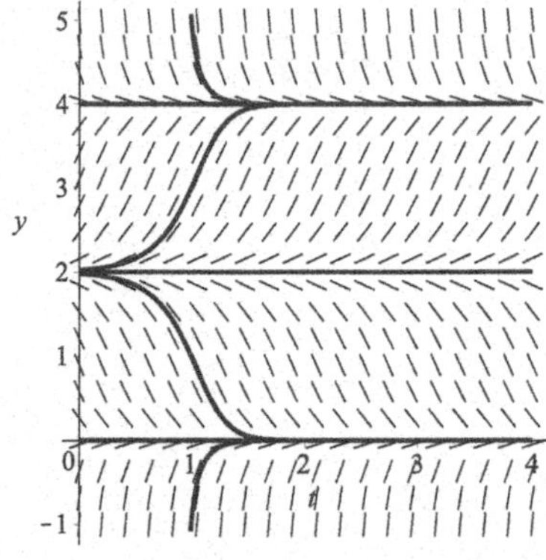


**15.**

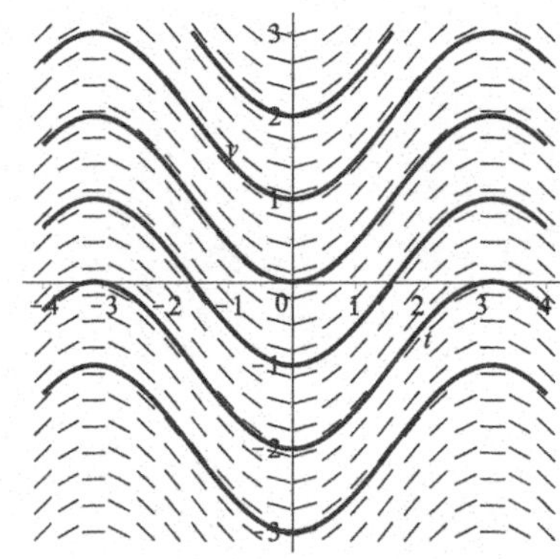

**17.**

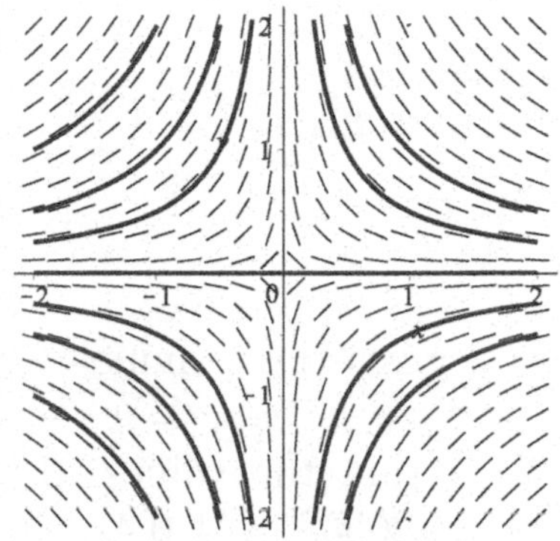

**19.**

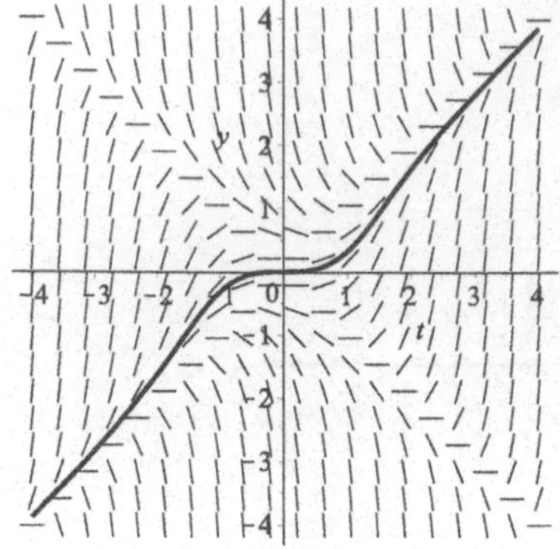


**21.**

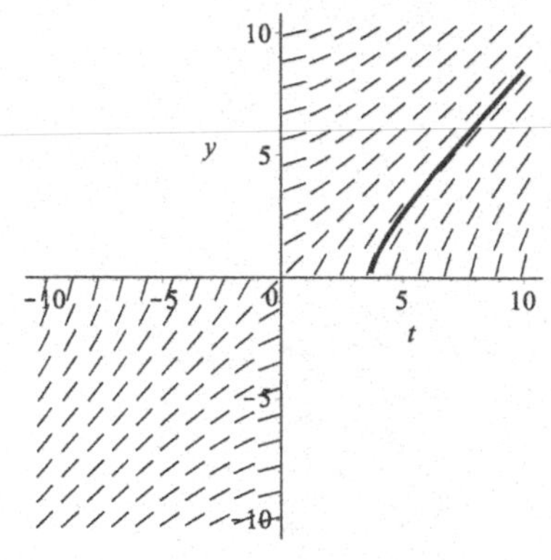


**23.**

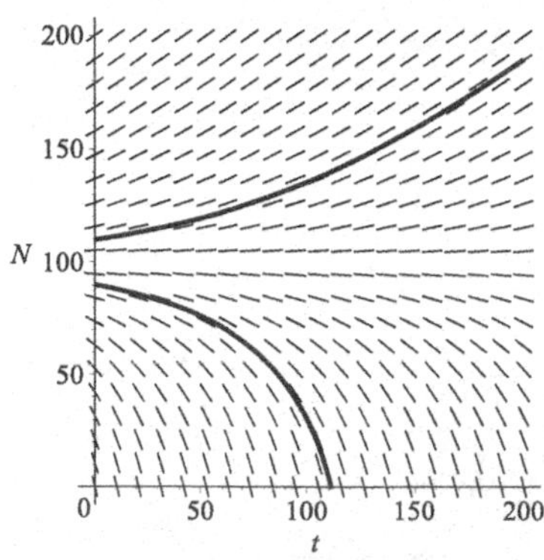


**25.** Euler's method gives the formulas $t_0 = 0$, $y_0 = 4$, $t_{n+1} = t_n + 1$, $y_{n+1} = y_n + 1 \cdot (t_n/y_n - t_n)$. We obtain the values $(0, 4)$, $(1, 4)$, $(2, 3.25)$, $(3, 1.86538)$, $(4, 0.473632)$, $(5, 4.91901)$, $(6, 0.935472)$, $(7, 1.34935)$.

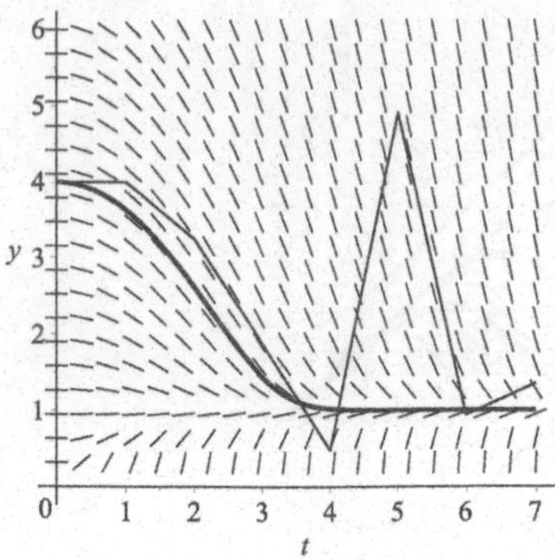


**27.** Euler's method gives the formulas $t_0 = 0$, $y_0 = 1$, $t_{n+1} = t_n + 0.5$, $y_{n+1} = y_n + 0.5 \cdot (2t_n(y_n - t_n^2))$. We obtain the values $(0, 1)$, $(0.5, 1)$, $(1, 1.375)$, $(1.5, 1.75)$, $(2, 1)$, $(2.5, -5)$, $(3, -33.125)$.

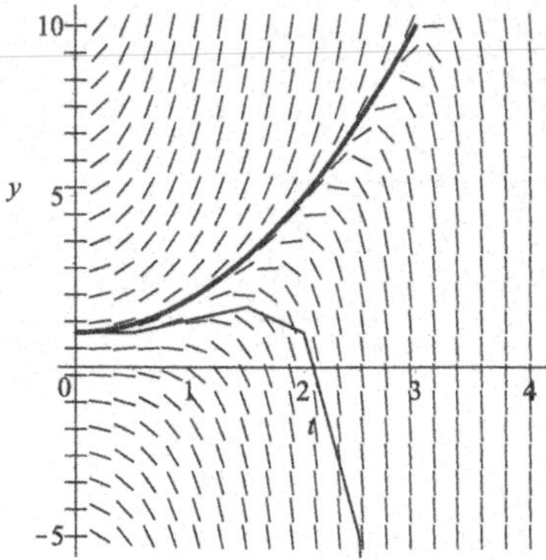


**29.** Euler's method gives the formulas $t_0 = 0$, $y_0 = 0.1$, $t_{n+1} = t_n + 0.5$, $y_{n+1} = y_n + 0.5 \cdot (4 - y_n)(y_n + 2)$. We obtain the values $(0, 0.1)$, $(0.5, 4.195)$, $(1, 3.59099)$, $(1.5, 4.73438)$, $(2, 2.26158)$.

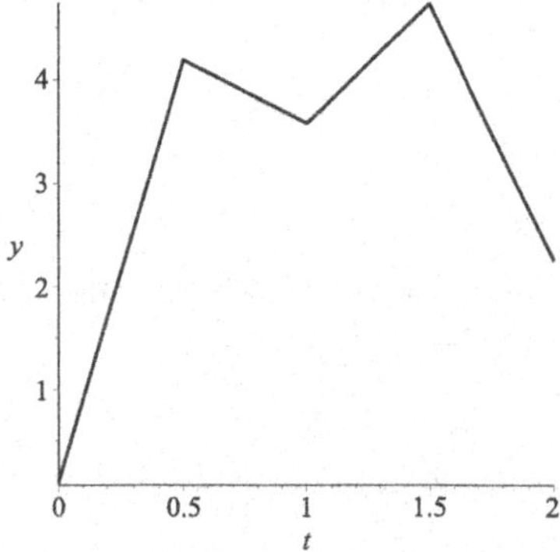


**31.** Euler's method gives the formulas $t_0 = 0$, $y_0 = 0$, $t_{n+1} = t_n + 0.5$, $y_{n+1} = y_n + 0.5 \cdot$

$(\sin \pi t_n - 2y_n)$. We obtain the values $(0,0)$, $(0.5,0)$, $(1,0.5)$, $(1.5,0)$, $(2,-0.5)$, $(2.5,0)$, $(3,0.5)$, $(3.5,0)$.

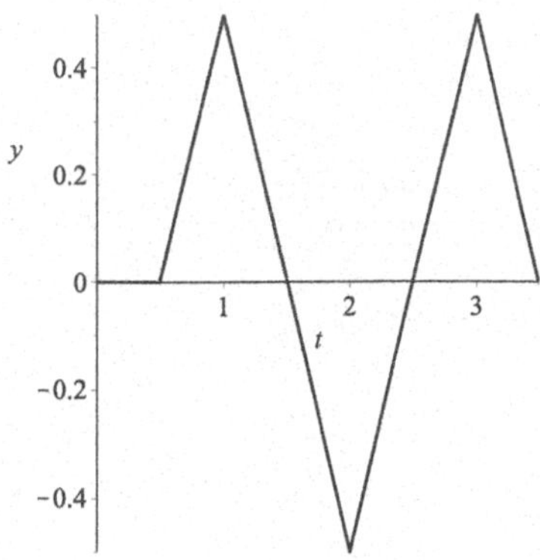


**33. a.** $\ln 1 = 0$, $dy/dt = 1/t$.

**b.** Euler's method gives the formulas $t_0 = 1$, $y_0 = 0$, $t_{n+1} = t_n + 0.5$, $y_{n+1} = y_n + 0.5 \cdot (1/t_n)$. We obtain the values $(1,0)$, $(1.5,05)$, $(2,0.833333)$.

**35. a.** Let $y$ denote the amount of drug in the patient's body. The model is $y' = a - by$, where $a = 100$ and $b = (\ln 2)/2$.

**b.**

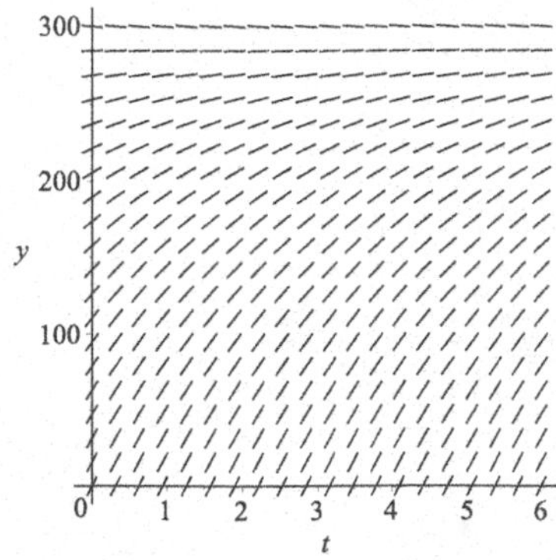


**c.** The limiting amount is $a/b = 200/\ln 2 \approx 288.5$.

**37. a.b.c.**

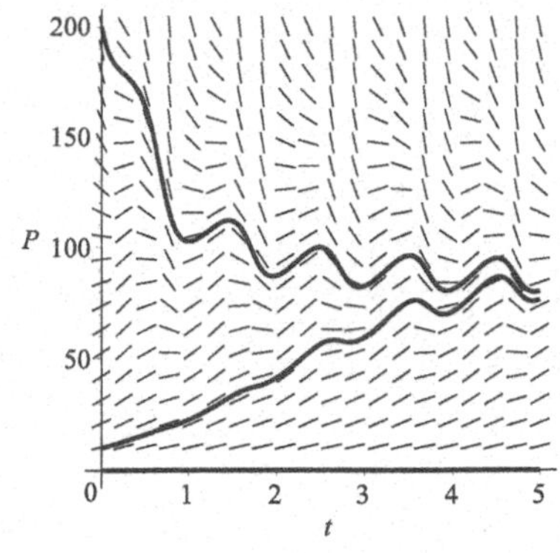


**39. a.**

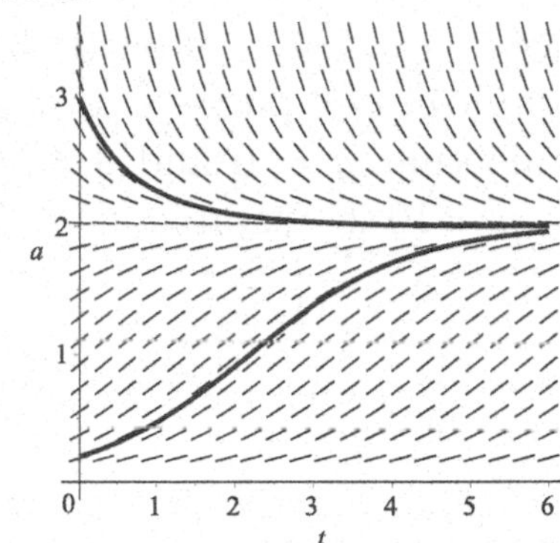


**b.** The limiting value is $a = k_1 b/k_2 = 2$.

## Problem Set 6.5 - Phase Lines and Classifying Equilibria

**1.** The equilibria are at the solutions of $1 - y^2 = 0$, i.e. at $y = 1$ and $y = -1$. We obtain that $f'(y) = -2y$, thus $y = 1$ is stable and $y = -1$ is unstable.

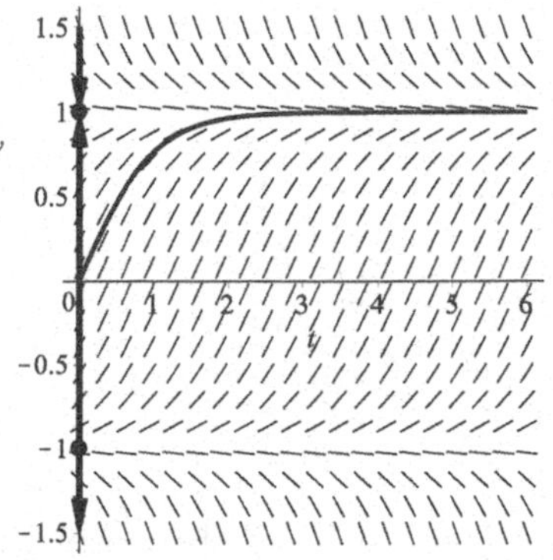

**3.** There are no equilibria.

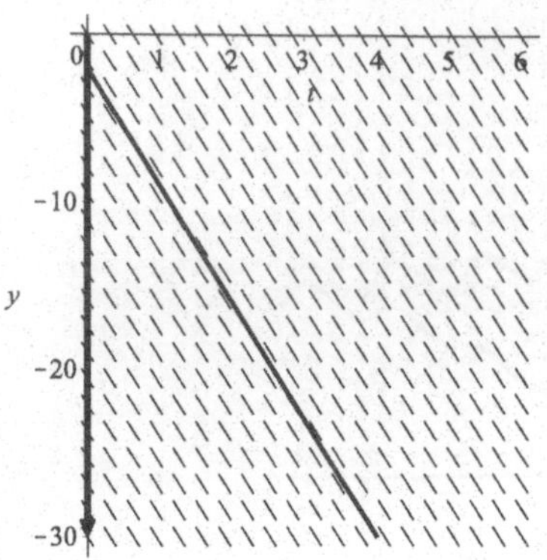

**5.** The equilibria are at $y = 0$, $y = 10$ and $y = 20$. Because $dy/dt < 0$ on $(0, 10)$ and $(20, \infty)$, and $dy/dt > 0$ on $(-\infty, 0)$ and $(10, 20)$, we obtain that $y = 0$ and $y = 20$ are stable, while $y = 10$ is unstable.

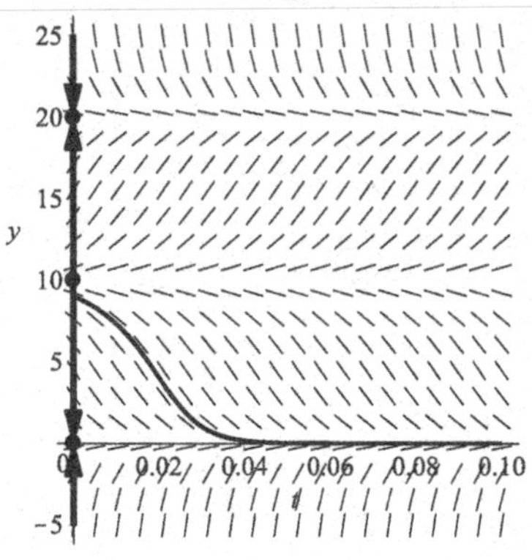

**7.** The equilibria are at $y = n\pi$, $n = 0, \pm 1, \pm 2, \ldots$. Because $f'(y) = \cos y$, we obtain that $y = n\pi$, $n = \pm 1, \pm 3, \ldots$ are stable, while $y = n\pi$, $n = 0, \pm 2, \pm 4, \ldots$ are unstable.

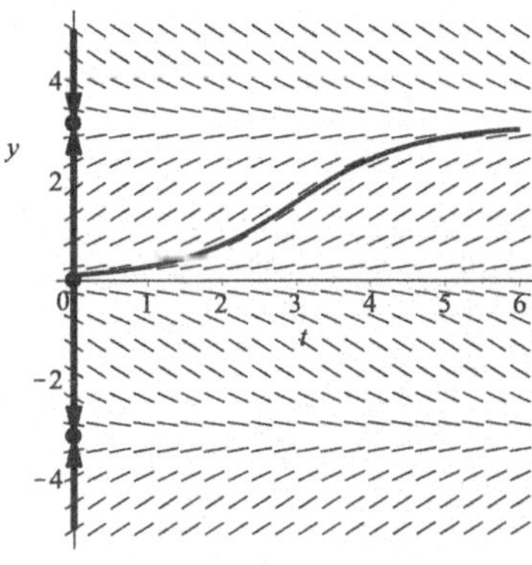

**9.** The only equilibrium is at the solution of $y^2 - 2y + 1 = (y-1)^2 = 0$, i.e. at $y = 1$. We obtain that $dy/dt = (y-1)^2 > 0$, for all values of $y$ except at the equilibrium, so it is semistable.

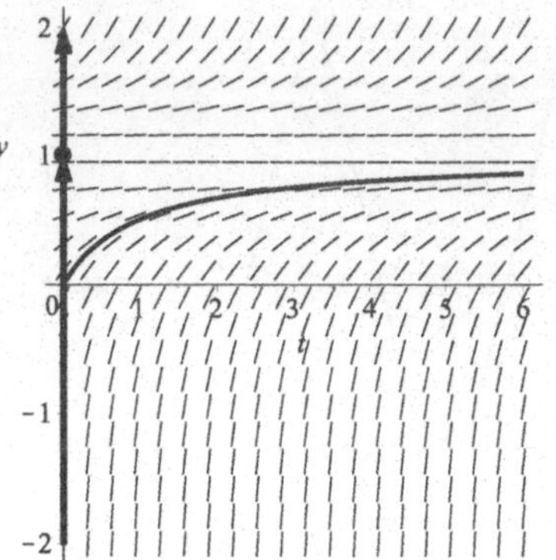

**11.**

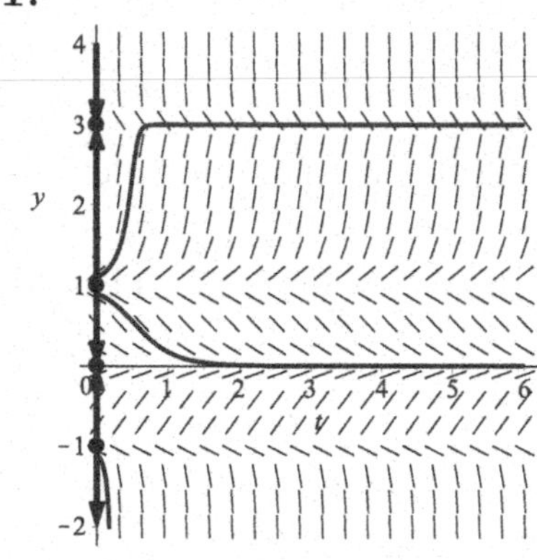

**13.**

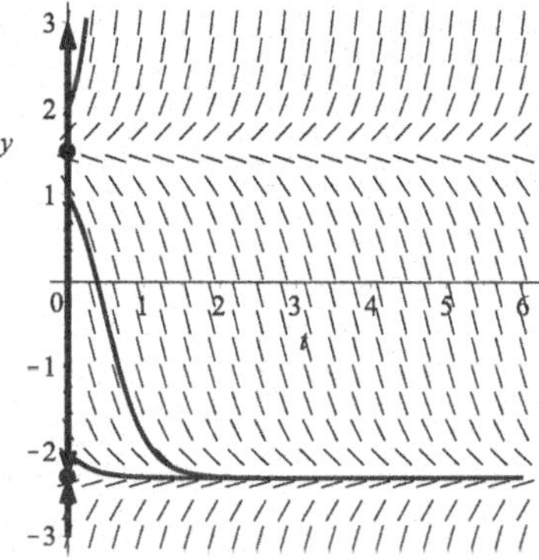

**15.** $f'(y^*) = -2y^* = -4$, thus $dy/dt \approx -4(y-2)$. The equilibrium is stable.

**17.** $f'(y^*) = \sin(y^*) = \sqrt{2}/2$, thus $dy/dt \approx (\sqrt{2}/2)(y - \pi/4)$. The equilibrium is unstable.

**19.** $f'(y^*) = -1$, thus the linearization is the original equation, $dy/dt = -1(y-3)$. The equilibrium is stable.

**21.** We get $y' = y(1-y)(0/2 - 2y/2) = -y^2(1-y)$. The equilibria are $y = 0$, which is semistable, and $y = 1$, which is unstable. The phase line shows the physically meaningful values $0 \le y \le 1$.

**23.** We get $y' = y(1-y)(-1/2 - 2y/2) = y(1-y)(-1/2-y)$. The equilibria are $y = 0$, which is stable, and $y = 1$ and $y = -1/2$, which are unstable. The phase line shows the physically meaningful values $0 \le y \le 1$.

**25.** The equation is $y' = y(1-y)((1-2) + y(4-2)) = y(1-y)(-1+2y)$. The equilibria are $y = 0$ and $y = 1$, which are stable, and $y = 1/2$, which is unstable. The phase line shows the physically meaningful values $0 \le y \le 1$.

**27.** The equation is $y' = y(1-y)((2-(-1)) + y(-2-3)) = y(1-y)(3-5y)$. The equilibria are $y = 0$ and $y = 1$, which are unstable, and $y = 3/5$, which is stable. The phase line shows the physically meaningful values $0 \le y \le 1$.

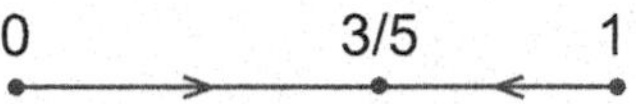


**29. a.** We obtain $y' = y(1-y)((4-5) + y(12.5-11)) = y(1-y)(-1+1.5y)$. The equilibria are $y = 0$ and $y = 1$, which are stable, and $y = 2/3$, which is unstable. The phase line shows the physically meaningful values $0 \le y \le 1$.

**b.**

**31. a.** Using the payoff matrix, the replicator equation is $y' = y(1-y)((-C-0) + y((B-C) - B + C)) = -Cy(1-y)$.

**b.** If $C > 0$, the equilibrium $y = 0$ is unstable, while the equilibrium $y = 1$ is stable. The phase line shows the physically meaningful values $0 \le y \le 1$.

**33. a.** The figure shows the graph of $f(N) = 0.1N(1 - N/1000) - 10N/(1+N)$. There are three equilibria, $N = 0$, $N \approx 111.6$ and $N \approx 887.4$. The phase line can be drawn by checking the sign of $f(N)$ between the equilibria. We obtain that $N = 0$ and $N = 887$ are stable, while $N = 112$ is unstable.

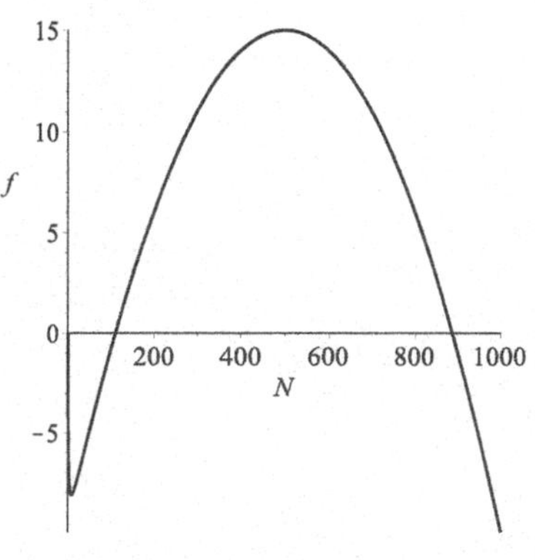


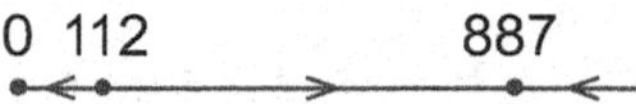

**b.** If the initial population abundance is less than $N = 112$, the population dies out; if it is greater than $N = 112$, it approaches $N = 887$.

**35.** The model is given by the equation $dV/dt = -3(V+65)(V-40)^2$. There are two equilibria, $V = -65$ and $V = 40$. Checking the sign of the right side, we conclude that $V = -65$ is stable, while $V = 40$ is semistable. This means that the membrane can switch, but if the current drops below 40 mV, the membrane switches back to $-65$ mV.

**37.** If $y(t) = (y_0 - y^*)e^{at} + y^*$, then $y(0) = (y_0 - y^*) + y^* = y_0$; also, we find that $dy/dt = a(y_0 - y^*)e^{at} = a(y - y^*)$.

**39.** Using the quotient rule, we obtain that

$$\begin{aligned}\frac{dy}{dt} &= \frac{(dN_a/dt)(N_a+N_A) - N_a(dN_a/dt + dN_A/dt)}{(N_a+N_A)^2} = \frac{(dN_a/dt)N_A - N_a(dN_A/dt)}{(N_a+N_A)^2} = \\ &= \frac{(r_aN_a)N_A - N_a(r_AN_A)}{(N_a+N_A)^2} = \frac{(r_a - r_A)N_aN_A}{(N_a+N_A)^2} = (r_a - r_A)\left(\frac{N_a}{N_a+N_A}\right)\left(\frac{N_A}{N_a+N_A}\right) = \\ &= (r_a - r_A)\left(\frac{N_a}{N_a+N_A}\right)\left(\frac{N_a+N_A-N_a}{N_a+N_A}\right) = (r_a - r_A)y(1-y).\end{aligned}$$

**41. a.** $A(y) = ay^b/(k^b+y^b) = a/((k/y)^b+1)$; thus $\lim_{y\to\infty} A(y) = a$ because $b \geq 1$. Also, $A(k) = ak^b/(k^b+k^b) = a/2$.

**b.** Let $f(y) = A(y) - cy$. Clearly, $f(0) = 0$, thus $y = 0$ is an equilibrium. We differentiate to obtain $f'(y) = A'(y) - c = (aby^{b-1}(k^b+y^b) - ay^b(by^{b-1}))/(k^b+y^b)^2 - c = abk^by^{b-1}/(k^b+y^b)^2 - c$. If $b > 1$, then $f'(0) = -c < 0$, thus the zero equilibrium is stable. If $b = 1$, then $f'(0) = a/k - c$, so if $a/k < c$, the zero equilibrium is stable.

**43. a.** $a = 1$, $b = 2$, $k = 1$, $c = 1$. There is only one nonnegative equilibrium, $y = 0$, which is stable.

**b.** $a = 2$, $b = 2$, $k = 1$, $c = 1$. There are two nonnegative equilibria, $y = 0$, which is stable, and $y = 1$, which is semistable.

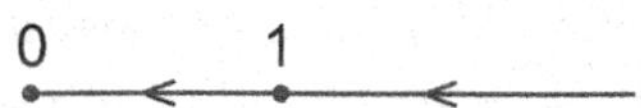


**c.** $a = 2.5$, $b = 2$, $k = 1$, $c = 1$. There are three nonnegative equilibria, $y = 0$, which is stable, $y = 1/2$, which is unstable, and $y = 2$, which is stable.

**Problem Set 6.6 - Bifurcations**

**1.** The equilibrium solutions can be found

by solving $f(y) = y(1 - y/100) - a = 0$; for $a = 0$, $y^* = 0$ and $y^* = 100$, for $a = 9$, $y^* = 10$ and $y^* = 90$, and for $a = 25$, $y^* = 50$.

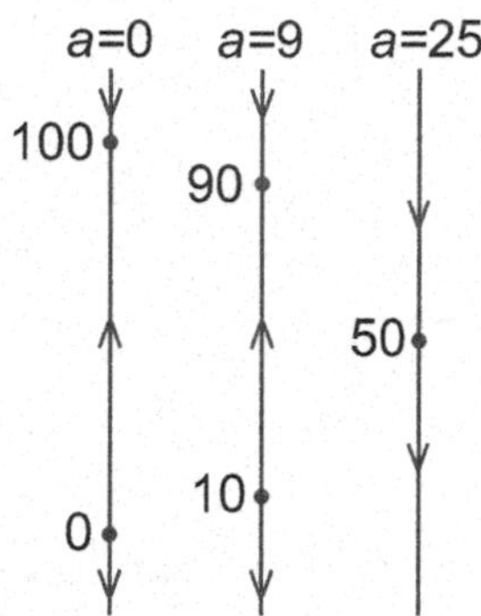


**3.** The equilibrium solutions can be found by solving $f(y) = 450 - ay = 0$; for $a = -10$, $y^* = -45$, for $a = 0$, there is no equilibrium, and for $a = 100$, $y^* = 45$.

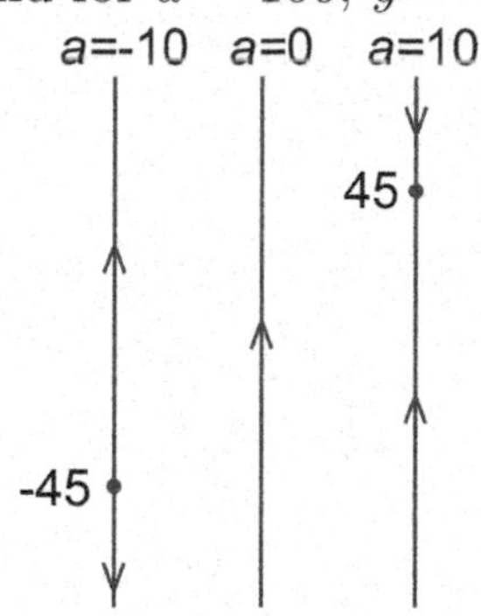


**5.** The equilibrium solutions can be found by solving $f(y) = y^2 - ay + 1 = 0$; for $a = 0$, there is no equilibrium, for $a = 2$, $y^* = 1$, and for $a = 4$, $y^* = 2 - \sqrt{3}$ and $y^* = 2 + \sqrt{3}$.

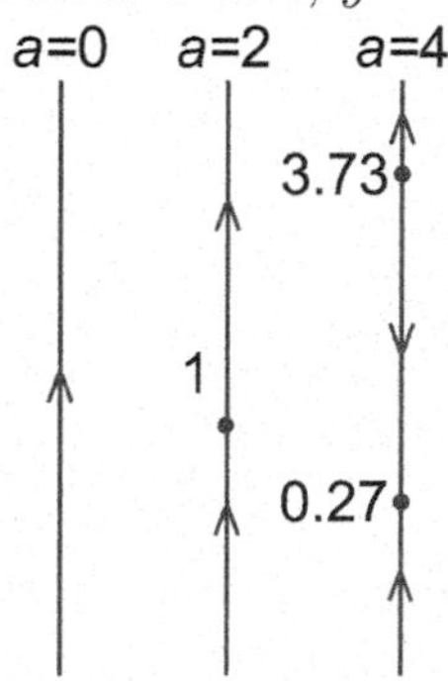


**7.** The equilibrium solutions are given by $ay - y^2 = y(a - y) = 0$; i.e. by $y = 0$ and $y = a$.

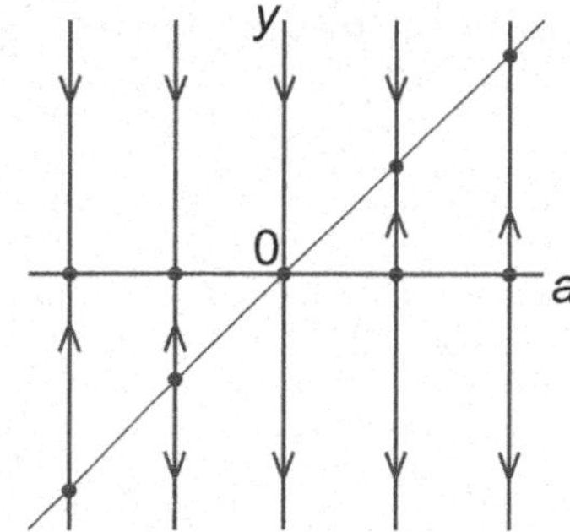


**9.** The equilibrium solutions are given by $1 + ay = 0$; i.e. by $y = -1/a$.

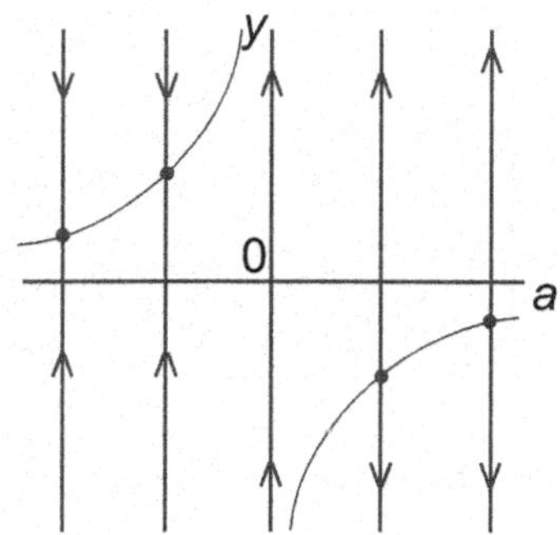


**11.** The equilibrium solutions are given by $\sin y + a = 0$.

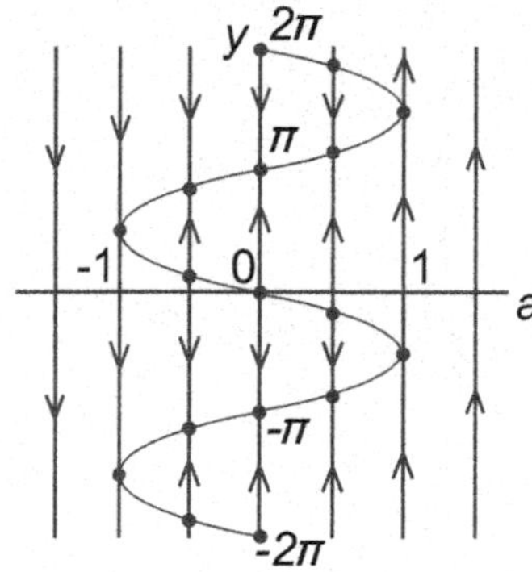


**13.** The equilibrium solutions are given by $y = 0$ and $a = 2(1 + y^2)/y$.

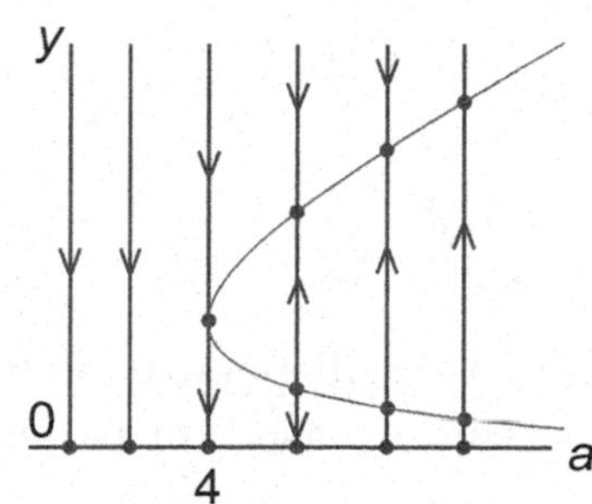

**15.** The equilibrium solutions are given by $y = 0$ and $c = 10y/(1 + y^2)$.

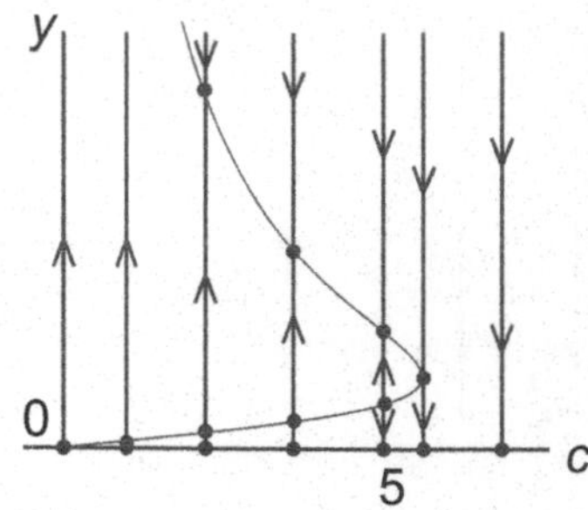


**17.** $b = 5$, $c = 1$.

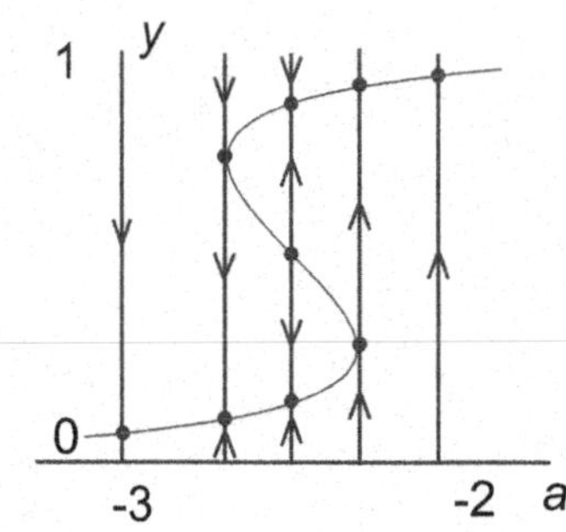


**19.** $b = 8$, $c = 1$.

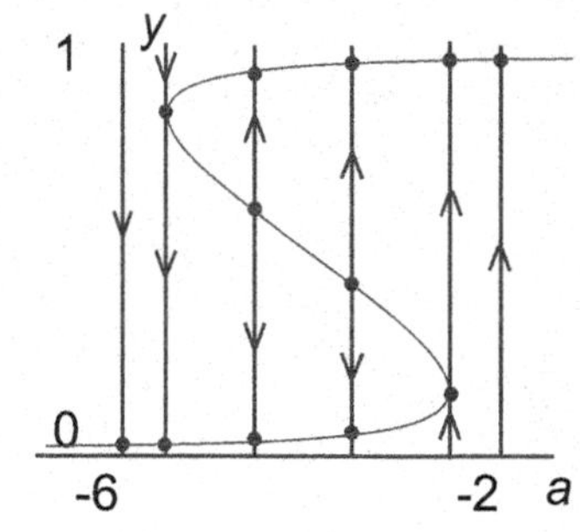


**21.** $b = 8$, $c = 2$.

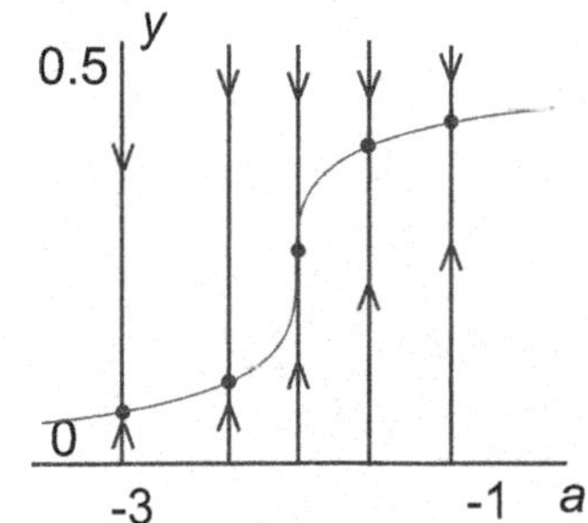


**23.** If $b = 1$, the disease will persist for all $r < 1000$; if $r = 1$, the disease will persist for $b > 0.001$.

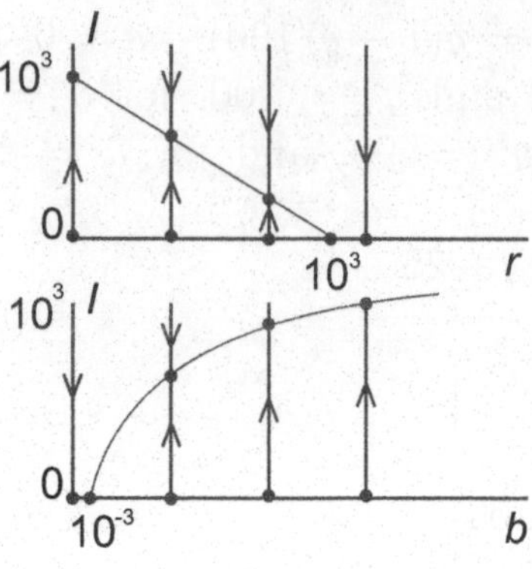


**25.** If $b = 1$, the disease will persist for all $r < 10^5$; if $r = 1$, the disease will persist for $b > 10^{-5}$.

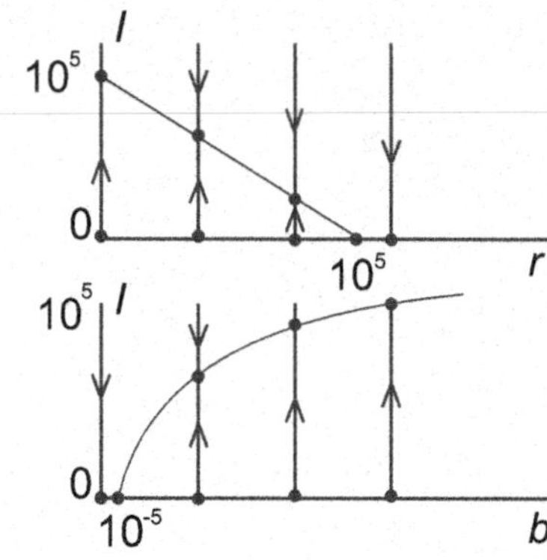


**27.** If $c = 1$, the population will persist for all $d < 1$; if $d = 0$, the population will persist for all $c > 0$.

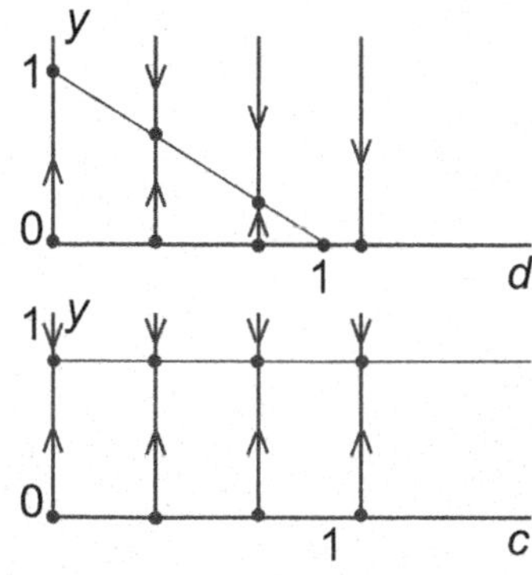


**29.** If $c = 1$, the population will persist for all $d < 1/2$; if $d = 1/2$, the population will persist for $c > 1$.

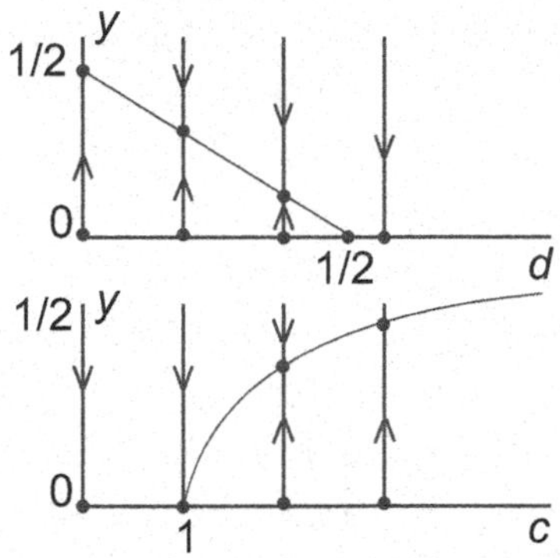


**31.** If $d = 1/2$, the population will persist for all $D < 2/3$; if $D = 0$, the population will persist for $d < 3/2$.

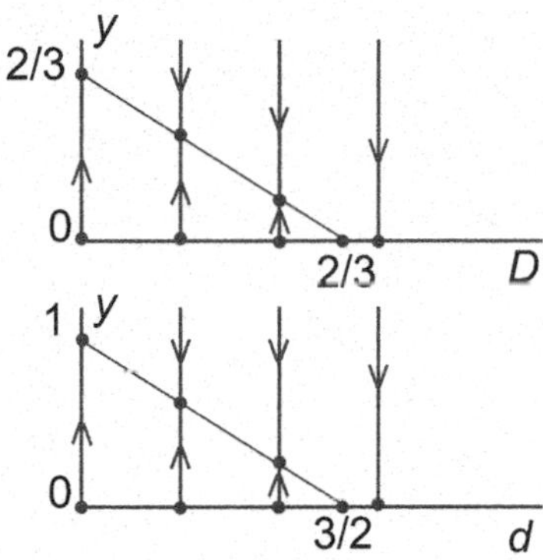


**33.** We obtain $dx/dt = 0.5x(1-x/3)-xy = x(0.5(1-x/3)-y)$. Thus the equilibria are $x = 0$ and $x = 3(1-2y)$. For $y < 1/2$, the population persists, for $y \geq 1/2$ it does not.

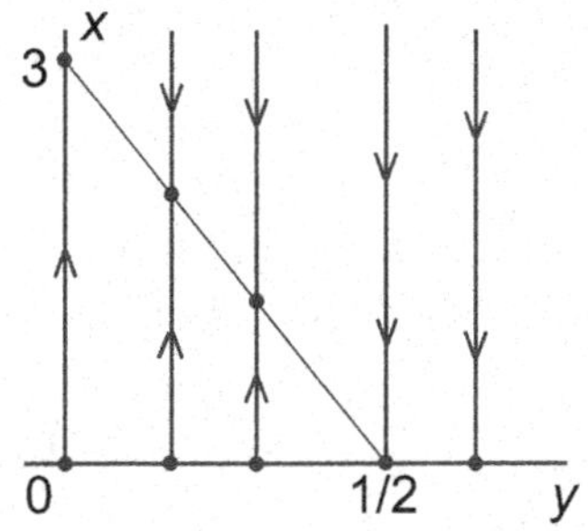


**35.** $dx/dt = 0.5x(1-x/3) - yx/(x+2) = x(0.5(1-x/3)-y/(x+2))$. Thus the equilibria are $x = 0$ and $y = 0.5(1-x/3)(x+2)$. For $y \leq 1$, there are two equilibria and the population persists; for $1 < y < 25/24$ there are three equilibria and depending on the initial condition, the population either persists or it dies out; for $y = 25/24$ there are two equilibria and depending on the initial condition, the population either persists or it dies out; for $y > 25/24$ the only equilibrium is 0 and the population dies out.

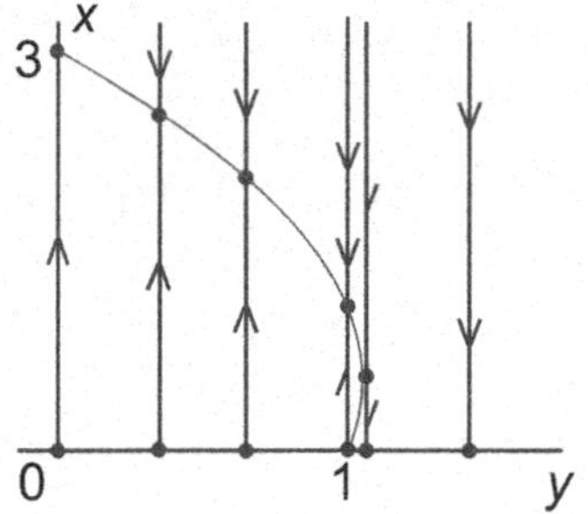


**37.** $dy/dt = 0.2yx - y^2 = y(0.2x-y)$. Thus the equilibria are $y = 0$ and $y = 0.2x$; the population persists for all $x > 0$.

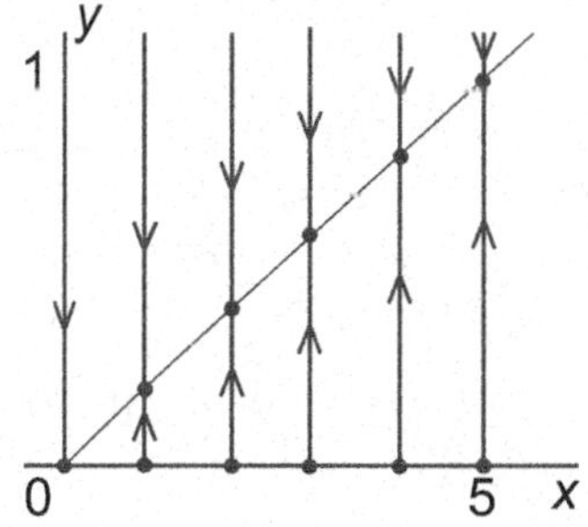


**39. a.** Using the payoff matrix, we get $y' = y(1-y)((-C-0)+y(n(B-C)-B+C)) = -y(1-y)(C-(n-1)(B-C)y)$.

**b.** For $B = 4$, $C = 3$, the equation is $y' = -y(1-y)(3-(n-1)y)$. The equilibria are $y = 0$, $y = 1$ and $y = 3/(n-1)$.

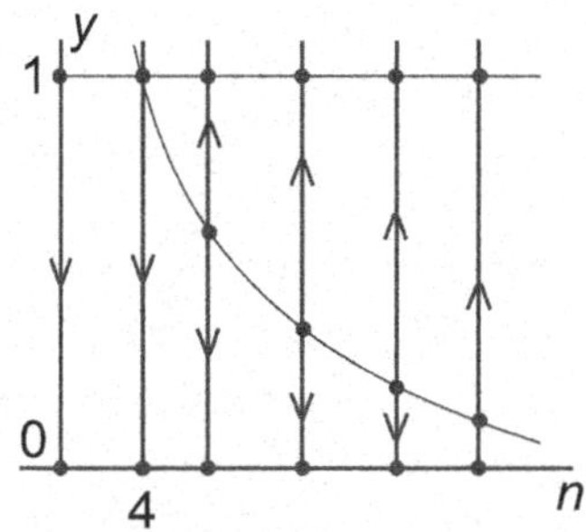


**c.** When $n > 4$, $y = 1$ becomes stable, so a stable cooperative equilibrium exists.

## Review Questions

**1.** We obtain $y' = e^t + 2at + b$, thus $t^2 + y' = y$ means $t^2 + e^t + 2at + b = e^t + at^2 + bt + c$. This gives $a = 1$, $b = 2$, and $c = 2$.

**3.** Separation of variables gives the equation $dL/(L_\infty - L) = kdt$, thus $-\ln(L_\infty - L) = kt + \tilde{C}$ and we obtain $L = L_\infty + Ce^{-kt}$. Now $L_\infty = 26.1$, so $12.5 = L(1) = 26.1 + Ce^k$, and $15.6 = L(2) = 26.1 + Ce^{2k}$. From the first equation, $Ce^k = -13.6$, and from the second, $Ce^{2k} = Ce^k e^k = -10.5$. Thus $e^k = 10.5/13.6$, and then $C = -13.6(13.6/10.5)$. These give that $L(t) = 26.1 - (13.6^2/10.5)(10.5/13.6)^t$.

**5.** Separating the variables we get $ye^{-2y}dy = e^t \sin t dt$, so integration gives $-e^{-2y}/4 - ye^{-2y}/2 = e^t(\sin t - \cos t)/2 + C$. Plugging in $t = 0$ and $y = 0$ we get $-1/4 = -1/2 + C$, thus $C = 1/4$ and the solution is $-e^{-2y} - 2ye^{-2y} = 2e^t(\sin t - \cos t) + 1$.

**7.**

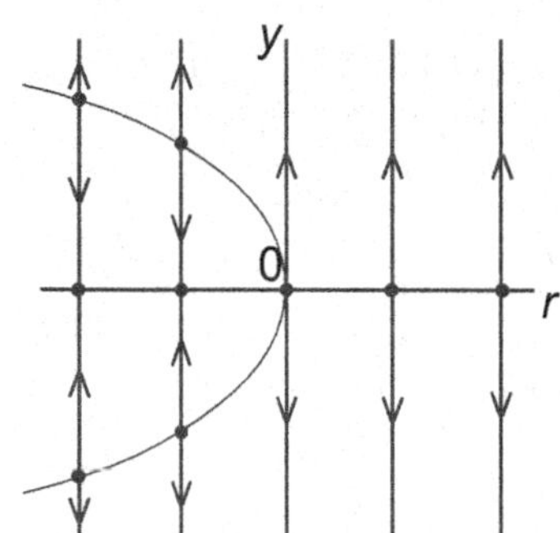


**9.** The amount of gallium is given by $y = y_0 e^{-kt}$, where $y_0 = 100$ mg, and $k = \ln 2/46.6$ per hour. The percentage lost between the 30th and 35th hour is $1 - y(35)/y(30) = 1 - e^{-5k} \approx 0.07167$, about 7.167%. The value $y(T+5)/y(T) = e^{-5k}$ does not depend on $T$, thus the percentage loss is the same for any 5-hour period.

**11.** Separation of variables and partial fraction decomposition give the equation $(1/N - 1/(N-K))dN = rdt$, so after integration, $\ln N - \ln(N-K) = rt + C$, which means $N/(N-K) = Ce^{rt}$, and then $N = KCe^{rt}/(Ce^{rt} - 1)$. Let $t = 0$ correspond to the year 1980; $K = 5000$, thus $1800 = N(0) = 5000C/(C-1)$, and this means $C = -9/16$. Also, $2000 = N(6) = 5000(-9/16)e^{6r}/(-9/16e^{6r} - 1)$, which gives $e^{6r} = 32/27$, or $r = \ln(32/27)/6 \approx 0.0283$. Then $N(20) = 5000(-9/16)e^{20r}/(-9/16e^{20r} - 1) \approx 2488.7$.

**13.** Let the amount of pollutant in the lake be $y(t)$; its initial amount is $6000 \cdot 0.0022 = 13.2$. The equation describing the rate of change is $y' = 350 \cdot 0.0006 - 350y/6000$. Separation of variables gives $dy/(0.21 - 7y/120) = dt$, thus $-(120/7)\ln(0.21 - 7y/120) = t + C$ and then $0.21 - 7y/120 = Ce^{-7t/120}$. The initial condition $y(0) = 13.2$ gives $0.21 - 7 \cdot 13.2/120 = C$, and then $y = (120/7)(0.21 - (0.21 - 7 \cdot 13.2/120)e^{-7t/120})$. We have to solve the equation $0.0015 \cdot 6000 = y(T)$, which gives $T \approx 9.86$ days.

**15.** $dy/dt = ay - y^2 = f(y)$. Clearly, $f(0) = 0$, thus 0 is an equilibrium. $f'(y) = a - 2y$, thus $f'(0) = a$ and it is stable when $a < 0$ and unstable when $a > 0$. We can check that in case $a = 0$, $y' = -y^2$ and $y = 0$ is semistable.

**17.** The equilibrium points are given by $N = 0$ and by the expression $r(1 - N/K) = bN/(a^2 + N^2)$; this gives the curve below on the $r$-$N$ plane for the values $a = b = 1$ and $K = 15$.

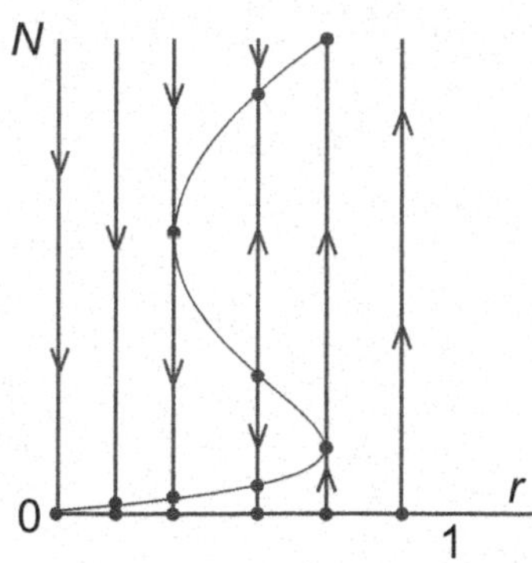


**b.** As $r$ slowly increases, the spruce budworm population slowly increases, then there is a value of $r$ where there is big jump in the population size of the budworm; an outbreak occurs.

**19. a.**

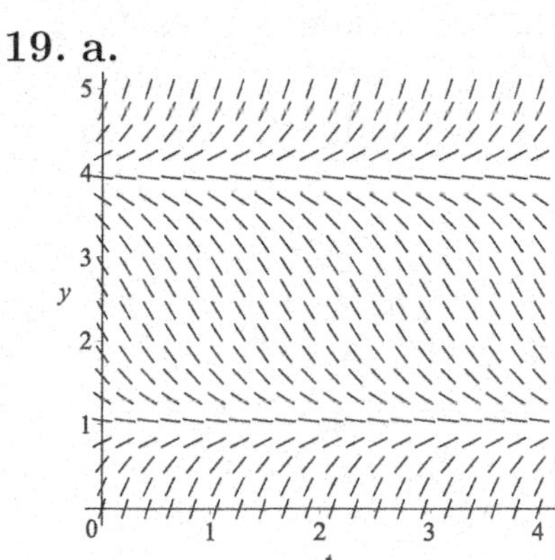


**b.** Euler's method gives $t_0 = 0$, $y_0 = 0$, $t_{n+1} = t_n + 0.5$, $y_{n+1} = y_n + 0.5 \cdot (y_n - 1)(y_n - 4)$. We obtain the values $(0,0)$, $(0.5,2)$, $(1,1)$, $(1.5,1)$.

**c.** The approximation is not reasonable, because the actual solution approaches $y = 1$ in a monotone increasing way, and never actually reaches this value.

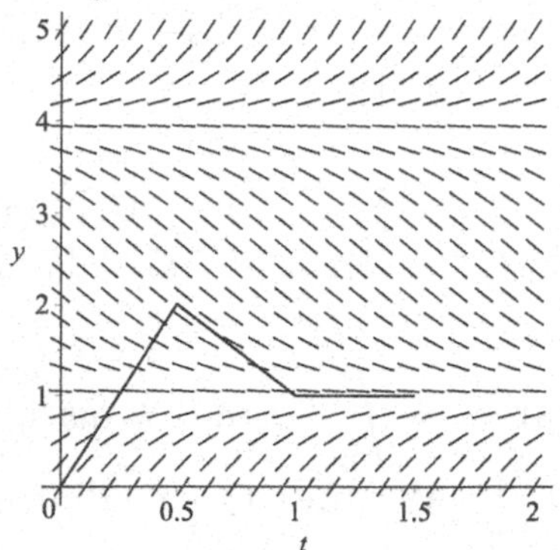


## Problem Set 7.1 - Histograms, PDFs and CDFs

**1.**

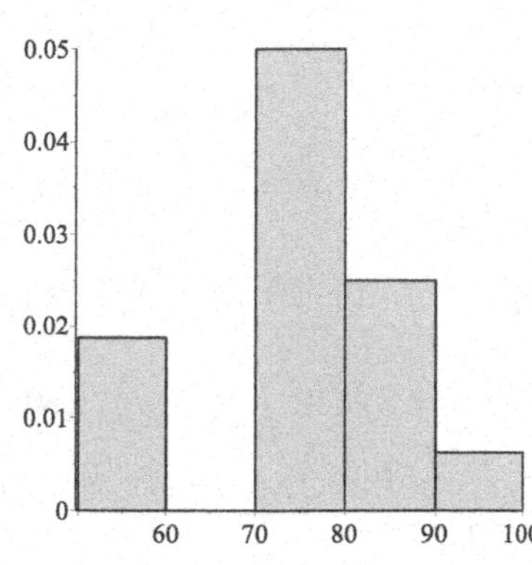

**3.**

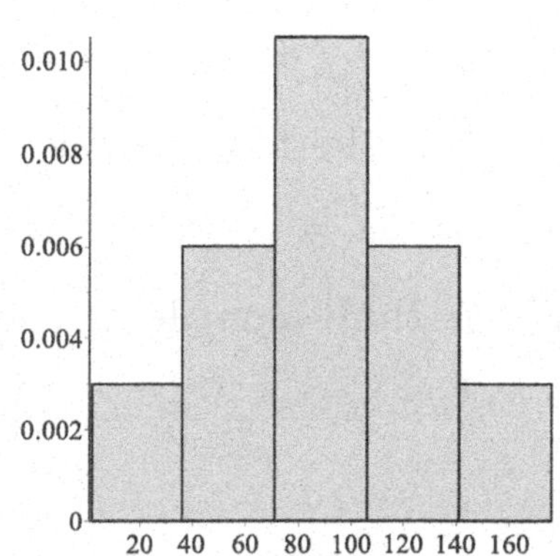

**5. a.** $P(50 \leq X \leq 59) = 3/(3+0+8+4+1) = 3/16$.

**b.** $P(50 \leq X < 69) = (3+0)/(3+0+8+4+1) = 3/16$.

**c.** $P(70 \leq X \leq 89) = (8+4)/(3+0+8+4+1) = 12/16 = 3/4$.

**d.** $P(90 \leq X < 100) = 1/(3+0+8+4+1) = 1/16$.

**7. a.** $P(X < 71) = (10+20)/(10+20+35+20+10) = 30/95 = 6/19$.

**b.** $P(1 \leq X < 141) = (10+20+35+20)/(10+20+35+20+10) = 85/95 = 17/19$.

**9.** We need that $1 = \int_0^2 2ax\,dx = ax^2|_0^2 = 4a$, thus $a = 1/4$. We find the median by solving $1/2 = \int_0^m x/2\,dx = x^2/4|_0^m = m^2/4$, thus $m = \sqrt{2}$; the first quartile by solving $1/4 = \int_0^q x/2\,dx = x^2/4|_0^q = q^2/4$, thus $q = 1$; and the third quartile by solving $3/4 = \int_0^t x/2\,dx = t^2/4|_0^t = t^2/4$, thus $t = \sqrt{3}$.

**11.** We need that $1 = \int_0^1 ax^2\,dx = ax^3/3|_0^1 = a/3$, thus $a = 3$. We find the median by solving $1/2 = \int_0^m 3x^2\,dx = x^3|_0^m = m^3$, thus $m = \sqrt[3]{1/2}$; the first quartile by solving $1/4 = \int_0^q 3x^2\,dx = x^3|_0^q = q^3$, thus $q = \sqrt[3]{1/4}$; and the third quartile by solving $3/4 = \int_0^t 3x^2\,dx = x^3|_0^t = t^3$, thus $t = \sqrt[3]{3/4}$.

**13.** 75th percentile for weight and 50th percentile for height.

**15.** 95th percentile for weight and above the 98th percentile for height.

**17.** $F(x) = \begin{cases} 0 & x \leq 0 \\ x/20 & 0 < x \leq 20 \\ 1 & x > 20 \end{cases}$

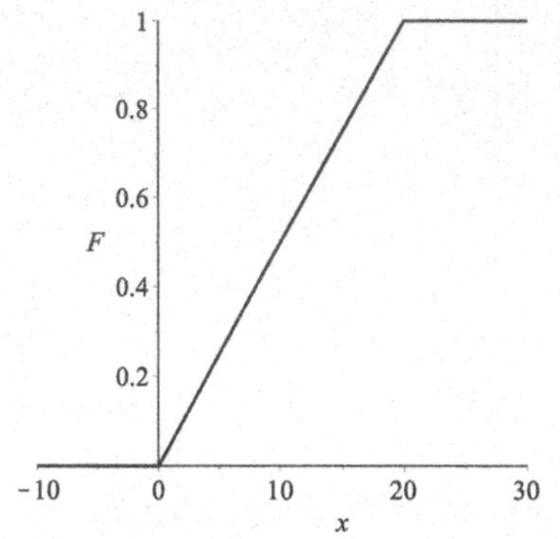


**19. a.** Separating the variables gives $dy/y = -0.25dt$, thus $\ln y = -t/4 + C$ and $y = Ce^{-t/4}$; the initial condition gives $y = y_0e^{-t/4}$.

**b.** $F(t) = 1 - y(t)/y_0 = 1 - e^{-t/4}$ is the exponential CDF with $c = 1/4$.

**c.** $P(0 < X \leq 1) = F(1) - F(0) = 1 - e^{-1/4} - 0 \approx 0.221$.

**21. a.** The area under the function is $1/2 + 1 + 1/2 = 2$, thus for $c = 1/2$ it is the graph of a PDF.

**b.** $P(2 \leq X \leq 3) = (1/2)(1/2) = 1/4$.

**23.** For $x \leq 0$, $F(x) = 0$. For $x > 0$, the CDF can be found by $F(x) = \int_0^x f(t)\,dt$, thus it is $F(x) = \int_0^x t/2\,dt = x^2/4$ for $0 < x < 1$, $F(x) = 1/4 + \int_1^x (1/2)\,dt = x/2 - 1/4$ for $1 < x < 2$, $F(x) = 3/4 + \int_2^x (3-t)/2\,dt = -5/4 + 3x/2 - x^2/4$ for $2 < x < 3$, and $F(x) = 1$ for $x > 3$.

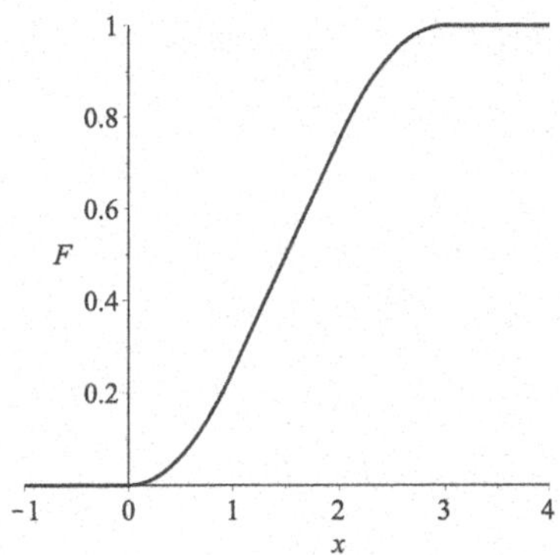


**25. a.** $F$ is non-decreasing, because for $x > 0$, $F' = 1/(1+x)^2 > 0$; $\lim_{x \to -\infty} F(x) = 0$ and $\lim_{x \to \infty} F(x) = 1$; and $F$ is right-continuous, because it is continuous everywhere.

**b.** $P(0 \le X \le 1) = F(1) - F(0) = 1/2$; $P(2 \le X \le 10) = F(10) - F(2) = 10/11 - 2/3 = 8/33$.

**27. a.** $P(60 \le X \le 72) = \int_{60}^{72} f(t)\,dt \approx 0.939$.

**b.** $1 - P(60 \le X \le 84) = 1 - \int_{60}^{84} f(t)\,dt \approx 0.0028$.

**29. a.**

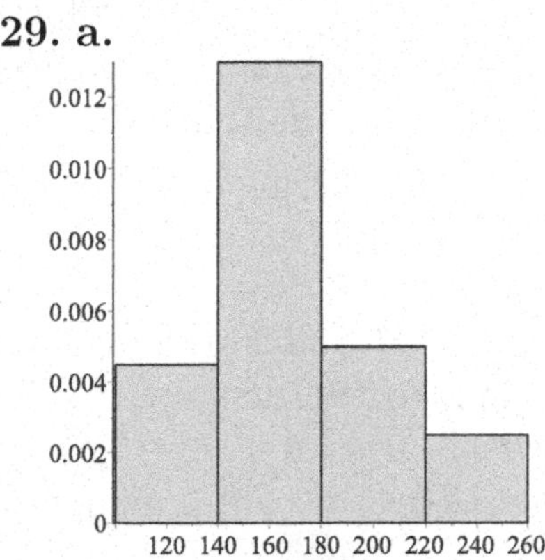


**b.** The probability is $(52+20+10)/(18+52+20+10) = 82/100 = 0.82$.

**c.** The probability is $(18+52+20)/(18+52+20+10) = 90/100 = 0.9$.

**31. a.**

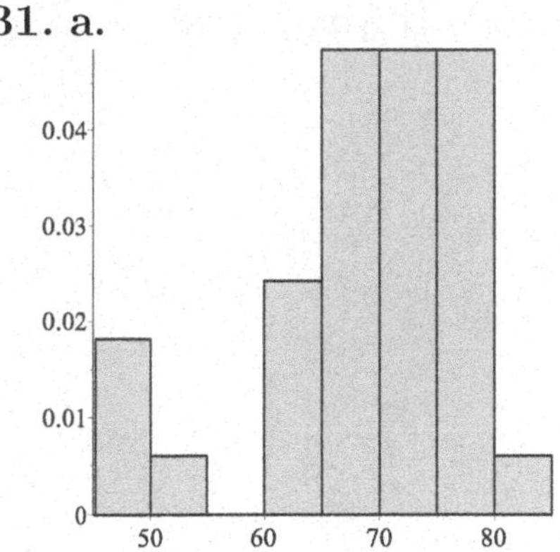


**b.** i. $4/33$ ii. $17/33$

**33. a.** $1 - P(13 \le X \le 14) = 1 - \int_{13}^{14} f(x)\,dx$.

**b.** $P(12 \le X \le 15) - P(13 \le X \le 14) = \int_{12}^{15} f(x)\,dx - \int_{13}^{14} f(x)\,dx$.

**35. a.** $f(x)$ is nonnegative, and we get $\int_0^2 (1 - x/2)\,dx = x - x^2/4|_0^2 = 1$.

**b.** $P(1 \le X \le 2) = \int_1^2 (1 - x/2)\,dx = x - x^2/4|_1^2 = 1 - 3/4 = 1/4$.

**c.** We obtain $P(1/4 \le X \le 1/2) = \int_{1/4}^{1/2} (1 - x/2)\,dx = x - x^2/4|_{1/4}^{1/2} = 13/64$.

**37. a.** Separating the variables gives $dy/y = -cdt$, thus $\ln y = -ct + C$ and $y = Ce^{-ct}$; the initial condition gives $y = y_0 e^{-ct}$.

**b.** $F(t) = (y_0 - y(t))/y_0 = 1 - y(t)/y_0 = 1 - e^{-ct}$.

**c.** $F(t)$ is the exponential CDF.

**d.** $P(r < X \le s) = F(s) - F(r) = 1 - e^{-cs} - (1 - e^{-cr}) = e^{-cr} - e^{-cs}$.

**Problem Set 7.2 - Improper Integrals**

**1.** $\int_4^\infty (1/x^2)dx = \lim_{t\to\infty} \int_4^t (1/x^2)dx = \lim_{t\to\infty} (-1/x|_4^t) = \lim_{t\to\infty} (-1/t + 1/4) = 1/4.$

**3.** We obtain that $\int_{-\infty}^0 1/(1-x)dx = \lim_{t\to-\infty} \int_t^0 1/(1-x)dx = \lim_{t\to-\infty} (-\ln(1-x)|_t^0) = \lim_{t\to-\infty} (\ln(1-t))$ does not exist, thus this is divergent.

**5.** $\int_0^\infty e^x dx = \lim_{t\to\infty} \int_0^t e^x dx = \lim_{t\to\infty} (e^x|_0^t) = \lim_{t\to\infty} (e^t - 1)$ does not exist, thus this is divergent.

**7.** $\int_0^\infty x^2e^{-x}dx = \lim_{t\to\infty} \int_0^t x^2e^{-x}dx = \lim_{t\to\infty} (-(x^2+2x+2)e^{-x}|_0^t) = \lim_{t\to\infty} (-(t^2+2t+2)e^{-t} + 2) = 2.$

**9.** We get $\int_{-\infty}^\infty x^2e^{-x}dx = \int_{-\infty}^0 x^2e^{-x}dx + \int_0^\infty x^2e^{-x}dx,$ and $\int_{-\infty}^0 x^2e^{-x}dx = \lim_{t\to-\infty} \int_t^0 x^2e^{-x}dx = \lim_{t\to-\infty} (-(x^2+2x+2)e^{-x}|_t^0) = \lim_{t\to-\infty} (-2+(t^2+2t+2)e^{-t})$ does not exist, thus this is divergent.

**11.** On $[1,\infty)$, $1/(1+e^x) < e^{-x}$ and we obtain that the integral is convergent.

**13.** On $[1,\infty)$, $\cos^2 x/(1+x^2) < 1/x^2$ and we obtain that the integral is convergent.

**15.** The CDF is $\int_{-\infty}^x e^s/(1+e^s)^2 ds = \lim_{t\to-\infty} \int_t^x e^s/(1+e^s)^2 ds = \lim_{t\to-\infty} (-1/(1+e^s)|_t^x) = \lim_{t\to-\infty} (-1/(1+e^x) + 1/(1+e^t)) = 1 - 1/(1+e^x) = e^x/(1+e^x).$

**17.** For $x < 1$, the CDF is 0; for $x \geq 1$, the CDF is $\int_1^x 1/s^2\, ds = -1/s|_1^x = 1 - 1/x.$

**19.** The PDF is given by $f(x) = F'(x) = e^{-x}/(1+e^{-x})^2.$

**21.** The PDF is $f(x) = F'(x) = e^x$ for $x < 0$ and $f(x) = F'(x) = 0$ for $x \geq 0$.

**23.** $F(x) = 1 - e^{-bx}$, thus the age of death corresponding to the first quartile is at $e^{-bx} = 3/4$, which gives $x = \ln(4/3)/b \approx 0.40$ years; the age of death corresponding to the second quartile is at $e^{-bx} = 1/2$, which gives $x = \ln(2)/b \approx 0.96$ years; and the age of death corresponding to the third quartile is at $e^{-bx} = 1/4$, which gives $x = \ln(4)/b \approx 1.93$ years.

**25.** $F(x) = 1 - e^{-bx}$, thus the age of death corresponding to the first quartile is at $e^{-bx} = 3/4$, which gives $x = \ln(4/3)/b \approx 0.93$ years; the age of death corresponding to the second quartile is at $e^{-bx} = 1/2$, which gives $x = \ln(2)/b \approx 2.24$ years; and the age of death corresponding to the third quartile is at $e^{-bx} = 1/4$, which gives $x = \ln(4)/b \approx 4.47$ years.

**27.** $F(x) = 1 - e^{-bx}$, thus the age of death corresponding to the first quartile is at $e^{-bx} = 3/4$, which gives $x = \ln(4/3)/b \approx 2.88$ years; the age of death corresponding to the second quartile is at $e^{-bx} = 1/2$, which gives $x = \ln(2)/b \approx 6.93$ years; and the age of death corresponding to the third quartile is at $e^{-bx} = 1/4$, which gives $x = \ln(4)/b \approx 13.86$ years.

**29.** Simpson's rule for the integral

$\int_0^4 e^{-x^2}\,dx$ with $n = 8$ gives $(1/6)(e^0 + 4e^{-1/4} + 2e^{-1} + 4e^{-9/4} + 2e^{-4} + 4e^{-25/4} + 2e^{-9} + 4e^{-49/4} + e^{-16}) \approx 0.886196$. Also, $\int_4^\infty e^{-x^2}\,dx \le \int_4^\infty e^{-4x}\,dx = -e^{-4x}/4|_0^\infty = e^{-16}/4 \approx 2.813 \cdot 10^{-8}$.

**31.** By definition, $\int_{-\infty}^\infty f(x)dx$ is convergent if and only if for any $a$ value $\int_{-\infty}^a f(x)dx$ and $\int_a^\infty f(x)dx$ are convergent. The statement of the problem is trivial for $a = 0$. Assume $a < 0$. Using the splitting property of the integral, we obtain that $\int_{-\infty}^0 f(x)dx = \int_{\infty}^a f(x)dx + \int_a^0 f(x)dx$ and $\int_a^\infty f(x)dx = \int_a^0 f(x)dx + \int_0^\infty f(x)dx$. Subtracting the second identity from the first one and rearranging the terms give the result. The proof for $a > 0$ is analogous.

**33. a.** For this Laplace distribution, $b = 2$. The fraction is given by $\int_2^\infty e^{-2x}\,dx = -e^{-2x}/2|_2^\infty = e^{-4}/2 \approx 0.00916$.

**b.** The fraction is given by $\int_{-\infty}^{-2} e^{2x}\,dx = e^{2x}/2|_{-\infty}^{-2} = e^{-4}/2 \approx 0.00916$.

**c.** The fraction is given by $1 - \int_2^\infty e^{-2x}\,dx = 1 + (e^{-2x}/2|_2^\infty) = 1 - e^{-4}/2 \approx 0.99084$.

**35.** Using the Pareto distribution with $p = 2.6$, we obtain that the proportion of trips between 5 and 20 hours is $F(20) - F(5) = 5^{-1.6} - 20^{-1.6} \approx 0.0679$.

**37.** Paint is a material, thus it is composed of atoms. They have a minimum positive size; however, Torricelli's trumpet's diameter will be smaller than that after a while. Thus the paint will get stuck at that point, and we cannot "fill it".

## Problem Set 7.3 - Mean and Variance

**1.** The mean is $\mu = (1{+}1{+}0{+}1{+}1)/5 = 4/5$, the variance is $\sigma^2 = ((1-4/5)^2+(1-4/5)^2+(0-4/5)^2+(1-4/5)^2+(1-4/5)^2)/5 = 4/25$, and the standard deviation is $\sigma = \sqrt{4/25} = 2/5$.

**3.** The mean is $\mu = (1{+}1{+}1{+}1{+}1)/5 = 1$, the variance is $\sigma^2 = 0$, and the standard deviation is $\sigma = 0$.

**5.** The mean is $\mu = (1{+}5{+}7)/3 = 13/3$, the variance is $\sigma^2 = ((1-13/3)^2 + (5-13/3)^2 + (7-13/3)^2)/3 = 56/9$, and the standard deviation is $\sigma = \sqrt{56/9} = \sqrt{56}/3$.

**7.** The mean is $\mu = (2 \cdot 0 + 6 \cdot 1 + 17 \cdot 2 + 8 \cdot 3)/33 = 64/33$, the variance is $\sigma^2 = (2(0 - 64/33)^2 + 6(1 - 64/33)^2 + 17(2 - 64/33)^2 + 8(3-64/33)^2)/33 = 722/1089$, and the standard deviation is $\sigma = \sqrt{722/1089}$.

**9.** The mean is $\mu = \int_0^2 x(1/2)\,dx = x^2/4|_0^2 = 1$.

**11.** The PDF is symmetric on the origin, thus $\mu = 0$.

**13.** The PDF is symmetric on the origin, thus $\mu = 0$.

**15.** The mean is $\mu = \int_0^\infty x^2 e^{-x}\,dx = -(x^2 + 2x + 2)e^{-x}|_0^\infty = 2$.

**17.** The mean is $\mu = 1$ from Problem 9, thus the variance is $\sigma^2 = \int_0^2 (x-1)^2(1/2)\,dx = x^3/6 - x^2/2 + x/2|_0^2 = 1/3$.

**19.** The mean is $\mu = 0$ from Problem 11, thus (because the PDF is an even function) the variance is $\sigma^2 = 2\int_1^\infty x^2((3/2)/x^4)\,dx = -3/x|_1^\infty = 3$.

**21. a.** The mean is $\mu = (-1+7\cdot 0+1)/9 = 0$, the variance is $\sigma^2 = ((-1-0)^2 + 7(0-0)^2 + (1-0)^2)/9 = 2/9$, and the standard deviation is $\sigma = \sqrt{2/9} = \sqrt{2}/3$.

**b.** $1 - 1/2^2 = 3/4$ of the data has to lie on the interval $[-2\sqrt{2}/3, 2\sqrt{2}/3]$; the actual fraction is 7/9.

**23. a.** We obtain that $P(0.2 \le X \le 0.6) = \int_{0.2}^{0.6} (1/2+x)\,dx = (1/2)x + x^2/2|_{0.2}^{0.6} = 0.36$.

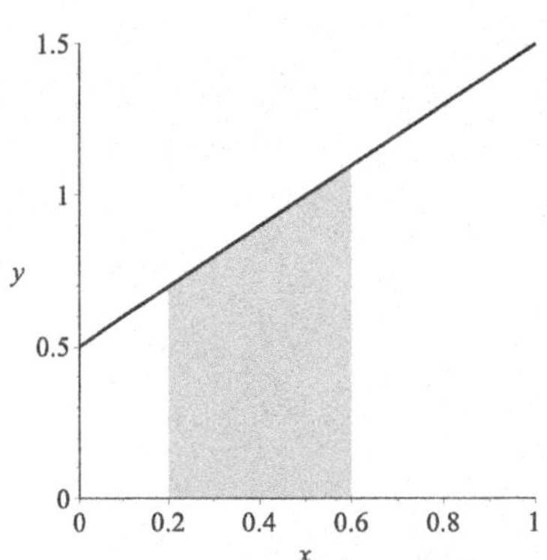


**b.** For $x \le 0$, $F(x) = 0$. For $x > 0$, the CDF can be found by $F(x) = \int_0^x f(t)\,dt$, thus it is $F(x) = \int_0^x 1/2 + t\,dt = x^2/2 + x/2$ for $0 < x < 1$, and $F(x) = 1$ for $x > 1$.

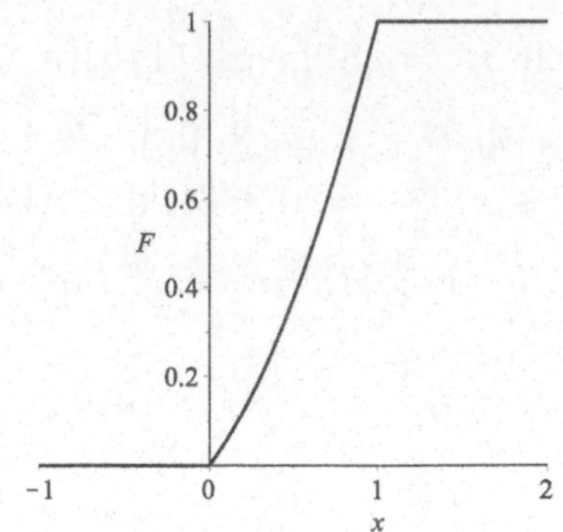


**c.** The mean is $\mu = \int_0^1 x(1/2+x)\,dx = (x^2/4 + x^3/3)|_0^1 = 7/12$.

**25. a.** The area of the center region with radius $x$ is $x^2\pi$, and the area of the whole board is $4\pi$. Thus the probability that the dart lands in the inside region with radius $x$ is $x^2/4$, and then the PDF is $(x^2/4)' = x/2$. (As long as $0 \le x \le 2$.)

**b.** The mean is $\mu = \int_0^2 x(x/2)\,dx = x^3/6|_0^2 = 4/3$, and then we obtain that $\sigma^2 = \int_0^2 (x - 4/3)^2(x/2)\,dx = (x^4/8 - 4x^3/9 + 4x^2/9)|_0^2 = 2/9$.

**27. a.** The mean is $\mu = (223+176+548)/6 = 947/6 \approx 157.8$.

**b.** The variance is $\sigma^2 = (3(0-947/6)^2 + (223-947/6)^2 + (176-947/6)^2 + (548-947/6)^2)/6$, and the standard deviation is $\sigma \approx 196.4$.

**29.** The mean is $\mu = \int_0^\infty tce^{-ct}\,dt = -e^{-ct}(1/c+t)|_0^\infty = 1/c$, the variance is $\sigma^2 = \int_0^\infty t^2ce^{-ct}\,dt - \mu^2 = -e^{-ct}(2/c^2 + 2t/c + t^2)|_0^\infty - 1/c^2 = 1/c^2$, and the standard deviation is $\sigma = 1/c$, so $\sigma = \mu$.

**31. a.** For mussel and clam species, the extinction rate is $20/297 \approx 0.0673$ over hundred years, thus $1 - e^{-100c} = 0.0673$, and then $c \approx 0.0006967$, and $\mu = 1/c \approx 1435$ years. For fish species, the extinction rate is $40/950 \approx 0.0421$ over hundred years, thus $1 - e^{-100c} = 0.0421$, and then $c \approx 0.0004301$, and $\mu = 1/c \approx 2325$ years.

**b.** We obtain $1 - e^{-100c} \approx 0.067$ for mussel and clam species, and $1 - e^{-100c} \approx 0.042$ for fish species.

**33. a.** We have $0.85 = P(X \geq 5) = 1 - (1 - e^{-5c})$, thus $c = -\ln(0.85)/5$. Then $\mu = -5/\ln(0.85) \approx 30.77$ years.

**b.** We obtain that $P(X \geq 10) = 1 - (1 - e^{-10c}) = e^{-10c} = 0.85^2 \approx 0.72$.

**35. a.** The mean is given by $7.5 \cdot 0.35 + 22.5 \cdot 0.25 + 37.5 \cdot 0.16 + 52.5 \cdot 0.12 + 67.5 \cdot 0.08 + 82.5 \cdot 0.02 + 142.5 \cdot 0.02 = 30.45$ minutes.

**b.** If $\mu = 1/c = 33.26$, then $P(X \leq 30) = 1 - e^{-30c} \approx 0.594$. According to the histogram, $P(X \leq 30) = 0.35 + 0.25 = 0.6$.

**c.** If $\mu = 1/c = 33.26$, then $P(X > 75) = e^{-75c} \approx 0.1049$. According to the histogram, $P(X > 80) = 0.02 + 0.02 = 0.04$.

**37.** The CDF of $Y = X - a$ is given by $G(x) = P(Y \leq x) = P(X - a \leq x) = P(X \leq x + a) = F(x + a)$, thus $g(x) = G'(x) = (F(x+a))' = F'(x+a) = f(x+a)$.

**Problem Set 7.4 - Bell-Shaped Distributions**

**1.** Using Table 7.3, $P(0 \leq X < 0.85) = 0.3023$.

**3.** Using Table 7.3, $P(X \geq 0.55) = 0.5 - 0.2088 = 0.2912$.

**5.** The first value is $0.4713 - 0.3849 = 0.0864$; the value at $z = 0.7$ is $0.2580$.

**7.** $P(X > 0) = P((X-1)/2 \geq -1/2) = P(Z \geq -1/2) = 1 - (0.5 - 0.1915) = 0.6915$.

**9.** $P(e^{-3} < X < e^{-1}) = P(-3 < \ln X < -1) = P(-1/2 < (\ln X + 2)/2 < 1/2) = P(-1/2 < Z < 1/2) = 2 \cdot 0.1915 = 0.3830$.

**11.** $P(0 < X < 0.5) = P(-\infty < \ln X < -\ln 2) = P(-\infty < Z < -\ln 2) = 0.2441$.

**13. a.** $y(0) = 1/(1 + e^4) \approx 0.0180$.

**b.** $y(1) = 1/(1 + e^{2.3}) \approx 0.0911$.

**c.** We have to solve $0.95 = 1/(1 + e^{4-1.7t})$; we obtain $t \approx 4.085$, i.e. about 408.5 days.

**15. a.** Separating the variables, we get $dy/(y(1-y)) = (1/y - 1/(y-1))dy = r dt$, thus $\ln y - \ln|y-1| = \ln(y/(1-y)) = rt + C$. Then $y/(1-y) = Ce^{rt}$, and $y = Ce^{rt}/(1 + Ce^{rt}) = 1/(1 + Ce^{-rt})$. Now $y_0 = y(0) = 1/(1+C)$, thus $C = 1/y_0 - 1$ and then $y = y_0/(y_0 + (1 - y_0)e^{-rt})$. For the given values $r = 0.1$ and $y_0 = 0.5$, we get $y = 0.5/(0.5 + 0.5e^{-t/10}) = 1/(1 + e^{-t/10})$.

**b.** $y(t)$ is monotone increasing, the limit at $-\infty$ is 0 and at $\infty$ is 1, and $y(t)$ is continuous; thus it is a CDF.

**c.** $y(2) - 0.5 \approx 0.0498$.

**17. a.** Separating the variables, we get $dy/(y(1-y)) = (1/y - 1/(y-1))dy = r dt$, thus $\ln y - \ln|y-1| = \ln(y/(1-y)) = rt + C$. Then $y/(1-y) = Ce^{rt}$, and $y = Ce^{rt}/(1 + Ce^{rt}) = 1/(1 + Ce^{-rt})$. Now $y_0 = y(0) = 1/(1+C)$, thus $C = 1/y_0 - 1$

and then $y = y_0/(y_0 + (1 - y_0)e^{-rt})$. For the given values $r = 1$ and $y_0 = 0.1$, we get $y = 0.1/(0.1 + 0.9e^{-t}) = 1/(1 + 9e^{-t})$.

**b.** $y(t)$ is monotone increasing, the limit at $-\infty$ is 0 and at $\infty$ is 1, and $y(t)$ is continuous; thus it is a CDF.

**c.** $y(1) - 0.1 \approx 0.1320$.

**19.** Using the logistic transformation $y = \ln(p/(1-p))$ and then linear regression, we obtain $y = 0.6707t - 2.8907$.

**21.** Using the logistic transformation $y = \ln(p/(1-p))$ and then linear regression, we obtain $y = 0.3605t - 5.4381$.

**23. a.** $P(X \le 128) = P((X - 120)/18 \le (128 - 120)/18) = P(Z \le 4/9) \approx 0.6716$.

**b.** $P(96 \le X \le 128) = P((96 - 120)/18 \le (X - 120)/18 \le (128 - 120)/18) = P(-4/3 \le Z \le 4/9) \approx 0.5804$.

**c.** $P(X \ge 144) = P((X - 120)/18 \ge (144 - 120)/18) = P(Z \ge 4/3) \approx 0.0912$.

**25. a.** $P(263 \le X \le 295) = P((263 - 279)/16 \le (X - 279)/16 \le (295 - 279)/16) = P(-1 \le Z \le 1) \approx 0.6827$.

**b.** $P(X \ge 303) = P((X - 279)/16 \ge (303 - 279)/16) = P(Z \ge 3/2) \approx 0.0668$.

**27. a.** The fraction in question is given by $P(X \le 3) = P(\ln X \le \ln 3) = P((\ln X - \ln 2.4)/\ln 1.47 \le (\ln 3 - \ln 2.4)/\ln 1.47) = P(Z \le (\ln 3 - \ln 2.4)/\ln 1.47) \approx 0.7188$.

**b.** We obtain that $P(X \ge 4) = P(\ln X \ge \ln 4) = P((\ln X - \ln 2.4)/\ln 1.47 \ge (\ln 4 - \ln 2.4)/\ln 1.47) = P(Z \ge (\ln 4 - \ln 2.4)/\ln 1.47) \approx 0.0924$.

**c.** $P(1 \le X \le 3) = P(\ln 1 \le \ln X \le \ln 3) = P((\ln 1 - \ln 2.4)/\ln 1.47 \le (\ln X - \ln 2.4)/\ln 1.47 \le (\ln 3 - \ln 2.4)/\ln 1.47) = P((\ln 1 - \ln 2.4)/\ln 1.47 \le Z \le (\ln 3 - \ln 2.4)/\ln 1.47) \approx 0.7072$.

**29. a.** We get $P(X \le 12) = P(\ln X \le \ln 12) = P((\ln X - \ln 17.2)/\ln 3.21 \le (\ln 12 - \ln 17.2)/\ln 3.21) = P(Z \le (\ln 12 - \ln 17.2)/\ln 3.21) \approx 0.3787$.

**b.** We get $P(X \ge 24) = P(\ln X \ge \ln 24) = P((\ln X - \ln 17.2)/\ln 3.21 \ge (\ln 24 - \ln 17.2)/\ln 3.21) = P(Z \ge (\ln 24 - \ln 17.2)/\ln 3.21) \approx 0.3876$.

**c.** We get $P(12 \le X \le 18) = P(\ln 12 \le \ln X \le \ln 18) = P((\ln 12 - \ln 17.2)/\ln 3.21 \le (\ln X - \ln 17.2)/\ln 3.21 \le (\ln 18 - \ln 17.2)/\ln 3.21) = P((\ln 12 - \ln 17.2)/\ln 3.21 \le Z \le (\ln 18 - \ln 17.2)/\ln 3.21) \approx 0.1369$.

**31.** Using technology, $m \approx 8.42$ and $v \approx 25.32$. Then $\mu = 2\ln m - (1/2)\ln(m^2 + v) \approx 1.98$ and $\sigma^2 = -2\ln m + \ln(m^2 + v) \approx 0.305$.

**33. a.** Separating the variables, we get $dy/(y \ln y) = -dt$, thus $\ln(\ln y) = -t + C$. Then $y = e^{Ce^{-t}}$. Now $1/e = y(0) = e^C$, thus $C = -1$ and then $y = e^{-e^{-t}}$.

**b.** $y(t)$ is monotone increasing, the limit at $-\infty$ is 0 and at $\infty$ is 1, and $y(t)$ is continuous; thus it is a CDF.

**c.** The PDF is $y'(t) = e^{-e^{-t}}e^{-t}$.

**d.** $y(2) = e^{-e^{-2}} \approx 0.873$.

**35.** The location of the maximum and inflection points does not depend on constant multipliers, thus we can consider the function $f(x) = e^{-(x-\mu)^2/2\sigma^2}$. Now $f'(x) = f(x)(-2(x - \mu)/2\sigma^2) = -f(x)((x - \mu)/\sigma^2)$.

$f(x) \neq 0$, thus the only critical point is at $x = \mu$, and the sign of the derivative changes from positive to negative there, thus it is the only maximum. From the above equation $f''(x) = -f'(x)((x-\mu)/\sigma^2) - f(x)/\sigma^2 = f(x)((x-\mu)/\sigma^2)((x-\mu)/\sigma^2) - f(x)/\sigma^2 = f(x)((x-\mu)^2/\sigma^4 - 1/\sigma^2)$ and the second derivative is zero when $(x-\mu)^2 = \sigma^2$, i.e. when $x = \mu \pm \sigma$. We can also check from the form of the second derivative that the sign changes sign there, thus these are inflection points.

**37.** Using the substitution $u = \ln x$, $du = (1/x)dx$, and $dx = xdu = e^u du$, thus the mean is $m = \int_0^\infty xf(x)dx = \int_0^\infty e^{-(\ln x-\mu)^2/2\sigma^2}/(\sqrt{2\pi}\sigma)dx =$

$= \int_{-\infty}^\infty e^{-(u-\mu)^2/2\sigma^2+u}/(\sqrt{2\pi}\sigma)du$; now let $y = (u-\mu)/\sigma$, then $dy = (1/\sigma)du$ and then completing the square in the exponential we get that $m = \int_{-\infty}^\infty e^{-y^2/2+\sigma y+\mu}/(\sqrt{2\pi})dy = e^{\mu+\sigma^2/2}\int_{-\infty}^\infty e^{-(y-\sigma)^2/2}/(\sqrt{2\pi})dy$, but the last integral is just 1 because it is the integral of the PDF of the normal distribution with mean $\sigma$ and variance 1. Now we compute $\int_0^\infty x^2 f(x)dx$ similarly; we get that $v + m^2 = \int_0^\infty x^2 f(x)dx = \int_0^\infty xe^{-(\ln x-\mu)^2/2\sigma^2}/(\sqrt{2\pi}\sigma)dx =$

$= \int_{-\infty}^\infty e^{-(u-\mu)^2/2\sigma^2+2u}/(\sqrt{2\pi}\sigma)du = \int_{-\infty}^\infty e^{-y^2/2+2\sigma y+2\mu}/(\sqrt{2\pi})dy =$

$= e^{2\mu+2\sigma^2}\int_{-\infty}^\infty e^{-(y-2\sigma)^2/2}/(\sqrt{2\pi})dy = e^{2\mu+2\sigma^2}$, because the last integral is the integral of the PDF of the normal distribution with mean $2\sigma$ and variance 1. Then $v = e^{2\mu+2\sigma^2} - m^2 = e^{2\mu+2\sigma^2} - e^{2\mu+\sigma^2} = (e^{\sigma^2}-1)e^{2\mu+\sigma^2}$ as claimed.

## Problem Set 7.5 - Life Tables

**1.** From Table 7.6, $l(14) - l(20) = 0.28$.

**3.** From Table 7.6, $l(6) = 0.54$.

**5.** From Table 7.7, $1 - l(24) = 0.41$.

**7.** From Table 7.7, $l(24) - l(48) = 0.41$.

**9.** We obtain $l_W(75) \approx 0.724$.

**11.** We obtain $l_W(25) - l_W(75) \approx 0.263$.

**13.** $-(\ln l(t))' = -l'(t)/l(t) = m(t) = a + bt$, thus $-\ln l(t) = at + bt^2/2 + c$, and then $l(t) = Ce^{-at-bt^2/2}$; also, $l(0) = 1$, thus $C = 1$ and $l(t) = e^{-at-bt^2/2}$.

**15.** The mortality rates are given by
$(1/8)(0.83 - 0.73)/0.83 \approx 0.015$,
$(1/8)(0.73 - 0.59)/0.73 \approx 0.024$,
$(1/8)(0.59 - 0.43)/0.59 \approx 0.034$,
$(1/8)(0.43 - 0.29)/0.43 \approx 0.041$,
$(1/8)(0.29 - 0.18)/0.29 \approx 0.047$,
$(1/8)(0.18 - 0.10)/0.18 \approx 0.056$,
$(1/8)(0.10 - 0.05)/0.10 \approx 0.063$,
$(1/8)(0.05 - 0.03)/0.05 \approx 0.050$.

**17.** The life expectancy is $\int_0^\infty e^{-t}\,dt = -e^{-t}|_0^\infty = 1$.

**19.** The life expectancy is $\int_0^\infty 1/(1+t)^2\,dt = -1/(1+t)|_0^\infty = 1$.

**21.** The life expectancy is $\int_0^\infty 1/(1+t)\,dt = -\ln(1+t)|_0^\infty = \infty$.

**23.** $R_0 = \int_1^\infty 2e^{-t}\,dt = -2e^{-t}|_1^\infty = 2.$

**25.** $R_0 = \int_5^\infty 5/(1+t)^2\,dt = -5/(1+t)|_5^\infty = 5/6.$

**27.** $R_0 = \int_0^\infty 5/(1+t)^2\,dt = -5/(1+t)|_0^\infty = 5.$

**29.** Using the chain rule, $-d\ln[l(t)]/dt = -(1/l(t))l'(t) = -l'(t)/l(t) = m(t)$.

**31.** The mortality function is given by $m(t) = -l'(t)/l(t) = 0.039 \cdot 0.187e^{0.187t}$.

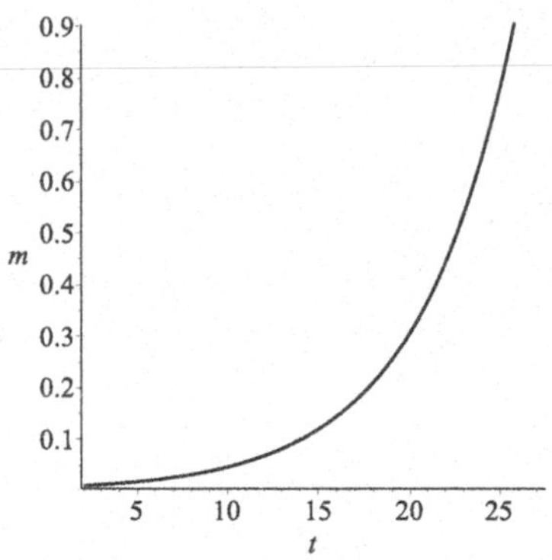

**33.** We have $-(\ln l(t))' = -l'(t)/l(t) = m(t) = 0.0059e^{0.2072t}$, thus $-\ln l(t) = (0.0059/0.2072)e^{0.2072t} + c$, and then $l(t) = Ce^{-(0.0059/0.2072)e^{0.2072t}}$; also, $l(0) = 1$, thus $C = e^{0.0059/0.2072}$ and $l(t) = e^{(0.0059/0.2072)(1-e^{0.2072t})}$.

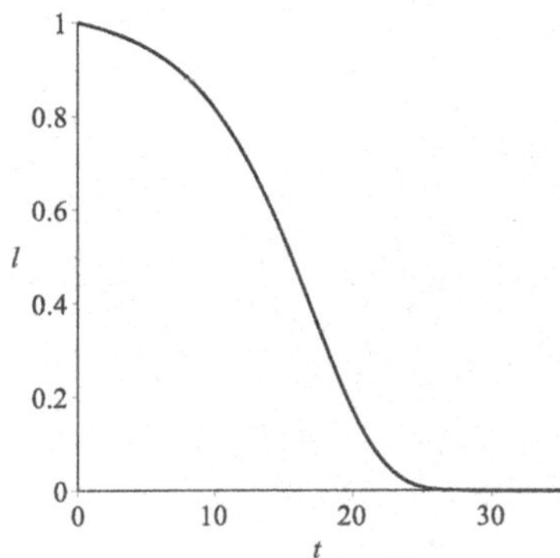

**35.** We obtain that $m(t) = -l'(t)/l(t) = 0.009 \cdot 0.2214e^{0.2214t}$.

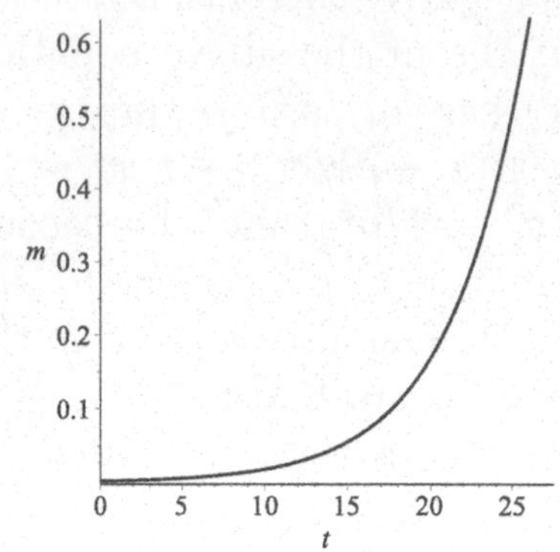

**37. a.** $R_0 = 0.2+0.98\cdot0.3+0.86\cdot0.33+0.74\cdot0.38+0.62\cdot0.43+0.52\cdot0.5+0.42\cdot0.6+0.4\cdot0.6+0.26\cdot0.6+0.25\cdot0.6+0.12\cdot0.6 = 2.4556 > 1$, thus the answer is yes.

**b.** We need $R_0(1-y) < 1$, thus when $y > 1 - 1/R_0 \approx 0.59$, then the disease will not spread.

**39. a.** For the untreated population, $R_0 = 0.5+0.2+0.2+0.2+0.95\cdot0.2+0.9\cdot0.2+0.8\cdot0.2+0.65\cdot0.1+0.45\cdot0.1+0.2\cdot0.1 = 1.76$; the analogous value for the treated population is $\sum_{i=1}^{20} l(i)b(i) = 2.02$. Thus the treatment exacerbates the epidemic.

**b.** We need $R_0(1-y) < 1$, i.e. $y > 1 - 1/R_0$. For the untreated population, we obtain $y \approx 0.43$, for the treated population, $y \approx 0.51$.

## Review Questions

**1. a.**

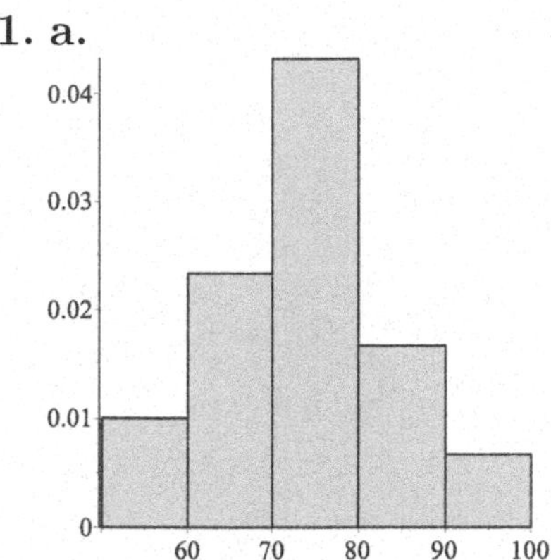


**b.** $P(0 \le X \le 89) = (6+14+26+10)/(6+14+26+10+4) = 14/15$.

**c.** $P(X > 79) = (10+4)/(6+14+26+10+4) = 7/30$.

**3. a.** $F' = k/(k+x)^2 > 0$, thus $F$ is nondecreasing, the limit of the function at $-\infty$ is 0 and at $\infty$ is 1, and it is continuous on $(-\infty, \infty)$.

**b.** $P(1 \le X \le 2) = F(2) - F(1) = 2/(k+2) - 1/(k+1) = k/(2+3k+k^2)$.

**5.** Using the substitution $u = \ln x$, we obtain that $\int_2^\infty (1/(x(\ln x)^p))\,dx = \int_{\ln 2}^\infty 1/u^p\,du$, thus we need that $p > 1$.

**7.** $\int_1^2 2/x^2\,dx = -2/x|_1^2 = 1$. The CDF is 0 if $x < 1$ and 1 if $x > 2$; if $1 \le x \le 2$, $F(x) = \int_1^x 2/t^2 dt = -2/t|_1^x = 2 - 2/x$.

**9.** The mean is $\int_1^\infty x(4/x^5)dx = -(4/3)x^{-3}|_1^\infty = 4/3$; the variance is $\sigma^2 = \int_1^\infty x^2(4/x^5)dx - \mu^2 = -2x^{-2}|_1^\infty - \mu^2 = 2 - (4/3)^2 = 2/9$.

**11.** $P(\mu - k\sigma \le X \le \mu + k\sigma) \ge 1 - 1/k^2$, where $\mu = 68164$, $\sigma = 17408$ and $k = (94276 - 68164)/17408 = (68164 - 42052)/17408 = 3/2$ gives that a lower bound for the fraction is $1 - 1/(3/2)^2 = 5/9$.

**13.** Assuming that at $t = 12$, $l(t) = 0$, and using Simpson's rule, the life expectancy can be approximated by $1(1+4\cdot 0.845+2\cdot 0.824+4\cdot 0.795+2\cdot 0.755+4\cdot 0.699+2\cdot 0.626+4\cdot 0.532+2\cdot 0.418+4\cdot 0.289+2\cdot 0.162+4\cdot 0.060+0.000) \approx 6.48$ years.

**15. a.** We obtain $l_W(40) \approx 0.757$.

**b.** The mortality rate is $m = -l_W'/l_W = 0.0507e^{-3.307+0.0507t}$, thus $m(40) = 0.014$ per year.

**17.** The PDF is $-l'(t)$, the CDF is $1 - l(t)$.

**19.** $P(X \ge 29) = P((X - 21)/64 \ge (29 - 21)/8) = P(Z \ge 1) \approx 0.1587$.

## Problem Set 8.1 - Multivariate Modeling

**1.** $z = 2(1/2) - 3(1/3) = 0$, the domain is all $(x, y)$ pairs of real numbers, and the range is all real numbers.

**3.** $z = \sqrt{1^2 + 2(-1)^2} = \sqrt{3}$, the domain is all $(x, y)$ pairs of real numbers, and the range is all nonnegative real numbers.

**5.** $z = \sqrt{1+2(-1/2)}/1 = 0$, the domain is all those $(x, y)$ pairs of real numbers such that $y \geq -x/2$ and $x \neq 0$, and the range is all real numbers.

**7.** $z = \ln(1) - (1/2)^2 = -1/4$, the domain is all $(x, y)$ pairs of real numbers such that $y > 0$, and the range is all real numbers.

**9.** $z = x - 0.3x = 0.7x$. The prey growth rate increases linearly as the prey density increases.

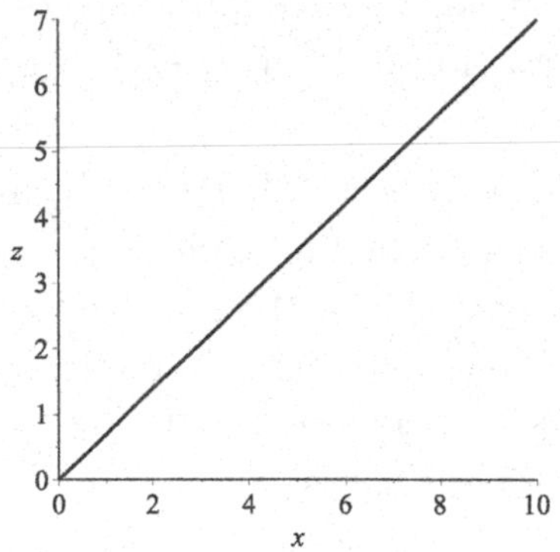


**11.** $z = 2.4 - y$. The prey growth rate decreases linearly as the predator density increases.

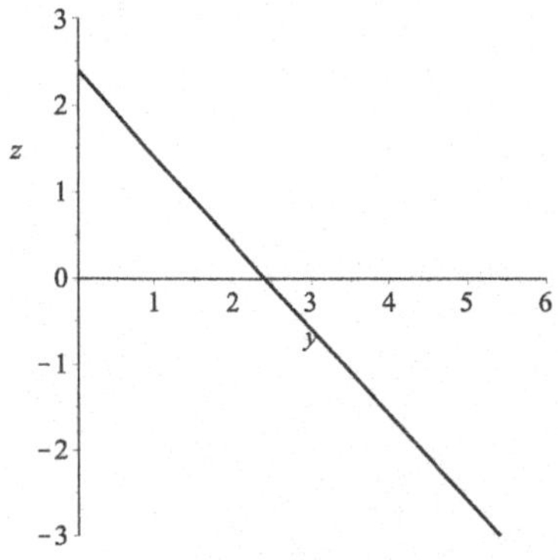


**13.** $z = 3/(1 + y)$. The rate of change of protein A decreases as the concentration of protein B increases.

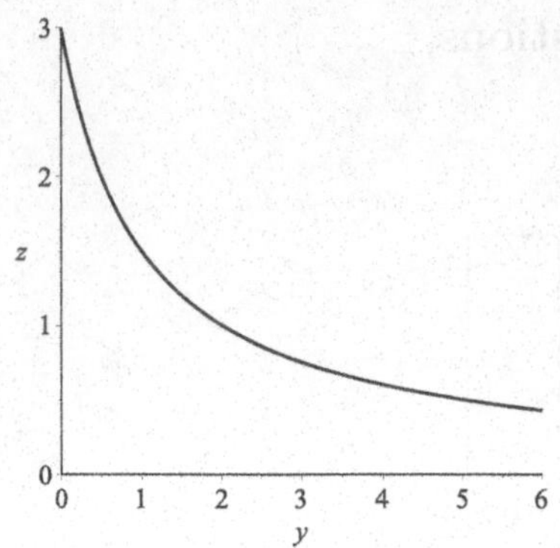


**15.**

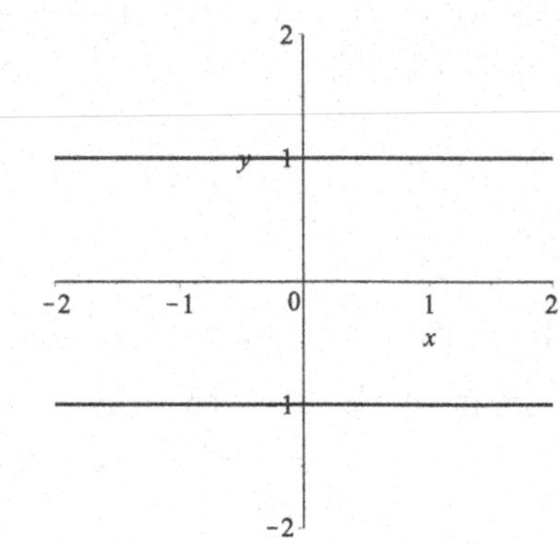


**17.**

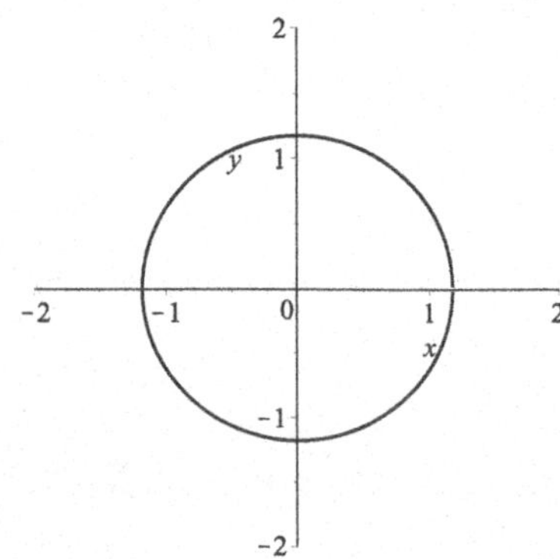


**19.** The contours correspond to $z = 1/2$; the function never takes the value $z = 2$.

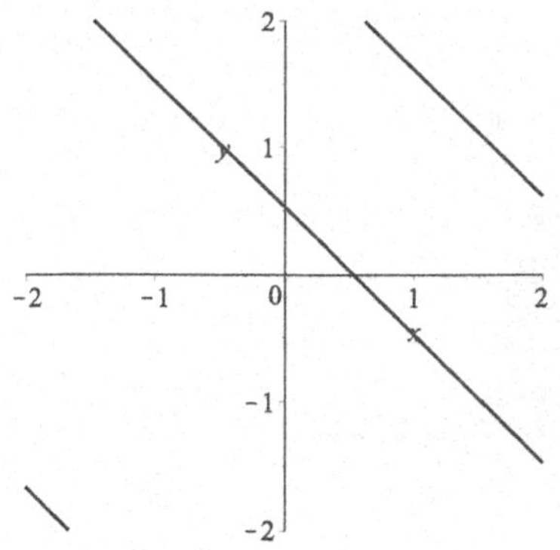


**21.** The surface has a minimum.

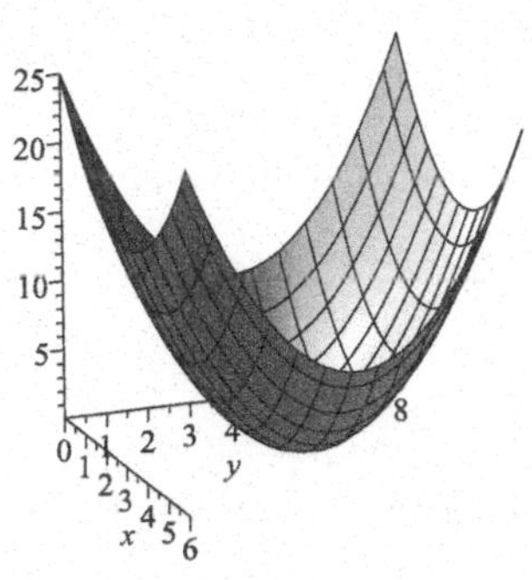


**23.** The surface has a maximum.

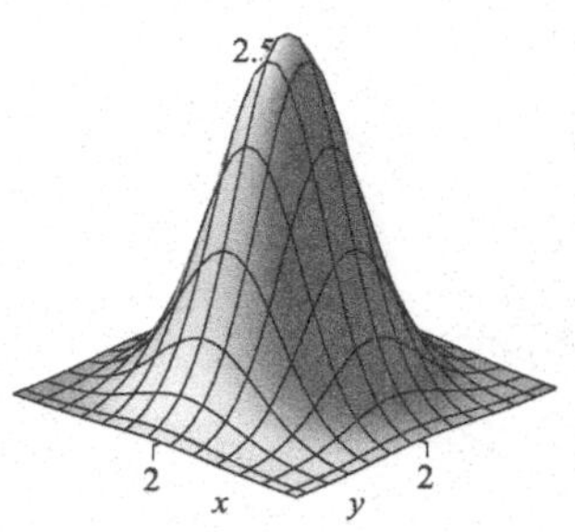


**25.** The surface is between $z = -2$ and $z = 2$, periodic.

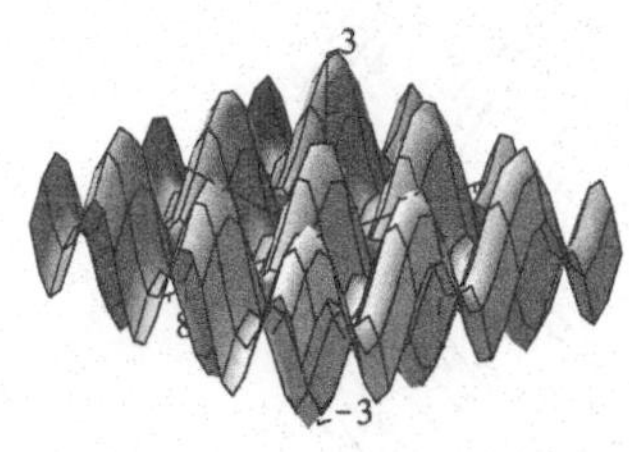


**27.** The surface is saddle-shaped.

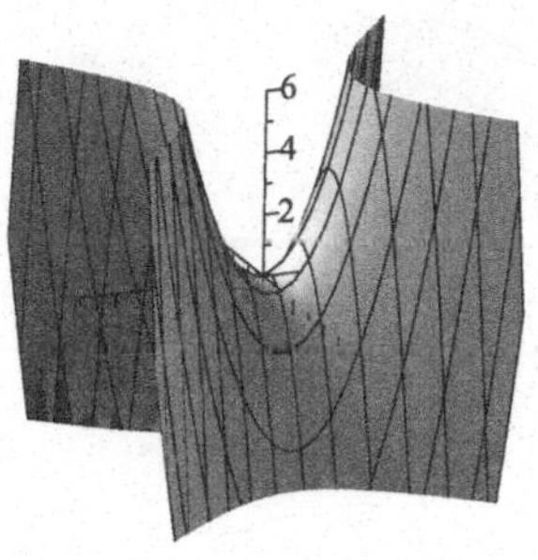


**29.** Maximum at the point $(0, 0)$. (Also see Problem 23.)

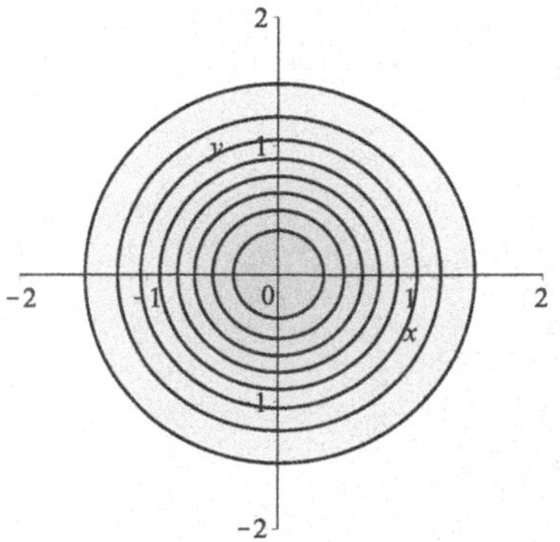


**31.** Minimum at $(0, 0)$; minimum values on white circles, maximum values on dark circles. (Also see Problem 28.)

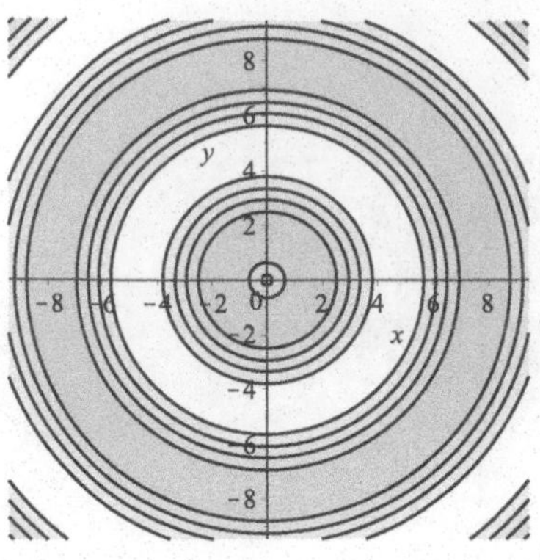


**33.** Each species has a higher growth rate when its competitor is at low densities.

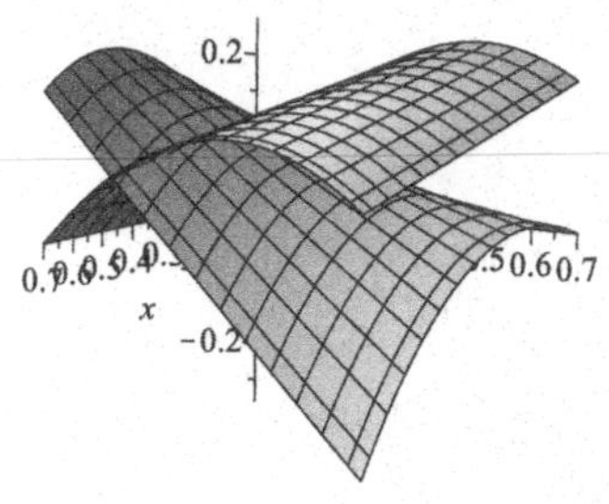


**35.** $f(x, y) = cx/(k + ax + by)$ satisfies all three properties.

**37.** The graph shows that increasing $N$ decreases the per-capita rate of resource extraction, for larger $d$ values this decrease is faster.

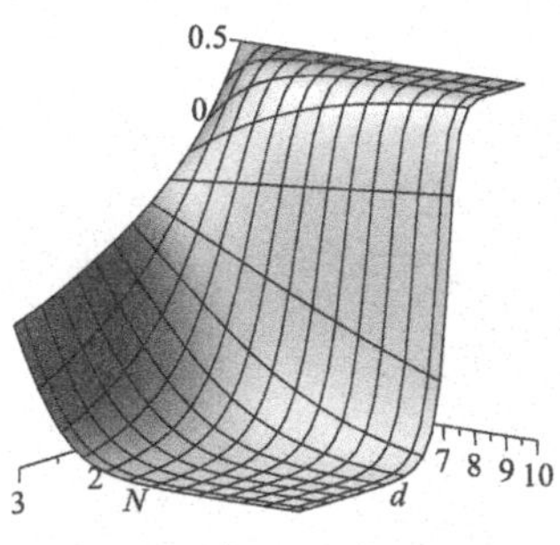


**39.** $c = 0.5$:

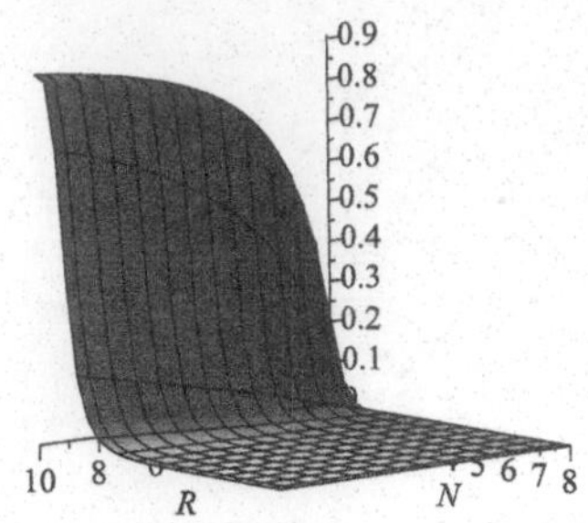


$c = 1$:

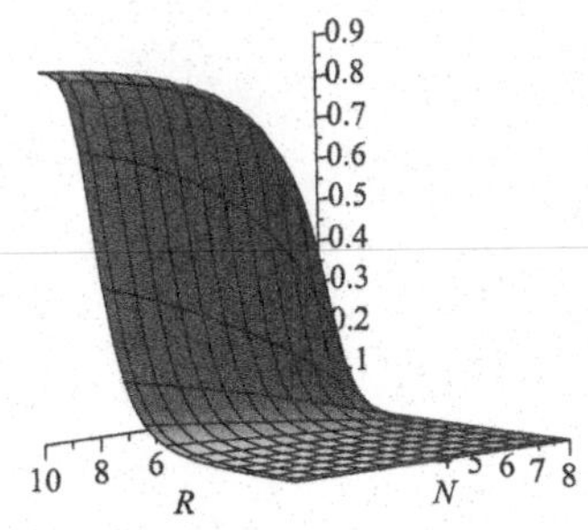


The parameter $c$ acts as a scaling factor in the $N$-direction.

**41. a.** We obtain $f(x, y) = a/(1 + y^c) - x$.

**b.**

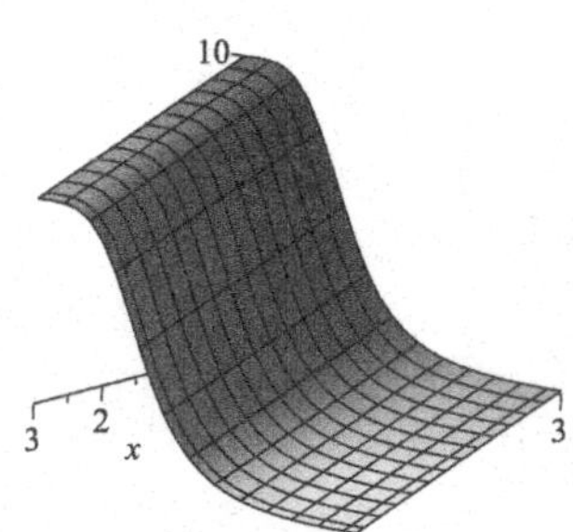


**43. a.** Using technology, the best fit line is $y = 0.476x - 57.377$.

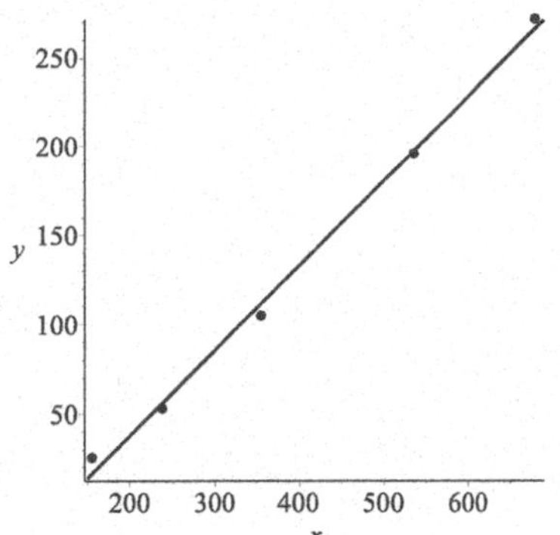


**b.** We obtain that $S(m,c) = (25.1 - 156.1m - c)^2 + (52.5 - 238.3m - c)^2 + (104.5 - 355.2m - c)^2 + (195.6 - 535.7m - c)^2 + (271.6 - 680.6m - c)^2 = 126332 - 1298.6c + 5c^2 - 686362m + 3931.8cm + 957512m^2$.

**c.**

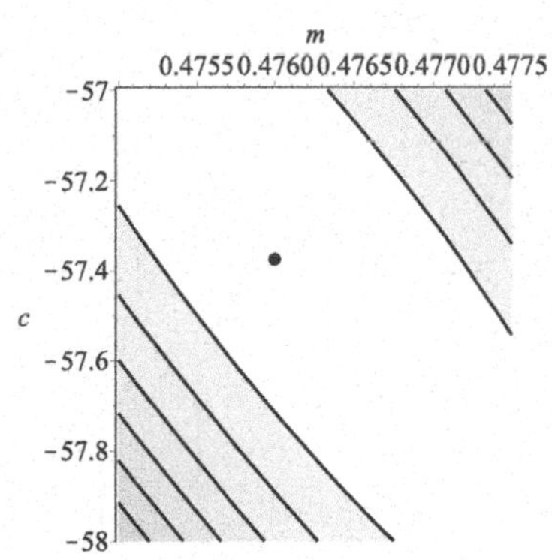


**45.**

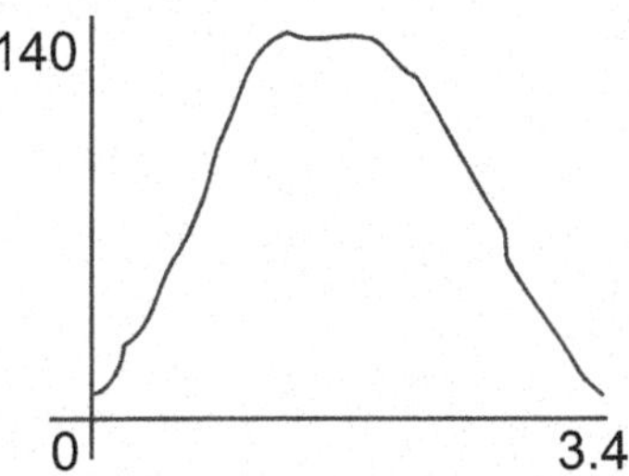


**Problem Set 8.2 - Matrices and Vectors**

**1.** $x = 2$, $y = 4$.

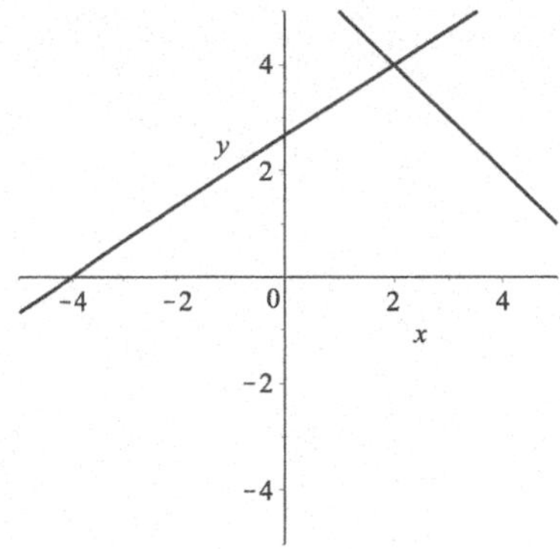


**3.** The two lines are the same, thus we have infinitely many solutions.

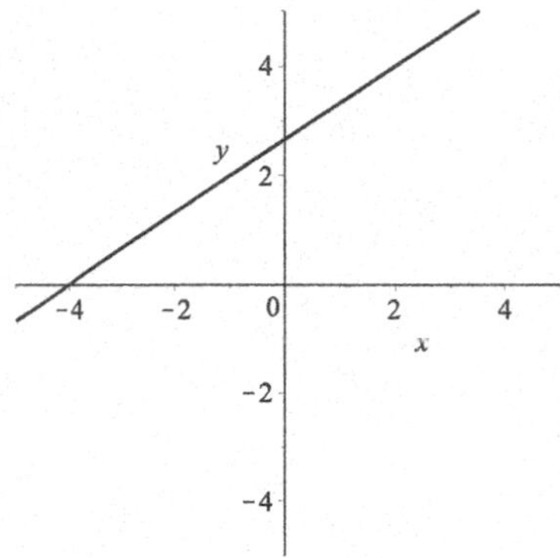


**5.** $x = -6/5$, $y = -23/5$.

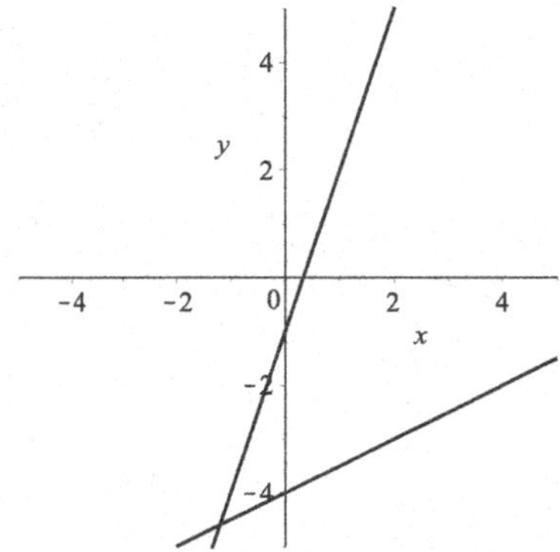


**7.** $\begin{bmatrix} 1 & 1 \\ 1 & -1 \end{bmatrix} \begin{bmatrix} x \\ y \end{bmatrix} = \begin{bmatrix} 2 \\ 0 \end{bmatrix}$; the inverse is $A^{-1} = \begin{bmatrix} 1/2 & 1/2 \\ 1/2 & -1/2 \end{bmatrix}$ and then $\begin{bmatrix} x \\ y \end{bmatrix} = \begin{bmatrix} 1/2 & 1/2 \\ 1/2 & -1/2 \end{bmatrix} \begin{bmatrix} 2 \\ 0 \end{bmatrix} = \begin{bmatrix} 1 \\ 1 \end{bmatrix}$.

**9.** $\begin{bmatrix} 4 & 2 \\ 1 & -2 \end{bmatrix} \begin{bmatrix} x \\ y \end{bmatrix} = \begin{bmatrix} 6 \\ -1 \end{bmatrix}$; the inverse is $A^{-1} = \begin{bmatrix} 1/5 & 1/5 \\ 1/10 & -2/5 \end{bmatrix}$ and then $\begin{bmatrix} x \\ y \end{bmatrix} = \begin{bmatrix} 1/5 & 1/5 \\ 1/10 & -2/5 \end{bmatrix}\begin{bmatrix} 6 \\ -1 \end{bmatrix} = \begin{bmatrix} 1 \\ 1 \end{bmatrix}$.

**11.** $\begin{bmatrix} 2 & 3 \\ 3 & -4 \end{bmatrix}\begin{bmatrix} x \\ y \end{bmatrix} = \begin{bmatrix} 4 \\ -4 \end{bmatrix}$; the inverse is $A^{-1} = \begin{bmatrix} 4/17 & 3/17 \\ 3/17 & -2/17 \end{bmatrix}$ and then $\begin{bmatrix} x \\ y \end{bmatrix} = \begin{bmatrix} 4/17 & 3/17 \\ 3/17 & -2/17 \end{bmatrix}\begin{bmatrix} 4 \\ -4 \end{bmatrix} = \begin{bmatrix} 4/17 \\ 20/17 \end{bmatrix}$.

**13.** The determinant is $1 \cdot 6 - 2 \cdot 3 = 0$, thus there is no inverse.

**15.** The determinant is $2 \cdot 2 - 1 \cdot 1 = 3$, thus the inverse is $A^{-1} = \begin{bmatrix} 2/3 & -1/3 \\ -1/3 & 2/3 \end{bmatrix}$.

**17.** The determinant is $2 \cdot 3 - 1 \cdot (-2) = 8$, thus the inverse is $A^{-1} = \begin{bmatrix} 3/8 & 1/4 \\ -1/8 & 1/4 \end{bmatrix}$.

**19.** $\mathbf{u} = A^{-1}\mathbf{v}$, where $A^{-1} = \begin{bmatrix} 2 & 7 \\ 1 & 4 \end{bmatrix}$; thus $\mathbf{u} = \begin{bmatrix} 2 & 7 \\ 1 & 4 \end{bmatrix}\begin{bmatrix} -2 \\ 1 \end{bmatrix} = \begin{bmatrix} 3 \\ 2 \end{bmatrix}$.

**21.** We know that $\mathbf{u} = A^{-1}\mathbf{v}$, where $A^{-1} = \begin{bmatrix} 1/11 & -3/22 \\ 1/22 & 2/11 \end{bmatrix}$; we obtain that $\mathbf{u} = \begin{bmatrix} 1/11 & -3/22 \\ 1/22 & 2/11 \end{bmatrix}\begin{bmatrix} 12 \\ -14 \end{bmatrix} = \begin{bmatrix} 3 \\ -2 \end{bmatrix}$.

**23.** We know that $\mathbf{u} = A^{-1}\mathbf{v}$, where $A^{-1} = \begin{bmatrix} 2/5 & 1/5 \\ 1/15 & -2/15 \end{bmatrix}$; we obtain that $\mathbf{u} = \begin{bmatrix} 2/5 & 1/5 \\ 1/15 & -2/15 \end{bmatrix}\begin{bmatrix} 9 \\ -3 \end{bmatrix} = \begin{bmatrix} 3 \\ 1 \end{bmatrix}$.

**25.** We know that $\mathbf{u} = A^{-1}\mathbf{v}$, where $A^{-1} = \begin{bmatrix} 1/2 & 1 \\ 1/4 & -1/2 \end{bmatrix}$; we obtain that $\mathbf{u} = \begin{bmatrix} 1/2 & 1 \\ 1/4 & -1/2 \end{bmatrix}\begin{bmatrix} 2 \\ -1 \end{bmatrix} = \begin{bmatrix} 0 \\ 1 \end{bmatrix}$.

**27.** $\mathbf{u} = A^{-1}\mathbf{v}$, where $A^{-1} = \begin{bmatrix} 4 & -3 \\ -2 & 3 \end{bmatrix}$; we obtain that $\mathbf{u} = \begin{bmatrix} 4 & -3 \\ -2 & 3 \end{bmatrix}\begin{bmatrix} 2 \\ 2 \end{bmatrix} = \begin{bmatrix} 2 \\ 2 \end{bmatrix}$.

**29.** $\mathbf{u} = A^{-1}\mathbf{v}$, where $A^{-1} = \begin{bmatrix} -1/3 & 2 \\ -2/3 & 2 \end{bmatrix}$; we obtain that $\mathbf{u} = \begin{bmatrix} -1/3 & 2 \\ -2/3 & 2 \end{bmatrix}\begin{bmatrix} 5/2 \\ 1 \end{bmatrix} = \begin{bmatrix} 7/6 \\ 1/3 \end{bmatrix}$.

**31.** $\mathbf{u} + \mathbf{v} = \begin{bmatrix} 1+0 \\ 2+(-1) \\ 3+(-2) \end{bmatrix} = \begin{bmatrix} 1 \\ 1 \\ 1 \end{bmatrix}$.

**33.** $A\mathbf{u} = \begin{bmatrix} 0 \cdot 1 + 1 \cdot 2 + 0 \cdot 3 \\ 1 \cdot 1 + 2 \cdot 2 + 0 \cdot 3 \\ 1 \cdot 1 + 0 \cdot 2 + 0 \cdot 3 \end{bmatrix} = \begin{bmatrix} 2 \\ 5 \\ 1 \end{bmatrix}$.

**35.** Let the number of dozens of tulips bought by the English customer be $x$, the number of dozens of tulips bought by the German customer be $y$. Then the equations are $x + y = 325$ and $2x + 4y/3 = 500$. The matrix of the system is $A = \begin{bmatrix} 1 & 1 \\ 2 & 4/3 \end{bmatrix}$, with inverse $A^{-1} = \begin{bmatrix} -2 & 3/2 \\ 3 & -3/2 \end{bmatrix}$. The solution is $A^{-1}\begin{bmatrix} 325 \\ 500 \end{bmatrix} = \begin{bmatrix} 100 \\ 225 \end{bmatrix}$. So the English customer bought 100 dozen tulips and 200 dozen roses and the German customer 225 dozen tulips and 300 dozen roses.

**37.** Let the number of type A units of habitat covered be $x$, the number of type B units of habitat covered be $y$. The equations are $15x+12y=90$ and $9x+17y=121$. The matrix of the system is $A=\begin{bmatrix}15 & 12\\ 9 & 17\end{bmatrix}$, with inverse $A^{-1}=\begin{bmatrix}17/147 & -4/49\\ -3/49 & 5/49\end{bmatrix}$. The solution is $A^{-1}\begin{bmatrix}90\\ 121\end{bmatrix}=\begin{bmatrix}26/49\\ 335/49\end{bmatrix}$. We estimate a half A habitat and 7 B habitats.

**39.** Let the number of vitamin capsules produced for men be $x$, for women, $y$. Then the total amount of B12 vitamin needed is $80x+70y$ units, and the total amount of C vitamin needed is $50x+40y$. The ratio of these is $13/22=(50x+40y)/(80x+70y)=(50+40(y/x))/(80+70(y/x))$. The solution of this equation for $y/x$ is 2, thus the customer base consists of twice as many women.

**41.** The input-output matrix is $\begin{bmatrix}0.2 & 0.3\\ 0.4 & 0\end{bmatrix}$. Thus in order to solve $\mathbf{d}=\mathbf{u}-A\mathbf{u}$, we have to find the inverse of $B=\begin{bmatrix}0.8 & -0.3\\ -0.4 & 1\end{bmatrix}$; we obtain that $B^{-1}=\begin{bmatrix}25/17 & 15/34\\ 10/17 & 20/17\end{bmatrix}$, thus $B^{-1}\begin{bmatrix}10000\\ 0\end{bmatrix}=\begin{bmatrix}250000/17\\ 100000/17\end{bmatrix}$; we need $100000/17\approx 5882$ units of coal.

**43.** The input-output matrix is given by $\begin{bmatrix}0.05 & 0.2\\ 0.1 & 0.1\end{bmatrix}$. Thus in order to solve $\mathbf{d}=\mathbf{u}-A\mathbf{u}$, we have to find the inverse of $B=\begin{bmatrix}0.95 & -0.2\\ -0.1 & 0.9\end{bmatrix}$; we obtain that $B^{-1}=\begin{bmatrix}180/167 & 40/167\\ 20/167 & 190/167\end{bmatrix}$, thus $B^{-1}\begin{bmatrix}0\\ 10^6\end{bmatrix}=\begin{bmatrix}4\cdot 10^7/167\\ 19\cdot 10^7/167\end{bmatrix}$; we need $4\cdot 10^7/167\approx 239,521$ units of water.

**45. a.** The input-output matrix is given by $\begin{bmatrix}0.13 & 0.33 & 0.25\\ 0.50 & 0.17 & 0.25\\ 0.25 & 0.17 & 0.25\end{bmatrix}$.

**b.** The system we have to solve is $10=x-(0.13x+0.33y+0.25z)$, $28=y-(0.50x+0.17y+0.25z)$, and $14=z-(0.25x+0.17y+0.25z)$. The solution we obtain after substitution and elimination is $x\approx 62.6$, $y\approx 89.5$, and $z\approx 59.8$.

**47.** Assume that on a given day, the chance of a cloudy day is $x$, the chance of a rainy day is $y$ and the chance of a sunny day is $z$. Then the vector $\begin{bmatrix}x\\ y\\ z\end{bmatrix}$ represents these chances on a given day, and the subsequent day the chances are given by $A\begin{bmatrix}x\\ y\\ z\end{bmatrix}$, where $A=\begin{bmatrix}0.5 & 0.3 & 0.3\\ 0.2 & 0.2 & 0\\ 0.3 & 0.5 & 0.7\end{bmatrix}$. To find the proportions, we have to solve the system $\begin{bmatrix}x\\ y\\ z\end{bmatrix}=A\begin{bmatrix}x\\ y\\ z\end{bmatrix}$, and $x+y+z=1$. The solution we obtain after substitution and elimination is $x\approx 0.375$, $y\approx 0.094$, and $z\approx 0.531$.

## Problem Set 8.3 - Eigenvalues and Eigenvectors

**1.** First, $\begin{bmatrix}1 & 2\\ 2 & 1\end{bmatrix}\begin{bmatrix}1\\ 1\end{bmatrix}=\begin{bmatrix}3\\ 3\end{bmatrix}=3\begin{bmatrix}1\\ 1\end{bmatrix}$, so the eigenvalue is 3 for this eigenvector. Second, $\begin{bmatrix}1 & 2\\ 2 & 1\end{bmatrix}\begin{bmatrix}-1\\ 1\end{bmatrix}=\begin{bmatrix}1\\ -1\end{bmatrix}=-1\begin{bmatrix}-1\\ 1\end{bmatrix}$, so the eigenvalue is $-1$ for this eigenvector.

**3.** First, $\begin{bmatrix} 1 & 2 \\ 3 & 2 \end{bmatrix}\begin{bmatrix} 2 \\ 3 \end{bmatrix} = \begin{bmatrix} 8 \\ 12 \end{bmatrix} = 4\begin{bmatrix} 2 \\ 3 \end{bmatrix}$, so the eigenvalue is 4 for this eigenvector. Second, $\begin{bmatrix} 1 & 2 \\ 3 & 2 \end{bmatrix}\begin{bmatrix} -1 \\ 1 \end{bmatrix} = \begin{bmatrix} 1 \\ -1 \end{bmatrix} = -1\begin{bmatrix} -1 \\ 1 \end{bmatrix}$, so the eigenvalue is $-1$ for this eigenvector.

**5.** First, $\begin{bmatrix} 1/2 & 1 \\ 1 & 1/2 \end{bmatrix}\begin{bmatrix} 1 \\ 1 \end{bmatrix} = \begin{bmatrix} 3/2 \\ 3/2 \end{bmatrix} = \frac{3}{2}\begin{bmatrix} 1 \\ 1 \end{bmatrix}$, so the eigenvalue is 3/2 for this eigenvector. Second, $\begin{bmatrix} 1/2 & 1 \\ 1 & 1/2 \end{bmatrix}\begin{bmatrix} -1 \\ 1 \end{bmatrix} = \begin{bmatrix} 1/2 \\ -1/2 \end{bmatrix} = -\frac{1}{2}\begin{bmatrix} -1 \\ 1 \end{bmatrix}$, so the eigenvalue is $-1/2$ for this eigenvector.

**7.** For $\lambda = 6$, we have to solve $\begin{bmatrix} 2 & 4 \\ 4 & 2 \end{bmatrix}\begin{bmatrix} x \\ y \end{bmatrix} = 6\begin{bmatrix} x \\ y \end{bmatrix}$, i.e. $2x + 4y = 6x$ and $4x + 2y = 6y$. Let $x = 1$, then $y = 1$ and the eigenvector is $\begin{bmatrix} 1 \\ 1 \end{bmatrix}$. For $\lambda = -2$, we have to solve $\begin{bmatrix} 2 & 4 \\ 4 & 2 \end{bmatrix}\begin{bmatrix} x \\ y \end{bmatrix} = -2\begin{bmatrix} x \\ y \end{bmatrix}$, i.e. $2x + 4y = -2x$ and $4x + 2y = -2y$. Let $x = 1$, then $y = -1$ and the eigenvector is $\begin{bmatrix} 1 \\ -1 \end{bmatrix}$.

**9.** For $\lambda = 4/3$, we have to solve $\begin{bmatrix} 1/3 & 2/3 \\ 1 & 2/3 \end{bmatrix}\begin{bmatrix} x \\ y \end{bmatrix} = \frac{4}{3}\begin{bmatrix} x \\ y \end{bmatrix}$, i.e. $x/3 + 2y/3 = 4x/3$ and $x + 2y/3 = 4y/3$. Let $x = 1$, then $y = 3/2$ and the eigenvector is $\begin{bmatrix} 1 \\ 3/2 \end{bmatrix}$. For $\lambda = -1/3$, we have to solve $\begin{bmatrix} 1/3 & 2/3 \\ 1 & 2/3 \end{bmatrix}\begin{bmatrix} x \\ y \end{bmatrix} = -\frac{1}{3}\begin{bmatrix} x \\ y \end{bmatrix}$, i.e. $x/3 + 2y/3 = -x/3$ and $x + 2y/3 = -y/3$. Let $x = 1$, then $y = -1$ and the eigenvector is $\begin{bmatrix} 1 \\ -1 \end{bmatrix}$.

**11.** For $\lambda = -3$, we have to solve $\begin{bmatrix} -1 & -1 \\ 2 & -4 \end{bmatrix}\begin{bmatrix} x \\ y \end{bmatrix} = -3\begin{bmatrix} x \\ y \end{bmatrix}$, i.e. $-x - y = -3x$ and $2x - 4y = -3y$. Let $x = 1$, then $y = 2$ and the eigenvector is $\begin{bmatrix} 1 \\ 2 \end{bmatrix}$. For $\lambda = -2$, we have to solve $\begin{bmatrix} -1 & -1 \\ 2 & -4 \end{bmatrix}\begin{bmatrix} x \\ y \end{bmatrix} = -2\begin{bmatrix} x \\ y \end{bmatrix}$, i.e. $-x - y = -2x$ and $2x - 4y = -2y$. Let $x = 1$, then $y = 1$ and the eigenvector is $\begin{bmatrix} 1 \\ 1 \end{bmatrix}$.

**13.** Solving $\det(A - \lambda I) = (1-\lambda)(2-\lambda) = 0$, we obtain $\lambda_1 = 1$ and $\lambda_2 = 2$. For $\lambda_1 = 1$, we have to solve $\begin{bmatrix} 1 & 0 \\ 0 & 2 \end{bmatrix}\begin{bmatrix} x \\ y \end{bmatrix} = 1\begin{bmatrix} x \\ y \end{bmatrix}$, i.e. $x = x$ and $2y = y$. Let $x = 1$, then $y = 0$ and the eigenvector is $\begin{bmatrix} 1 \\ 0 \end{bmatrix}$. For $\lambda_2 = 2$, we have to solve $\begin{bmatrix} 1 & 0 \\ 0 & 2 \end{bmatrix}\begin{bmatrix} x \\ y \end{bmatrix} = 2\begin{bmatrix} x \\ y \end{bmatrix}$, i.e. $x = 2x$ and $2y = 2y$. Let $y = 1$, then $x = 0$ and the eigenvector is $\begin{bmatrix} 0 \\ 1 \end{bmatrix}$.

**15.** Solving $\det(A - \lambda I) = (1-\lambda)(2-\lambda) = 0$, we obtain $\lambda_1 = 1$ and $\lambda_2 = 2$. For $\lambda_1 = 1$, we have to solve $\begin{bmatrix} 1 & 2 \\ 0 & 2 \end{bmatrix}\begin{bmatrix} x \\ y \end{bmatrix} = 1\begin{bmatrix} x \\ y \end{bmatrix}$, i.e. $x + 2y = x$ and $2y = y$. Let $x = 1$, then $y = 0$ and the eigenvector is $\begin{bmatrix} 1 \\ 0 \end{bmatrix}$. For $\lambda_2 = 2$, we have to solve $\begin{bmatrix} 1 & 2 \\ 0 & 2 \end{bmatrix}\begin{bmatrix} x \\ y \end{bmatrix} = 2\begin{bmatrix} x \\ y \end{bmatrix}$, i.e. $x + 2y = 2x$ and $2y = 2y$. Let $y = 1$, then $x = 2$ and the eigenvector is $\begin{bmatrix} 2 \\ 1 \end{bmatrix}$.

**17.** Solving $\det(A - \lambda I) = (2-\lambda)(-10-\lambda) = 0$, we obtain $\lambda_1 = 2$ and $\lambda_2 = -10$. For $\lambda_1 =$

2, we have to solve $\begin{bmatrix} 2 & -10 \\ 0 & -10 \end{bmatrix} \begin{bmatrix} x \\ y \end{bmatrix} = 2 \begin{bmatrix} x \\ y \end{bmatrix}$, i.e. $2x - 10y = 2x$ and $-10y = 2y$. Let $x = 1$, then $y = 0$ and the eigenvector is $\begin{bmatrix} 1 \\ 0 \end{bmatrix}$. For $\lambda_2 = -10$, we have to solve $\begin{bmatrix} 2 & -10 \\ 0 & -10 \end{bmatrix} \begin{bmatrix} x \\ y \end{bmatrix} = -10 \begin{bmatrix} x \\ y \end{bmatrix}$, i.e. $2x - 10y = -10x$ and $-10y = -10y$. Let $y = 1$, then $x = 5/6$ and the eigenvector is $\begin{bmatrix} 5/6 \\ 1 \end{bmatrix}$.

**19.** Solving $\det(A-\lambda I) = (2-\lambda)(3-\lambda)-2 = \lambda^2 - 5\lambda + 4 = 0$, we obtain $\lambda_1 = 4$ and $\lambda_2 = 1$. For $\lambda_1 = 4$, we have to solve $\begin{bmatrix} 2 & 2 \\ 1 & 3 \end{bmatrix} \begin{bmatrix} x \\ y \end{bmatrix} = 4 \begin{bmatrix} x \\ y \end{bmatrix}$, i.e. $2x + 2y = 4x$ and $x + 3y = 4y$. Let $x = 1$, then $y = 1$ and the eigenvector is $\begin{bmatrix} 1 \\ 1 \end{bmatrix}$. For $\lambda_2 = 1$, we have to solve $\begin{bmatrix} 2 & 2 \\ 1 & 3 \end{bmatrix} \begin{bmatrix} x \\ y \end{bmatrix} = 1 \begin{bmatrix} x \\ y \end{bmatrix}$, i.e. $2x+2y = x$ and $x+3y = y$. Let $y = 1$, then $x = -2$ and the eigenvector is $\begin{bmatrix} -2 \\ 1 \end{bmatrix}$.

**21.** Solving the equation $\det(A - \lambda I) = (2 - \lambda)(-2 - \lambda) - 12 = \lambda^2 - 16 = 0$, we obtain $\lambda_1 = 4$ and $\lambda_2 = -4$. For $\lambda_1 = 4$, we have to solve $\begin{bmatrix} 2 & 6 \\ 2 & -2 \end{bmatrix} \begin{bmatrix} x \\ y \end{bmatrix} = 4 \begin{bmatrix} x \\ y \end{bmatrix}$, i.e. $2x + 6y = 4x$ and $2x - 2y = 4y$. Let $x = 1$, then $y = 1/3$ and the eigenvector is $\begin{bmatrix} 1 \\ 1/3 \end{bmatrix}$. For $\lambda_2 = -4$, we have to solve $\begin{bmatrix} 2 & 6 \\ 2 & -2 \end{bmatrix} \begin{bmatrix} x \\ y \end{bmatrix} = -4 \begin{bmatrix} x \\ y \end{bmatrix}$, i.e. $2x + 6y = -4x$ and $2x - 2y = -4y$. Let $y = 1$, then $x = -1$ and the eigenvector is $\begin{bmatrix} 1 \\ -1 \end{bmatrix}$.

**23.** Solving the equation $\det(A - \lambda I) = (-10-\lambda)(8-\lambda)+5 = \lambda^2+2\lambda-75 = 0$, we obtain $\lambda_1 = -1-2\sqrt{19}$ and $\lambda_2 = -1+2\sqrt{19}$. For $\lambda_1 = -1 - 2\sqrt{19}$, we have to solve $\begin{bmatrix} -10 & 1 \\ -5 & 8 \end{bmatrix} \begin{bmatrix} x \\ y \end{bmatrix} = (-1-2\sqrt{19}) \begin{bmatrix} x \\ y \end{bmatrix}$, i.e. $-10x + y = (-1 - 2\sqrt{19})x$ and $-5x + 8y = (-1 - 2\sqrt{19})y$. Let $x = 1$, then $y = 9 - 2\sqrt{19}$ and the eigenvector is $\begin{bmatrix} 1 \\ 9 - 2\sqrt{19} \end{bmatrix}$. For $\lambda_2 = -1 + 2\sqrt{19}$, we have to solve $\begin{bmatrix} -10 & 1 \\ -5 & 8 \end{bmatrix} \begin{bmatrix} x \\ y \end{bmatrix} = (-1+2\sqrt{19}) \begin{bmatrix} x \\ y \end{bmatrix}$, i.e. $-10x + y = (-1 + 2\sqrt{19})x$ and $-5x + 8y = (-1+2\sqrt{19})y$. Let $x = 1$, then $y = 9+2\sqrt{19}$ and the eigenvector is $\begin{bmatrix} 1 \\ 9 + 2\sqrt{19} \end{bmatrix}$.

**25.** We obtain $\mathbf{u}_1 = A\mathbf{u}_0 = \begin{bmatrix} 24 \\ 48 \end{bmatrix}$, $\mathbf{u}_2 = A\mathbf{u}_1 = \begin{bmatrix} 57.6 \\ 48 \end{bmatrix}$, $\mathbf{u}_3 = A\mathbf{u}_2 = \begin{bmatrix} 57.6 \\ 61.44 \end{bmatrix}$, $\mathbf{u}_4 = A\mathbf{u}_3 = \begin{bmatrix} 73.728 \\ 72.192 \end{bmatrix}$, $\mathbf{u}_5 = A\mathbf{u}_4 = \begin{bmatrix} 86.6304 \\ 87.2448 \end{bmatrix}$, $\mathbf{u}_6 = A\mathbf{u}_5 = \begin{bmatrix} 104.694 \\ 104.448 \end{bmatrix}$.

**27.** We obtain $\mathbf{u}_1 = A\mathbf{u}_0 = \begin{bmatrix} 0 \\ 25 \end{bmatrix}$, $\mathbf{u}_2 = A\mathbf{u}_1 = \begin{bmatrix} 50 \\ 15 \end{bmatrix}$, $\mathbf{u}_3 = A\mathbf{u}_2 = \begin{bmatrix} 30 \\ 21.5 \end{bmatrix}$, $\mathbf{u}_4 = A\mathbf{u}_3 = \begin{bmatrix} 43 \\ 20.4 \end{bmatrix}$, $\mathbf{u}_5 = A\mathbf{u}_4 = \begin{bmatrix} 40.8 \\ 22.99 \end{bmatrix}$, $\mathbf{u}_6 = A\mathbf{u}_5 = \begin{bmatrix} 45.98 \\ 23.994 \end{bmatrix}$.

**29.** We obtain $\mathbf{u}_1 = A\mathbf{u}_0 = \begin{bmatrix} 100 \\ 45 \end{bmatrix}$, $\mathbf{u}_2 = A\mathbf{u}_1 = \begin{bmatrix} 90 \\ 46 \end{bmatrix}$, $\mathbf{u}_3 = A\mathbf{u}_2 = \begin{bmatrix} 92 \\ 45.8 \end{bmatrix}$, $\mathbf{u}_4 = A\mathbf{u}_3 = \begin{bmatrix} 91.6 \\ 45.84 \end{bmatrix}$, $\mathbf{u}_5 = A\mathbf{u}_4 = \begin{bmatrix} 91.68 \\ 45.832 \end{bmatrix}$, $\mathbf{u}_6 = A\mathbf{u}_5 = \begin{bmatrix} 91.664 \\ 45.8336 \end{bmatrix}$.

**31.** $A = \begin{bmatrix} 0 & 2 \\ 0.5 & 0.5 \end{bmatrix}$; $\mathbf{u}_1 = A\mathbf{u}_0 = \begin{bmatrix} 100 \\ 50 \end{bmatrix}$, $\mathbf{u}_2 = A\mathbf{u}_1 = \begin{bmatrix} 100 \\ 75 \end{bmatrix}$, $\mathbf{u}_3 = A\mathbf{u}_2 = \begin{bmatrix} 150 \\ 87.5 \end{bmatrix}$.

**33.** $A = \begin{bmatrix} 0 & 1.6 \\ 0.4 & 0.5 \end{bmatrix}$; $\mathbf{u}_1 = A\mathbf{u}_0 = \begin{bmatrix} 80 \\ 45 \end{bmatrix}$, $\mathbf{u}_2 = A\mathbf{u}_1 = \begin{bmatrix} 72 \\ 54.5 \end{bmatrix}$, $\mathbf{u}_3 = A\mathbf{u}_2 = \begin{bmatrix} 87.2 \\ 56.05 \end{bmatrix}$.

**35.** $A = \begin{bmatrix} 0 & 1 \\ 0.3 & 0.6 \end{bmatrix}$; $\mathbf{u}_1 = A\mathbf{u}_0 = \begin{bmatrix} 50 \\ 45 \end{bmatrix}$, $\mathbf{u}_2 = A\mathbf{u}_1 = \begin{bmatrix} 45 \\ 42 \end{bmatrix}$, $\mathbf{u}_3 = A\mathbf{u}_2 = \begin{bmatrix} 42 \\ 38.7 \end{bmatrix}$.

**37.** $A = \begin{bmatrix} 0.4 & 2 \\ 0.5 & 0 \end{bmatrix}$; $\mathbf{u}_1 = A\mathbf{u}_0 = \begin{bmatrix} 88 \\ 35 \end{bmatrix}$, $\mathbf{u}_2 = A\mathbf{u}_1 \approx \begin{bmatrix} 105 \\ 44 \end{bmatrix}$, $\mathbf{u}_3 = A\mathbf{u}_2 \approx \begin{bmatrix} 130 \\ 53 \end{bmatrix}$. We obtain $p \approx 0.71$, and $r \approx 0.22$ (the largest eigenvalue is approximately 1.22).

**39.** $A = \begin{bmatrix} 0.8 & 1 \\ 0.5 & 0 \end{bmatrix}$; $\mathbf{u}_1 = A\mathbf{u}_0 = \begin{bmatrix} 86 \\ 35 \end{bmatrix}$, $\mathbf{u}_2 = A\mathbf{u}_1 \approx \begin{bmatrix} 104 \\ 43 \end{bmatrix}$, $\mathbf{u}_3 = A\mathbf{u}_2 \approx \begin{bmatrix} 126 \\ 52 \end{bmatrix}$. We obtain $p \approx 0.71$, and $r \approx 0.21$ (the largest eigenvalue is approximately 1.21).

**41.** $A = \begin{bmatrix} 0.4 & 0.8 \\ 0.5 & 0 \end{bmatrix}$; $\mathbf{u}_1 = A\mathbf{u}_0 = \begin{bmatrix} 52 \\ 35 \end{bmatrix}$, $\mathbf{u}_2 = A\mathbf{u}_1 \approx \begin{bmatrix} 49 \\ 26 \end{bmatrix}$, $\mathbf{u}_3 = A\mathbf{u}_2 \approx \begin{bmatrix} 40 \\ 24 \end{bmatrix}$. We obtain $p \approx 0.63$, and $r \approx -0.14$ (the largest eigenvalue is approximately 0.86).

**43.** Solving the equation $\det(A - \lambda I) = (1/3 - \lambda)(2/3 - \lambda) - 2/3 = \lambda^2 - \lambda - 4/9 = 0$, we obtain $\lambda_1 = 4/3$ and $\lambda_2 = -1/3$. For $\lambda_1 = 4/3$, we have to solve $\begin{bmatrix} 1/3 & 2/3 \\ 1 & 2/3 \end{bmatrix}\begin{bmatrix} x \\ y \end{bmatrix} = \frac{4}{3}\begin{bmatrix} x \\ y \end{bmatrix}$, i.e. $x/3 + 2y/3 = 4x/3$ and $x + 2y/3 = 4y/3$. Let $x = 1$, then $y = 3/2$ and the eigenvector is $\begin{bmatrix} 1 \\ 3/2 \end{bmatrix}$. For $\lambda_2 = -1/3$, we have to solve $\begin{bmatrix} 1/3 & 2/3 \\ 1 & 2/3 \end{bmatrix}\begin{bmatrix} x \\ y \end{bmatrix} = \frac{-1}{3}\begin{bmatrix} x \\ y \end{bmatrix}$, i.e. $x/3 + 2y/3 = -x/3$ and $x + 2y/3 = -y/3$. Let $y = 1$, then $x = -1$ and the eigenvector is $\begin{bmatrix} -1 \\ 1 \end{bmatrix}$. The general solution is $\mathbf{u}_n = a(4/3)^n \begin{bmatrix} 1 \\ 3/2 \end{bmatrix} + b(-1/3)^n \begin{bmatrix} -1 \\ 1 \end{bmatrix}$.

**45.** Solving the equation $\det(A - \lambda I) = (13/4 - \lambda)(7/4 - \lambda) - 27/16 = \lambda^2 - 5\lambda + 4 = 0$, we obtain $\lambda_1 = 1$ and $\lambda_2 = 4$. For $\lambda_1 = 1$, we have to solve $\begin{bmatrix} 13/4 & -3/4 \\ -9/4 & 7/4 \end{bmatrix}\begin{bmatrix} x \\ y \end{bmatrix} = 1\begin{bmatrix} x \\ y \end{bmatrix}$, i.e. $13x/4 - 3y/4 = x$ and $-9x/4 + 7y/4 = y$. Let $x = 1$, then $y = 3$ and the eigenvector is $\begin{bmatrix} 1 \\ 3 \end{bmatrix}$. For $\lambda_2 = 4$, we have to solve $\begin{bmatrix} 13/4 & -3/4 \\ -9/4 & 7/4 \end{bmatrix}\begin{bmatrix} x \\ y \end{bmatrix} = 4\begin{bmatrix} x \\ y \end{bmatrix}$, i.e. $13x/4 - 3y/4 = 4x$ and $-9x/4 + 7y/4 = 4y$. Let $y = 1$, then $x = -1$ and the eigenvector is $\begin{bmatrix} -1 \\ 1 \end{bmatrix}$. The general solution is $\mathbf{u}_n = a\begin{bmatrix} 1 \\ 3 \end{bmatrix} + b \cdot 4^n \begin{bmatrix} -1 \\ 1 \end{bmatrix}$.

**47.** From Problem 43, we obtain that $\mathbf{u}_n = a(4/3)^n \begin{bmatrix} 1 \\ 3/2 \end{bmatrix} + b(-1/3)^n \begin{bmatrix} -1 \\ 1 \end{bmatrix}$. Also, $\begin{bmatrix} 30 \\ 20 \end{bmatrix} = \mathbf{u}_0 = a\begin{bmatrix} 1 \\ 3/2 \end{bmatrix} + b\begin{bmatrix} -1 \\ 1 \end{bmatrix}$, thus $a = 20$, $b = -10$ and then $\mathbf{u}_n = 20(4/3)^n \begin{bmatrix} 1 \\ 3/2 \end{bmatrix} - 10(-1/3)^n \begin{bmatrix} -1 \\ 1 \end{bmatrix}$. The stable distribution is $\begin{bmatrix} 1/(1+3/2) \\ 3/2/(1+3/2) \end{bmatrix} = \begin{bmatrix} 0.4 \\ 0.6 \end{bmatrix}$.

**49.** From Problem 45, we obtain that $\mathbf{u}_n = a\begin{bmatrix}1\\3\end{bmatrix} + b\cdot 4^n\begin{bmatrix}-1\\1\end{bmatrix}$. Also, $\begin{bmatrix}3\\5\end{bmatrix} = \mathbf{u}_0 = a\begin{bmatrix}1\\3\end{bmatrix} + b\begin{bmatrix}-1\\1\end{bmatrix}$, thus $a = 2$, $b = -1$ and then $\mathbf{u}_n = 2\begin{bmatrix}1\\3\end{bmatrix} - 4^n\begin{bmatrix}-1\\1\end{bmatrix}$.

**51.** $x_{n+1} = 2.4y_n$; $y_{n+1} = 0.5x_n + 0.4y_n$. Thus $A = \begin{bmatrix}0 & 2.4\\0.5 & 0.4\end{bmatrix}$; $\mathbf{u}_1 = A\mathbf{u}_0 = \begin{bmatrix}120\\35\end{bmatrix}$, $\mathbf{u}_2 = A\mathbf{u}_1 = \begin{bmatrix}84\\74\end{bmatrix}$, $\mathbf{u}_3 = A\mathbf{u}_2 \approx \begin{bmatrix}178\\72\end{bmatrix}$. We obtain that $p \approx 0.65$, and $r \approx 0.31$ (the largest eigenvalue is approximately 1.31).

**53.** $x_{n+1} = y_n$; $y_{n+1} = 0.6x_n + sy_n$. Thus $A = \begin{bmatrix}0 & 1\\0.6 & s\end{bmatrix}$. If $s = 1$, we obtain that the largest eigenvalue is approximately 1.42, thus $r \approx 0.42$. For $s = 2/3$, the corresponding value is $r \approx 0.18$, and for $s = 1/3$, $r \approx -0.04$. The population will not decline if the largest eigenvalue of $A$ is at least 1; this happens when $s \geq 2/5 = 0.4$.

**55.** The eigenvalues of $A$ are given by $\det(A - \lambda I) = (a-\lambda)(c-\lambda) - b\cdot 0 = 0$, i.e. they are $a$ and $c$. The eigenvalues of $B$ are given by $\det(B-\lambda I) = (a-\lambda)(c-\lambda) - 0\cdot b = 0$, so they are also $a$ and $c$.

**57. a.** Let the number of small, medium and large plants in year $n$ be given by $x_n$, $y_n$ and $z_n$, respectively. Then the model is $x_{n+1} = 0.2(1-0.8)x_n + 30\cdot 0.1y_n + 50\cdot 0.1z_n$, $y_{n+1} = 0.2\cdot 0.8x_n + 0.5(1-0.6)y_n$, and $z_{n+1} = 0.5\cdot 0.6y_n + 0.5z_n$. Thus the matrix is $A = \begin{bmatrix}0.04 & 3 & 5\\0.16 & 0.2 & 0\\0 & 0.3 & 0.5\end{bmatrix}$.

**b.** $\mathbf{u}_0 = \begin{bmatrix}1000\\0\\0\end{bmatrix}$, $\mathbf{u}_1 = A\mathbf{u}_0 = \begin{bmatrix}40\\160\\0\end{bmatrix}$, and $\mathbf{u}_2 = A\mathbf{u}_1 \approx \begin{bmatrix}482\\38\\48\end{bmatrix}$.

**c.** We obtain that the growth rate is $r = 0.0719$ (approximately 7.2%), and the proportion of small plants is $0.9792/(0.9792 + 0.1797 + 0.0943) \approx 0.7814$ (i.e. 78.2%).

## Problem Set 8.4 - Linear Differential Equations

**1.** $A = \begin{bmatrix}2 & 0\\0 & 3\end{bmatrix}$; the eigenvalues are 2 and 3, with eigenvectors $\begin{bmatrix}1\\0\end{bmatrix}$ and $\begin{bmatrix}0\\1\end{bmatrix}$, respectively. This gives that the solution is $\mathbf{u}(t) = ae^{2t}\begin{bmatrix}1\\0\end{bmatrix} + be^{3t}\begin{bmatrix}0\\1\end{bmatrix} = \begin{bmatrix}ae^{2t}\\be^{3t}\end{bmatrix}$.

**3.** $A = \begin{bmatrix}-1 & 0\\3 & -2\end{bmatrix}$; the eigenvalues are $-1$ and $-2$, with eigenvectors $\begin{bmatrix}1\\3\end{bmatrix}$ and $\begin{bmatrix}0\\1\end{bmatrix}$, respectively. This gives that the solution is $\mathbf{u}(t) = ae^{-t}\begin{bmatrix}1\\3\end{bmatrix} + be^{-2t}\begin{bmatrix}0\\1\end{bmatrix} = \begin{bmatrix}ae^{-t}\\3a^{-t} + be^{-2t}\end{bmatrix}$.

**5.** $A = \begin{bmatrix}1 & 2\\2 & 1\end{bmatrix}$; the eigenvalues are 3 and $-1$, with eigenvectors $\begin{bmatrix}1\\1\end{bmatrix}$ and $\begin{bmatrix}-1\\1\end{bmatrix}$, respectively. This gives that the solution is $\mathbf{u}(t) = ae^{3t}\begin{bmatrix}1\\1\end{bmatrix} + be^{-t}\begin{bmatrix}-1\\1\end{bmatrix} =$

$\begin{bmatrix} ae^{3t} - be^{-t} \\ ae^{3t} + be^{-t} \end{bmatrix}$.

**7.** $A = \begin{bmatrix} 2 & -4 \\ -1 & -1 \end{bmatrix}$; the eigenvalues are 3 and $-2$, with eigenvectors $\begin{bmatrix} -4 \\ 1 \end{bmatrix}$ and $\begin{bmatrix} 1 \\ 1 \end{bmatrix}$, respectively. This gives that the solution is $\mathbf{u}(t) = ae^{3t}\begin{bmatrix} -4 \\ 1 \end{bmatrix} + be^{-2t}\begin{bmatrix} 1 \\ 1 \end{bmatrix} = \begin{bmatrix} -4ae^{3t} + be^{-2t} \\ ae^{3t} + be^{-2t} \end{bmatrix}$.

**9.** $A = \begin{bmatrix} 1 & 0 \\ 1 & -2 \end{bmatrix}$; the eigenvalues are 1 and $-2$, with eigenvectors $\begin{bmatrix} 3 \\ 1 \end{bmatrix}$ and $\begin{bmatrix} 0 \\ 1 \end{bmatrix}$, respectively. This gives that the general solution is $\mathbf{u}(t) = ae^{t}\begin{bmatrix} 3 \\ 1 \end{bmatrix} + be^{-2t}\begin{bmatrix} 0 \\ 1 \end{bmatrix} = \begin{bmatrix} 3ae^{t} \\ ae^{t} + be^{-2t} \end{bmatrix}$. Now $\begin{bmatrix} 1 \\ 1 \end{bmatrix} = \mathbf{u}(0) = \begin{bmatrix} 3a \\ a+b \end{bmatrix}$, thus $a = 1/3$ and then $b = 2/3$. The particular solution is $\mathbf{u}(t) = \begin{bmatrix} e^{t} \\ e^{t}/3 + 2e^{-2t}/3 \end{bmatrix}$.

**11.** $A = \begin{bmatrix} 1 & 3 \\ 1 & 1 \end{bmatrix}$; the eigenvalues are $1 + \sqrt{3}$ and $1 - \sqrt{3}$, with eigenvectors $\begin{bmatrix} \sqrt{3} \\ 1 \end{bmatrix}$ and $\begin{bmatrix} -\sqrt{3} \\ 1 \end{bmatrix}$, respectively. This gives that the general solution is $\mathbf{u}(t) = ae^{(1+\sqrt{3})t}\begin{bmatrix} \sqrt{3} \\ 1 \end{bmatrix} + be^{(1-\sqrt{3})t}\begin{bmatrix} -\sqrt{3} \\ 1 \end{bmatrix} = \begin{bmatrix} \sqrt{3}ae^{(1+\sqrt{3})t} - \sqrt{3}be^{(1-\sqrt{3})t} \\ ae^{(1+\sqrt{3})t} + be^{(1-\sqrt{3})t} \end{bmatrix}$. Now $\begin{bmatrix} 0 \\ 1 \end{bmatrix} = \mathbf{u}(0) = \begin{bmatrix} \sqrt{3}a - \sqrt{3}b \\ a+b \end{bmatrix}$, thus $a = b = 1/2$. The particular solution is $\mathbf{u}(t) = \begin{bmatrix} \sqrt{3}e^{(1+\sqrt{3})t}/2 - \sqrt{3}e^{(1-\sqrt{3})t}/2 \\ e^{(1+\sqrt{3})t}/2 + e^{(1-\sqrt{3})t}/2 \end{bmatrix}$.

**13.** $A = \begin{bmatrix} 1 & 0 \\ 1 & -2 \end{bmatrix}$, thus $\mathbf{u}^* = -A^{-1}\mathbf{b} = \begin{bmatrix} 1 & 0 \\ 1/2 & -1/2 \end{bmatrix}\begin{bmatrix} -1 \\ 1 \end{bmatrix} = \begin{bmatrix} 1 \\ 1 \end{bmatrix}$. The eigenvalues of $A$ are 1 and $-2$, with eigenvectors $\begin{bmatrix} 3 \\ 1 \end{bmatrix}$ and $\begin{bmatrix} 0 \\ 1 \end{bmatrix}$, respectively. This gives that the general solution is $\mathbf{u}(t) = ae^{t}\begin{bmatrix} 3 \\ 1 \end{bmatrix} + be^{-2t}\begin{bmatrix} 0 \\ 1 \end{bmatrix} + \mathbf{u}^* = \begin{bmatrix} 3ae^{t} + 1 \\ ae^{t} + be^{-2t} + 1 \end{bmatrix}$. We need that $\begin{bmatrix} 0 \\ 0 \end{bmatrix} = \mathbf{u}(0) = \begin{bmatrix} 3a+1 \\ a+b+1 \end{bmatrix}$, thus $a = -1/3$ and $b = -2/3$. The solution is $\mathbf{u}(t) = \begin{bmatrix} -e^{t} + 1 \\ -e^{t}/3 - 2e^{-2t}/3 + 1 \end{bmatrix}$.

**15.** $A = \begin{bmatrix} 1 & -2 \\ -2 & 1 \end{bmatrix}$, thus $\mathbf{u}^* = -A^{-1}\mathbf{b} = \begin{bmatrix} -1/3 & -2/3 \\ -2/3 & -1/3 \end{bmatrix}\begin{bmatrix} -3 \\ 1 \end{bmatrix} = \begin{bmatrix} -1/3 \\ -5/3 \end{bmatrix}$. The eigenvalues of $A$ are 3 and $-1$, with eigenvectors $\begin{bmatrix} -1 \\ 1 \end{bmatrix}$ and $\begin{bmatrix} 1 \\ 1 \end{bmatrix}$, respectively. This gives that the general solution is $\mathbf{u}(t) = ae^{3t}\begin{bmatrix} -1 \\ 1 \end{bmatrix} + be^{-t}\begin{bmatrix} 1 \\ 1 \end{bmatrix} + \mathbf{u}^* = \begin{bmatrix} -ae^{3t} + be^{-t} - 1/3 \\ ae^{3t} + be^{-t} - 5/3 \end{bmatrix}$. We need that $\begin{bmatrix} 1 \\ 0 \end{bmatrix} = \mathbf{u}(0) = \begin{bmatrix} -a+b-1/3 \\ a+b-5/3 \end{bmatrix}$, thus $a = 1/6$ and $b = 3/2$. The solution is $\mathbf{u}(t) = \begin{bmatrix} -e^{3t}/6 + 3e^{-t}/2 - 1/3 \\ e^{3t}/6 + 3e^{-t}/2 - 5/3 \end{bmatrix}$.

**17.** $A = \begin{bmatrix} 1 & 2 \\ 3 & 2 \end{bmatrix}$; the eigenvalues of $A$ are 4 and $-1$, with eigenvectors $\begin{bmatrix} 2 \\ 3 \end{bmatrix}$ and

$\begin{bmatrix} -1 \\ 1 \end{bmatrix}$, respectively. This gives that the general solution is $\mathbf{u}(t) = ae^{4t}\begin{bmatrix} 2 \\ 3 \end{bmatrix} + be^{-t}\begin{bmatrix} -1 \\ 1 \end{bmatrix} = \begin{bmatrix} 2ae^{4t} - be^{-t} \\ 3ae^{4t} + be^{-t} \end{bmatrix}$. We need that $\begin{bmatrix} 0 \\ -4 \end{bmatrix} = \mathbf{u}(0) = \begin{bmatrix} 2a - b \\ 3a + b \end{bmatrix}$, thus $a = -4/5$ and $b = -8/5$. The solution is $\mathbf{u}(t) = \begin{bmatrix} -8e^{4t}/5 + 8e^{-t}/5 \\ -12e^{4t}/5 - 8e^{-t}/5 \end{bmatrix}$.

**19.** Let $x(t)$ be the amount of drug in the blood, and $y(t)$ be the amount of drug in the cerebrospinal fluid. The equations describing the process are $x' = -x/2 + y/4 - x/5$ and $y' = x/2 - y/4 - y/10$. This corresponds to the system $\mathbf{u}' = A\mathbf{u}$, where $A = \begin{bmatrix} -7/10 & 1/4 \\ 1/2 & -7/20 \end{bmatrix}$. Also, $\mathbf{u}(0) = \begin{bmatrix} 100 \\ 50 \end{bmatrix}$. The equilibrium solution is given by $\mathbf{u}^* = \begin{bmatrix} 0 \\ 0 \end{bmatrix}$. The eigenvalues and corresponding eigenvectors are approximately $\lambda_1 = -0.919$, $\lambda_2 = -0.131$, and $\begin{bmatrix} -1.139 \\ 1 \end{bmatrix}$ and $\begin{bmatrix} 0.439 \\ 1 \end{bmatrix}$, respectively. The initial conditions give the solution $\begin{bmatrix} 56.34e^{-0.919t} + 43.66e^{-0.131t} \\ -49.46e^{-0.919t} + 99.46e^{-0.131t} \end{bmatrix}$. Thus after two hours, $\mathbf{u}(2) \approx \begin{bmatrix} 42.59 \\ 68.75 \end{bmatrix}$ (mg).

**21.** $a$ and $b$ represent the per capita growth rate in pools 1 and 2, respectively. $c$ and $k$ are the outside flow rates into the pools. $g$ and $h$ represent the per capita flow rates from pool 1 into pool 2 and vice versa.

**23.** The system is given by $x' = -0.15x - 0.1x + 0.1y + 10$ and $y' = 0.1x - 0.1y - 0.1y + 20$. This corresponds to the system $\mathbf{u}' = A\mathbf{u} + \mathbf{b}$, where $A = \begin{bmatrix} -1/4 & 1/10 \\ 1/10 & -2/10 \end{bmatrix}$ and $\mathbf{b} = \begin{bmatrix} 10 \\ 20 \end{bmatrix}$. Also, $\mathbf{u}(0) = \begin{bmatrix} 30 \\ 40 \end{bmatrix}$. The equilibrium solution is given by $\mathbf{u}^* = -A^{-1}\mathbf{b} = \begin{bmatrix} 100 \\ 150 \end{bmatrix}$. The eigenvalues and corresponding eigenvectors are approximately $\lambda_1 = -0.328$, $\lambda_2 = -0.122$, and $\begin{bmatrix} -1.281 \\ 1 \end{bmatrix}$ and $\begin{bmatrix} 0.781 \\ 1 \end{bmatrix}$, respectively. The initial conditions then give the solution $\begin{bmatrix} 100 + 9.87e^{-0.328t} - 79.87e^{-0.122t} \\ 150 - 7.71e^{-0.328t} - 102.29e^{-0.122t} \end{bmatrix}$. At $t = 10$, $\mathbf{u}(10) \approx \begin{bmatrix} 77 \\ 119 \end{bmatrix}$.

**25.** The system is given by $\mathbf{u}' = A\mathbf{u} + \mathbf{b}$, where $A = \begin{bmatrix} -1/8 & 0 & 0 \\ 1/8 & -1/2 & 0 \\ 0 & 1/2 & -1/10 \end{bmatrix}$, $\mathbf{b} = \begin{bmatrix} 0 \\ 1/8 \\ 0 \end{bmatrix}$, and $\mathbf{u}(0) = \begin{bmatrix} 3 \\ 0 \\ 0 \end{bmatrix}$. First, we find the equilibrium solution $\mathbf{u}^* = -A^{-1}\mathbf{b} = \begin{bmatrix} 0 \\ 1/4 \\ 5/4 \end{bmatrix}$. After this, the solution is found the same way as in Example 9. We get $\mathbf{u} = \begin{bmatrix} 3e^{-t/8} \\ 1/4 - 5e^{-t/2}/4 + e^{-t/8} \\ 5/4 + 25e^{-t/2}/16 - 20e^{-t/8} + 275e^{-t/10}/16 \end{bmatrix}$. The figure shows the morphine level in the brain for Example 9 and this exercise; we can see that the morphine level does not go as high, but sustains a steadier level later with the new initial conditions (and constant morphine input).

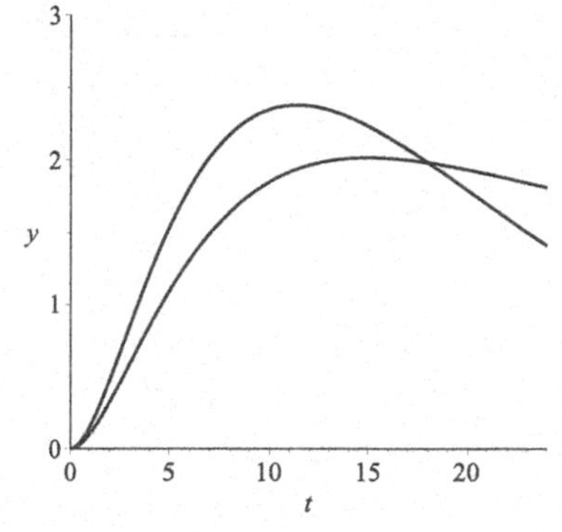

**27.** By differentiating the given $\mathbf{u}(t)$, a lengthy computation shows that $\mathbf{u}' = A\mathbf{u}$. By substituting $t = 0$ in $\mathbf{u}(t)$, the given initial condition equations are obtained. Thus the specified $\mathbf{u}$ is the solution of the initial value problem.

**29.** $\mathbf{u}^*$ is a constant vector solving the equation $A\mathbf{u}^* = -\mathbf{b}$, thus we compute: $d\mathbf{u}/dt = d\mathbf{z}/dt = A\mathbf{z} = A(\mathbf{u} - \mathbf{u}^*) = A\mathbf{u} - A\mathbf{u}^* = A\mathbf{u} + \mathbf{b}$.

## Problem Set 8.5 - Nonlinear Systems

**1.**

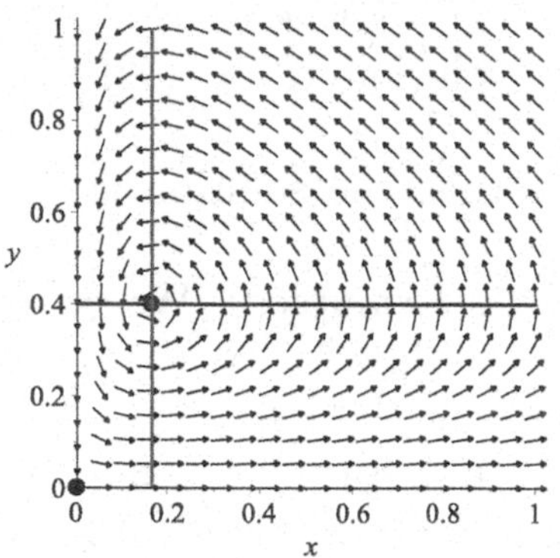

**3.**

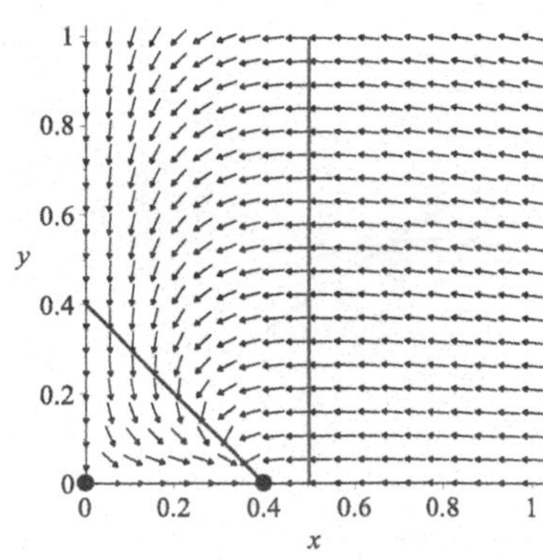

**5.**

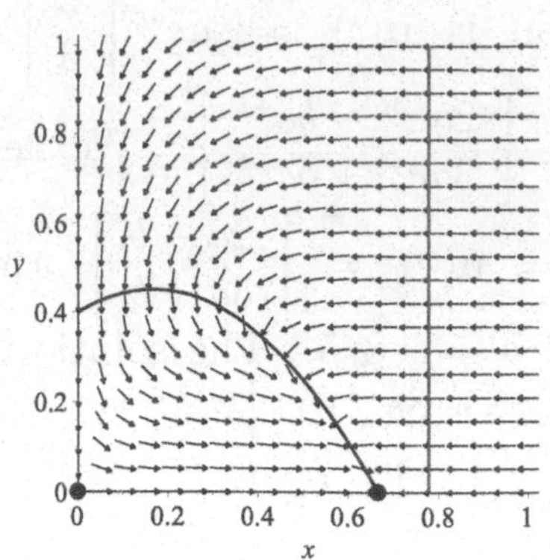

**7.**

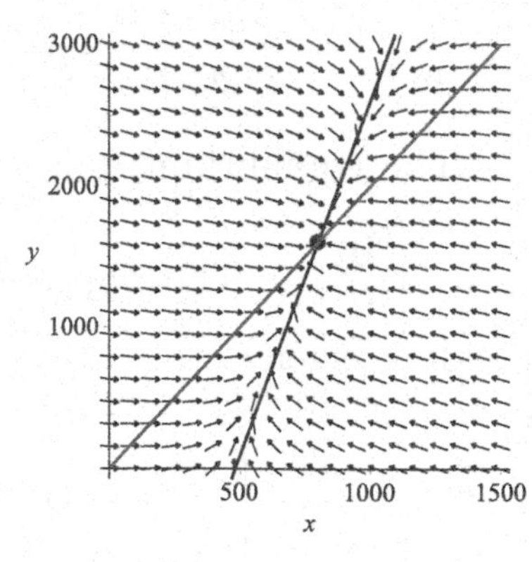

**9.**

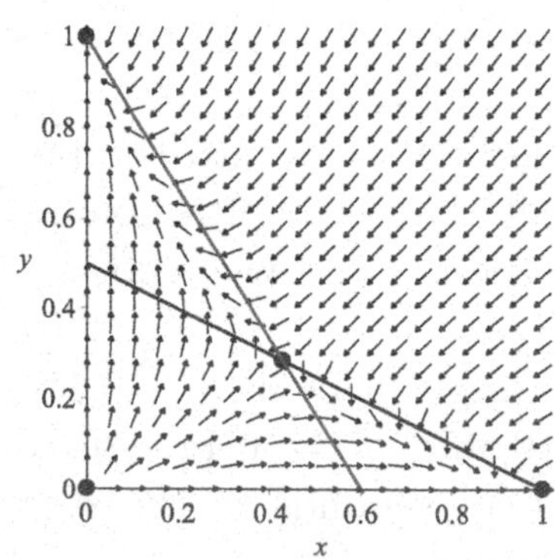

**11.**

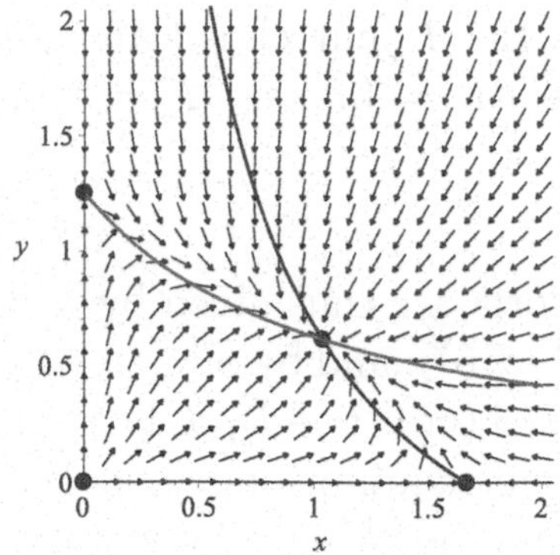

**13.**

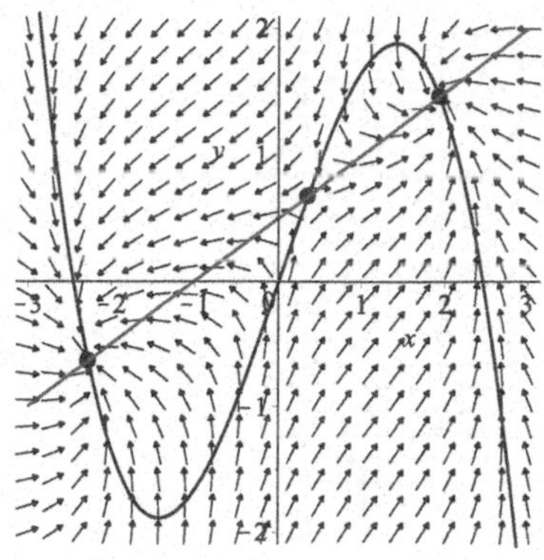

**15.**

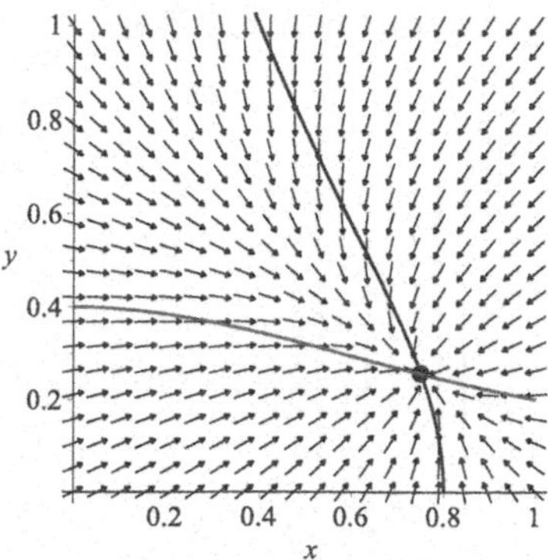

**17.** The model is $x' = 0.3x(1-x) - bxy$, $y' = 100bxy - cy$. The figure shows the phase plane diagram for $b = 0.5$, $c = 30$. The positive equilibrium is at $(0.6, 0.24)$.

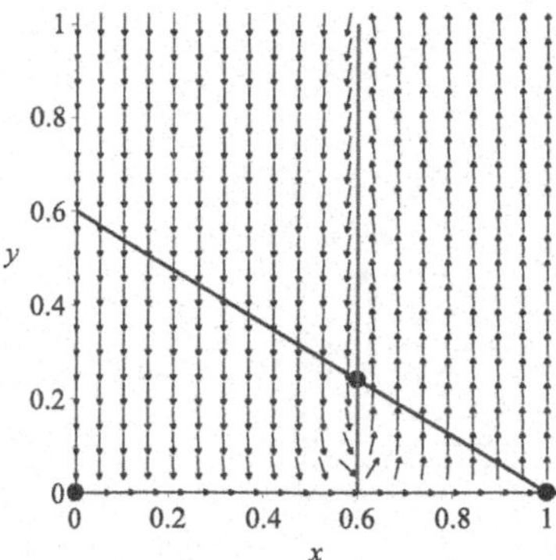

**19.** The model is given by $x' = ax(x/b - 1)(1 - x/K) - xy/10000$, $y' = xy/10000 - cy$. The phase diagrams are given for $c = 0.04$, $c = 0.1$ and $c = 0.2$. There are equilibria at $(0,0)$, $(500,0)$ and $(2000,0)$ for all $c$ values, and in case of $c = 0.1$, there is an equilibrium at $(1000, 500)$. In the first case, solutions seem to approach $(0,0)$. In the second case, depending on the initial condition, the solution either approaches $(0,0)$ or it rotates around $(1000, 500)$. In the third case, depending on the initial condition, we either approach $(0,0)$ or $(2000,0)$.

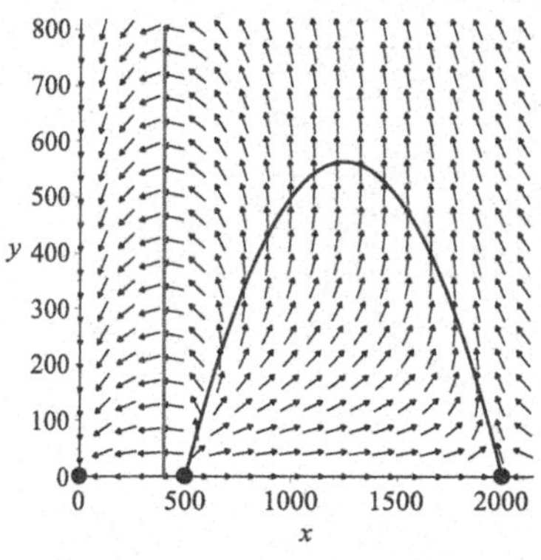

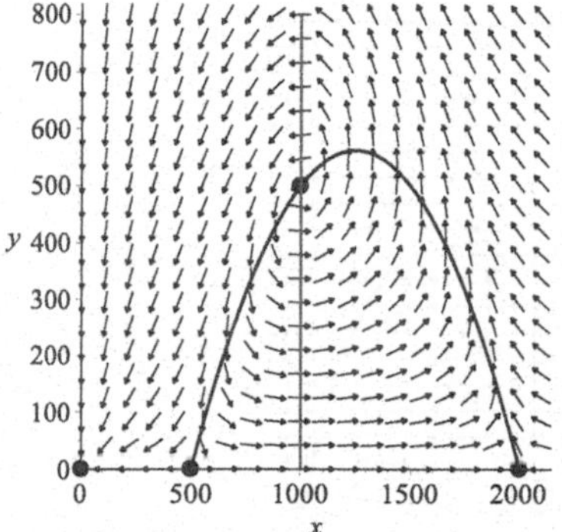

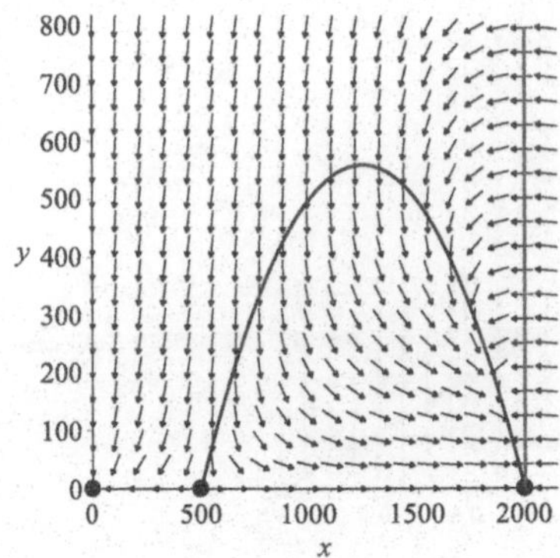

**21.** $a = 0.5$, $c = 1$:

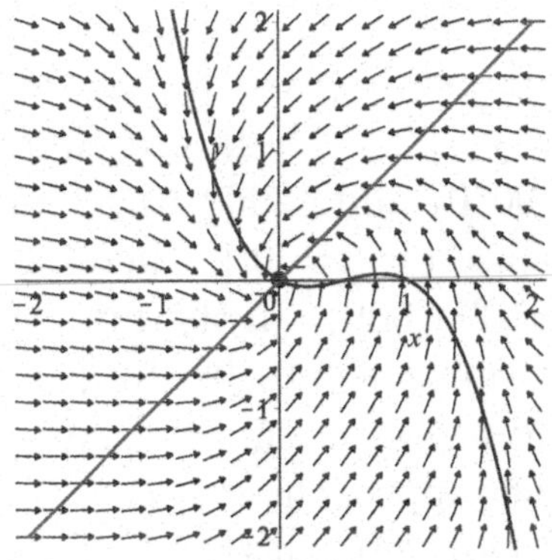

$b = 0.5$:

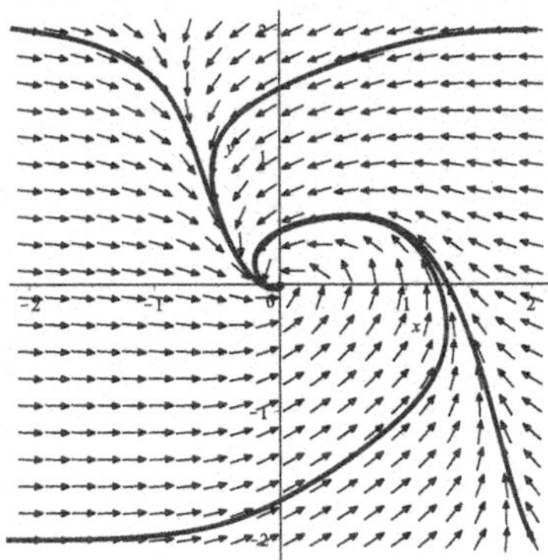

$b = 0.05$:

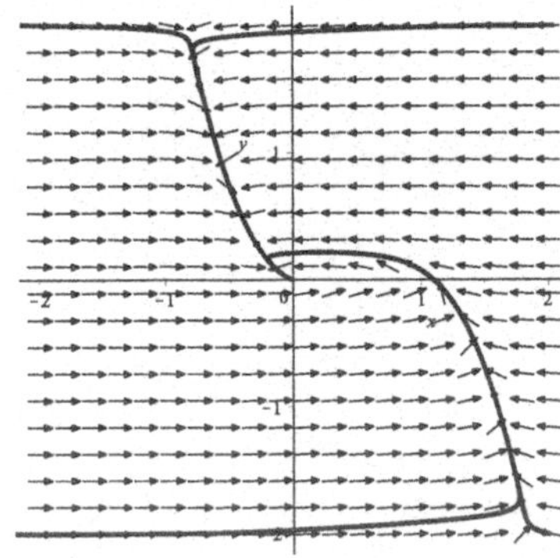

**23. a.** The $S$ null cline is given by $1 - \beta SI - dS = 0$, i.e. $I = (1 - dS)/\beta S = 1/(\beta S) - d/\beta$; the $I$ null clines are given by $\beta SI - rI - dI = 0$, i.e. $I = 0$ and $S = (r + d)/\beta$.

**b.** The equilibria are thus given by $I = 0$, $S = 1/d$ (i.e. $(1/d, 0)$); and by $S = (r+d)/\beta$, $I = 1/(r+d) - d/\beta$ (i.e. $((r+d)/\beta, 1/(r+d) - d/\beta)$). An endemic equilibrium exists if this second equilibrium point is in the first quadrant, i.e. if $1/(r + d) > d/\beta$ or $d(r + d) < \beta$.

**c.** $\beta = 0.1$, $d = 0.02$, $r = 0.3$.

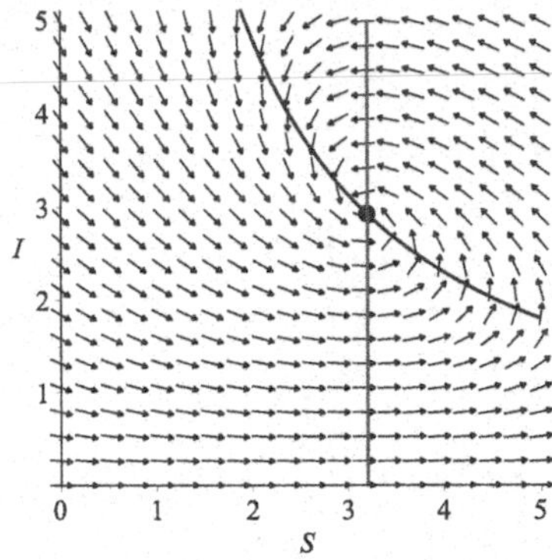

**25. a.** $d_1 = d_2 = 1$, $b_1 = 2$, and $b_2 = 2.25$. The equilibria are at $(0, 0)$, $(0, 5/9)$, and $(1/2, 0)$. Solutions approach $(1/2, 0)$.

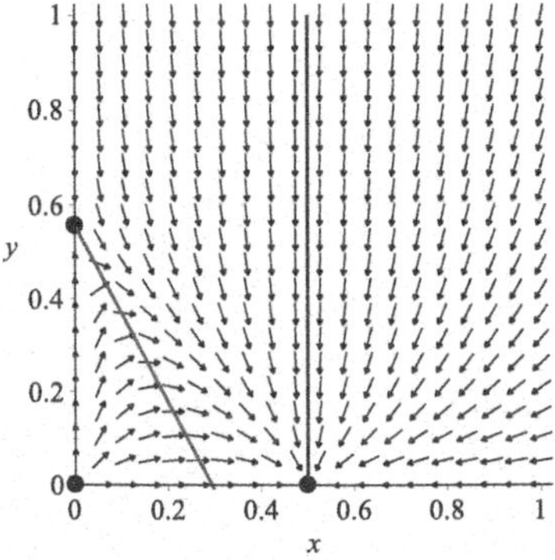

**b.** $d_1 = d_2 = 1$, $b_1 = 2$, and $b_2 = 5$. The equilibria are at $(0, 0)$, $(0, 4/5)$, $(1/2, 0)$, and $(1/2, 1/10)$. Solutions approach $(1/2, 1/10)$.

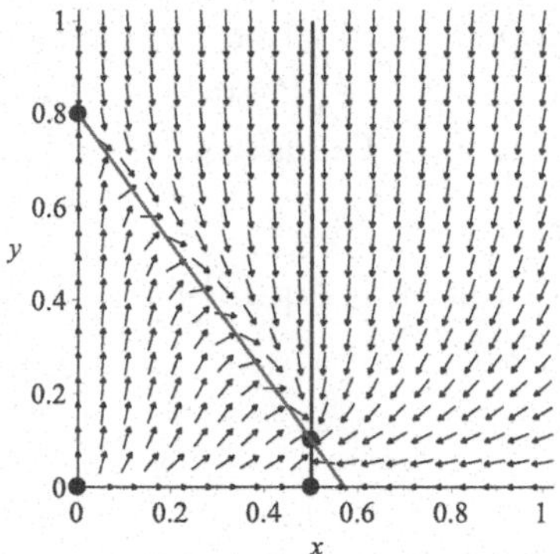


## Review Questions

**1. a.** $c = 0$:

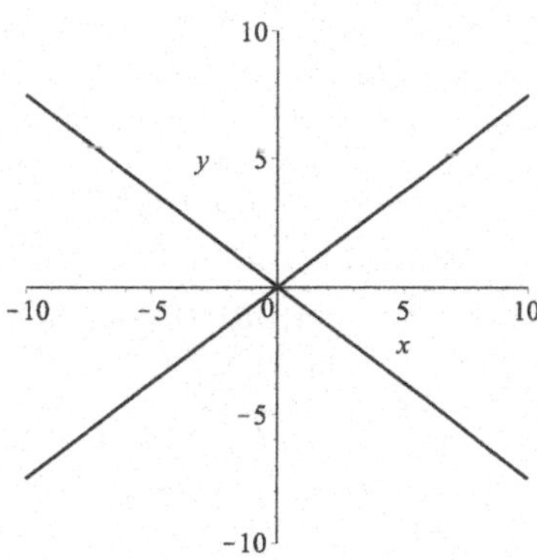

**b.**

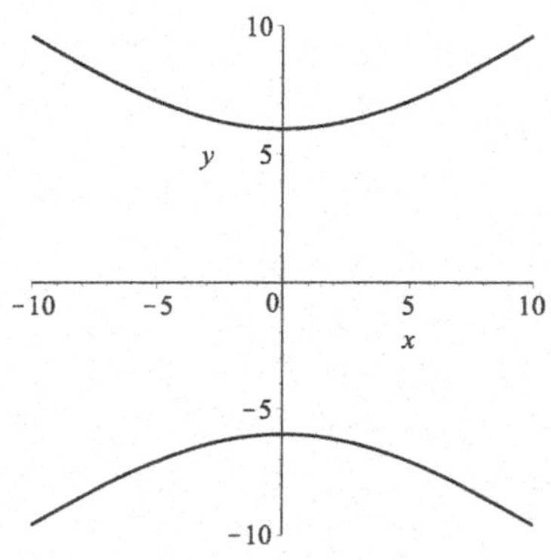

**c.** $f(2,3) = 3^2/9 - 2^2/16 = 3/4$, $f(4,-9) = (-9)^2/9 - 4^2/16 = 8$.

**d.**

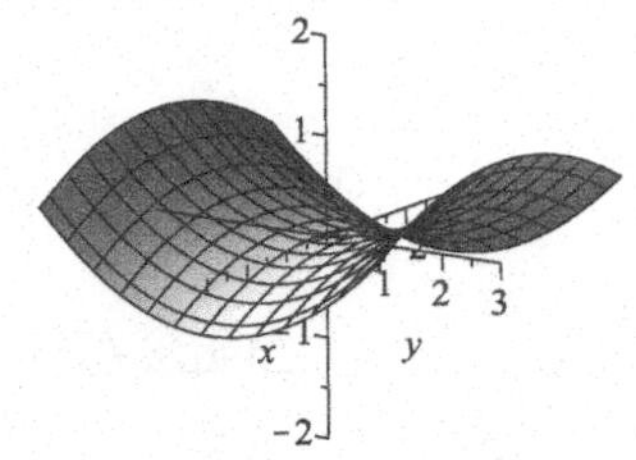

**3. a.** C **b.** D **c.** A **d.** B

**5. a.** $\begin{bmatrix} -2/3 & 1/3 \\ 1 & 1/3 \end{bmatrix}\begin{bmatrix} x \\ y \end{bmatrix} = \begin{bmatrix} -2/5 \\ 1 \end{bmatrix}$; the inverse is $A^{-1} = \begin{bmatrix} -3/5 & 3/5 \\ 9/5 & 6/5 \end{bmatrix}$ and then $\begin{bmatrix} x \\ y \end{bmatrix} = \begin{bmatrix} -3/5 & 3/5 \\ 9/5 & 6/5 \end{bmatrix}\begin{bmatrix} -2/5 \\ 1 \end{bmatrix} = \begin{bmatrix} 21/25 \\ 12/25 \end{bmatrix}$.

**b.** $\begin{bmatrix} 3 & -5/4 \\ 2/3 & 1 \end{bmatrix}\begin{bmatrix} x \\ y \end{bmatrix} = \begin{bmatrix} -1 \\ -2/3 \end{bmatrix}$; the inverse is $A^{-1} = \begin{bmatrix} 6/23 & 15/46 \\ -4/23 & 18/23 \end{bmatrix}$ and then $\begin{bmatrix} x \\ y \end{bmatrix} = \begin{bmatrix} 6/23 & 15/46 \\ -4/23 & 18/23 \end{bmatrix}\begin{bmatrix} -1 \\ -2/3 \end{bmatrix} = \begin{bmatrix} -11/23 \\ -8/23 \end{bmatrix}$.

**7.** The input-output matrix is $\begin{bmatrix} 0.5 & 0.2 \\ 0.4 & 0.1 \end{bmatrix}$. Thus in order to solve $\mathbf{d} = \mathbf{u} - A\mathbf{u}$, we have to find the inverse of $B = \begin{bmatrix} 0.5 & -0.2 \\ -0.4 & 0.9 \end{bmatrix}$; we obtain that $B^{-1} = \begin{bmatrix} 90/37 & 20/37 \\ 40/37 & 50/37 \end{bmatrix}$, thus $B^{-1}\begin{bmatrix} 7 \\ 4 \end{bmatrix} = \begin{bmatrix} 710/37 \\ 480/37 \end{bmatrix}$; we need $710/37 \approx 19.19$ units of good A and $480/37 \approx 12.97$ units of good B.

**9.** The determinant is $(1/2)(-1/3) - (1/4)(2/3) = -1/3$ and thus the inverse is $C^{-1} = \begin{bmatrix} 1 & 2 \\ 3/4 & -3/2 \end{bmatrix}$.

**11.** First, we obtain $\begin{bmatrix} 1 & 1 \\ 1 & 3 \end{bmatrix} \begin{bmatrix} -1+\sqrt{2} \\ 1 \end{bmatrix} = \begin{bmatrix} \sqrt{2} \\ 2+\sqrt{2} \end{bmatrix} = (2+\sqrt{2}) \begin{bmatrix} -1+\sqrt{2} \\ 1 \end{bmatrix}$, so the eigenvalue is $2+\sqrt{2}$ for this eigenvector. Second, we get $\begin{bmatrix} 1 & 1 \\ 1 & 3 \end{bmatrix} \begin{bmatrix} 1+\sqrt{2} \\ -1 \end{bmatrix} = \begin{bmatrix} \sqrt{2} \\ -2+\sqrt{2} \end{bmatrix} = (2-\sqrt{2}) \begin{bmatrix} 1+\sqrt{2} \\ -1 \end{bmatrix}$, so the eigenvalue is $2-\sqrt{2}$ for this eigenvector.

**13.** The eigenvalues can be found by solving $\det(A - \lambda I) = (1/2 - \lambda)(s - \lambda) - 2/25 = 0$, i.e. $\lambda^2 - (s + 1/2)\lambda + s/2 - 2/25 = 0$. The solutions are $\lambda = (s/2 + 1/4) + \sqrt{(s+1/2)^2/4 - (s/2 - 2/25)}$ and $\lambda = (s/2+1/4) - \sqrt{(s+1/2)^2/4 - (s/2 - 2/25)}$. We need that the bigger eigenvalue is at least 1; this happens if $s \geq 21/25 = 0.84$.

**15.** $A = \begin{bmatrix} 1 & 1/3 \\ 2/3 & 2/3 \end{bmatrix}$; the eigenvalues are 4/3 and 1/3, with eigenvectors $\begin{bmatrix} 1 \\ 1 \end{bmatrix}$ and $\begin{bmatrix} -1 \\ 2 \end{bmatrix}$, respectively. This gives that the general solution is $\mathbf{u}(t) = ae^{4t/3} \begin{bmatrix} 1 \\ 1 \end{bmatrix} + be^{t/3} \begin{bmatrix} -1 \\ 2 \end{bmatrix} = \begin{bmatrix} ae^{4t/3} - be^{t/3} \\ ae^{4t/3} + 2be^{t/3} \end{bmatrix}$. Now $\begin{bmatrix} 2 \\ -1 \end{bmatrix} = \mathbf{u}(0) = \begin{bmatrix} a-b \\ a+2b \end{bmatrix}$, thus $a = 1$ and $b = -1$. The particular solution is $\mathbf{u}(t) = \begin{bmatrix} e^{4t/3} + e^{t/3} \\ e^{4t/3} - 2e^{t/3} \end{bmatrix}$.

**17. a.** The null clines are given by (for $S$): $1 - 0.1SI - 0.05S = 0$, i.e. $I = (1 - 0.05S)/(0.1S) = 10/S - 1/2$ and by (for $I$): $0.1SI - 0.25I = 0$, i.e. $I = 0$ and $S = 2.5$.

**b.** Biologically relevant equilibria are thus $(20, 0)$ and $(2.5, 3.5)$.

**c.**

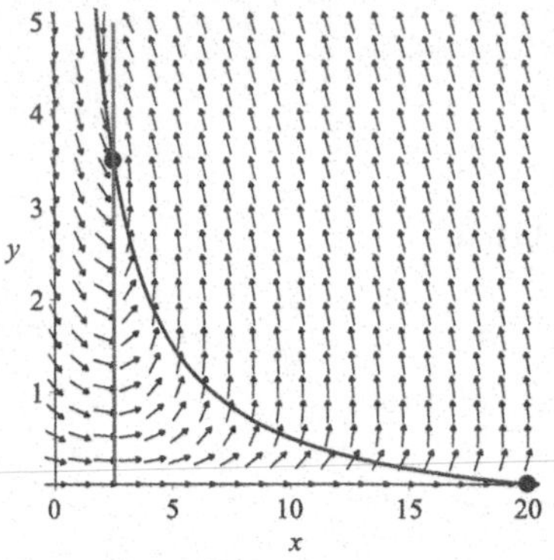

**19.** Let $x(t)$ be the amount of drug in the blood, and $y(t)$ be the amount of drug in the cerebrospinal fluid. The equations describing the process are $x' = -x/3 + y/4 - x/4 + 1$ and $y' = x/3 - y/4 - y/8$. This corresponds to the system $\mathbf{u}' = A\mathbf{u} + \mathbf{b}$, where $A = \begin{bmatrix} -7/12 & 1/4 \\ 1/3 & -3/8 \end{bmatrix}$ and $\mathbf{b} = \begin{bmatrix} 1 \\ 0 \end{bmatrix}$. Also, $\mathbf{u}(0) = \begin{bmatrix} 0 \\ 0 \end{bmatrix}$. The equilibrium solution is given by $\mathbf{u}^* = -A^{-1}\mathbf{b} = \begin{bmatrix} 36/13 \\ 32/13 \end{bmatrix} \approx \begin{bmatrix} 2.77 \\ 2.46 \end{bmatrix}$. We also obtain that the eigenvalues and corresponding eigenvectors are approximately $\lambda_1 = -0.786$, $\lambda_2 = -0.172$, and $\begin{bmatrix} -1.233 \\ 1 \end{bmatrix}$ and $\begin{bmatrix} 0.608 \\ 1 \end{bmatrix}$, respectively. The initial conditions then give the solution $\begin{bmatrix} 2.77 - 0.85e^{-0.786t} - 1.92e^{-0.172t} \\ 2.46 + 0.69e^{-0.786t} - 3.15e^{-0.172t} \end{bmatrix}$. Thus after two hours, $\mathbf{u}(1) \approx \begin{bmatrix} 0.77 \\ 0.12 \end{bmatrix}$ (mg).

CPSIA information can be obtained
at www.ICGtesting.com
Printed in the USA
BVOW04s0229031117
499272BV00004B/4/P

9 781118 645598